The Second Age of Computer Science

FROM ALGOL GENES TO NEURAL NETS

Subrata Dasgupta

OXFORD
UNIVERSITY PRESS

UNIVERSITY PRESS

Oxford University Press is a department of the University of Oxford. It furthers
the University's objective of excellence in research, scholarship, and education
by publishing worldwide. Oxford is a registered trade mark of Oxford University
Press in the UK and certain other countries.

Published in the United States of America by Oxford University Press
198 Madison Avenue, New York, NY 10016, United States of America.

Library of Congress Cataloging-in-Publication Data
Names: Dasgupta, Subrata, author.
Title: The second age of computer science : from ALGOL genes to neural nets / Subrata Dasgupta.
Description: New York, NY : Oxford University Press, 2018. |
Includes bibliographical references and index.
Identifiers: LCCN 2017057808 | ISBN 9780190843861
Subjects: LCSH: Computer science—History—20th century. | Genetic programming
(Computer science) | ALGOL (Computer program language) | Neural networks (Computer science)
Classification: LCC QA76.17 .D363 2018 | DDC 004.09—dc23
LC record available at https://lccn.loc.gov/2017057808

9 8 7 6 5 4 3 2 1

Printed by Sheridan Books, Inc., United States of America

To
Sarmistha (Mithu) Dasgupta
and
Heidi Holmquist

Contents

Acknowledgments

THANK YOU JEREMY LEWIS, my publisher at Oxford University Press, who embraced this project from the day he received the manuscript. As always, it has been a pleasure working with him.

Thank you Anna Langley, editor at OUP, for all your help.

Lincy Priya steered this project very professionally from final manuscript to print. I thank her and her production team at Newgen.

My thanks to Victoria Danahy for her careful, nuanced, and informed copyediting. Her comments as "asides" were always interesting.

Thank you Elman Bashar who took time off from his own work to create the images that appear in this book.

Four anonymous reviewers read drafts of some of the chapters on behalf of the publisher and offered thoughtful and invaluable comments. I thank them, though I would have preferred to acknowledge them by name.

Finally, as always, my thanks to Mithu, Deep, Heidi, and Shome.

Prologue

P.1 THE FIRST AGE OF COMPUTER SCIENCE

In the 1960s a science the likes of which had never been quite known before came into being. At its epicenter were an idea and an artifact. The idea was *automatic computation*, the notion that symbol structures representing information about things in the world could be mechanically transformed into other symbol structures representing other pieces of information. The artifact was the *digital computer*, a machine for effecting such symbol or information processing. And with the naming of this science as *computer science*,[1] its teachers, researchers, and practitioners chose to call themselves *computer scientists*.

The genesis of computer science has a long and tenuous history reaching back to the first quarter of the 19th century, to pre–Victorian England, in fact.[2] Charles Babbage with whom the story of the birth of automatic computation properly began was a product of the Romantic Age in Europe.[3] By the time computer science assumed an identity and a name of its own in the 1960s, its gestation had spanned the long cultural period called the Modern Age and, indeed, when the science emerged, the world was in the first throes of the Postmodern era.

At any rate, as the 1960s drew to a close the discipline was in place. The first batches of formally trained computer scientists were emerging. Universities were offering bachelor's, master's, and doctoral degrees in the discipline. By the end of the decade, computer science could boast a certain level of maturity. It could claim the status of what philosopher of science Thomas Kuhn called in his seminal book *The Structure* of *Scientific Revolutions* a scientific paradigm.[4] We may call this the *computational paradigm*.

A seemingly clear representation of this brand new, and quite unique, paradigm was offered in an influential document defining the contents of computer science circa 1968. *Curriculum 68* was a comprehensive report published under the auspices of the Association for Computing Machinery (ACM), founded in 1947 and the world's first professional society dedicated to computing. This document recommended a curriculum for academic programs in computer science, from "basic" through "intermediate" to "advanced" courses. The report was authored by a committee of 12 academics all drawn from North American universities—11 American and 1 Canadian—who had solicited and received input and opinions from over three-score consultants and reviewers.[5] It seems fair to say (despite its North American authorship) that publication of *Curriculum 68* implied a certain perceived maturity and stability of the computational paradigm as the '60s drew to a close. With historical hindsight let us call the state of the paradigm circa 1969, reflected in *Curriculum 68*, as marking the *first age of computer science*.

The subject matter of computer science, according to this report, consisted of three "major divisions." The first, *information structures and processes*, concerned the "representations and transformations of information structures and with theoretical models for such representations and transformations."[6] It included "data structures," "programming languages," and "models of computation."[7] The second division was named *information processing systems* and included such topics as "computer design and organization," "translators and interpreters," and "computer and operating systems."[8] The third division, called (somewhat confusingly) *methodologies*, consisted of "broad areas of application of computing which has common structures, processes and techniques" and included such topics as text processing, information retrieval, artificial intelligence (AI), and numerical mathematics.[9]

The message was clear: The basic stuff of the computational paradigm was *information*. Further support of this idea was given by the formation, in 1960, of the official world body on computing, called the International Federation for Information Processing (IFIP).

Unfortunately, the term "information" possesses two quite opposite meanings. Communication engineers use it to mean a commodity *devoid of meaning* that gets passed around in telephone and other electrical communication systems, rather as money is circulated in financial systems. The unit of meaningless information is the *bit*, allowing only two values, commonly denoted as "1" and "0." In contrast, in a commonsense or everyday connotation, information is an assertion about something in the world; here information has semantic content: It is *meaningful*. The only shared attribute of these oppositional notions of information is that information—meaningless or meaningful—reduces uncertainty.[10]

So which of these connotations was *Curriculum 68* referring to when they spoke of "information structures and processing" or "information processing systems"? The report does not explicitly clarify, but judging from the contents of the three major "divisions" it would seem that what the authors had in mind was *predominantly* meaningful or semantic information. So we can say that as the 1960s came to an end the basic stuff of the computational paradigm was taken to be meaningful information, with the caveat that any piece of meaningful information is "ultimately" *reducible* to patterns of meaningless information, that is, patterns of 1's and 0's.

P.2 RECOGNIZING COMPUTER SCIENCE AS A SCIENCE OF THE ARTIFICIAL

The authors of *Curriculum 68* understood clearly that, despite its originality and strangeness, computer science was not an island of its own. Recognizing that certain branches of mathematics and engineering were "related" to computer science they added two other auxiliary divisions of knowledge, calling them *mathematical sciences* and *physical and engineering sciences*, respectively. The former included standard topics such as linear algebra, combinatorics, number theory, statistics and probability, and differential equations. The latter consisted almost entirely of topics in electrical and communication engineering.

But what *Curriculum 68* did *not* state is what *kind* of science computer science was conceived to be. How did it differ, for example, from such sciences as physics and chemistry or from the social sciences such as economics? What (if any) were their commonalities? What was the relationship between the brand new science of computing and the venerable field of mathematics? And what distinguished (if at all) computer science from the traditional engineering sciences?

These are metascientific questions. They pertain to the essential nature of a science (its ontology), its methods of inquiry (methodology), and the nature of the knowledge it elicits (its epistemology). Every paradigm governing a science includes metascientific propositions that collectively articulate the philosophical foundation of the science. If we take *Curriculum 68* as a representation of the computational paradigm at the end of the 1960s, then such a metascientific foundation was noticeably missing.

Yet the founding figures of computer science were very much cognizant of this aspect of their discipline. Largely in defense against skeptics who questioned whether computer science was at all a science, Allen Newell, Alan Perlis, and Herbert Simon, three of the founding faculty of the computer science department in the Carnegie Institute of Technology (later Carnegie-Mellon University), published in 1967 a short article in the journal *Science* titled "What is Computer Science?."[11] Their answer was, quite baldly, that computer science is the study of computers and their associated phenomena. The disbelievers charged that science deals with natural phenomena whereas the computer is an *artifact*, an instrument; and that instruments per se do not have their own sciences (for example, the science underlying the electron microscope is part of physics). In response, Newell and his coauthors maintained that though the computer is an instrument and an artifact, its complexity, richness, and uniqueness set it apart from other instruments, and thus the phenomena surrounding it cannot be adequately parceled out to an existing science.

Herbert Simon was a polymath who had already contributed seminally to administrative decision making, economics, psychology, philosophy of science, and, most pertinently, to the creation of AI, a branch of computer science.[12] Given his particular range of interests, it was almost inevitable that he would want to explore their shared scientific identity and how they differed from or were similar to natural science. The fruits of this exploration was his highly original book *The Sciences of the Artificial* (1969). Here, Simon dwelt on the nature of the artificial world and the things that populate it—artifacts. The artificial world, he pointed out, is to be distinguished from the natural world. The problem that especially intrigued him was how empirical propositions could at all be made about artificial phenomena and subject to corroboration or falsification in the sense that natural scientists construct and test empirical propositions about natural phenomena. In other words, he was arguing for the possibility of the artificial having its own distinct sciences. Thus computers, being artifacts, the empirical phenomena surrounding them belonged to the artificial world. If Simon's argument is pursued then we have an answer to our metascientific question: *computer science is a science of the artificial.*

P.3 THREE CLASSES OF COMPUTATIONAL ARTIFACTS

If we accept the authors and other contributors to *Curriculum 68* as representatives of the *first generation* of computer scientists, then this generation clearly recognized that the richness of the phenomena surrounding the computer lay in that the computer was not a singular, monolithic artifact; nor was it the only artifact of interest to computer scientists. Rather, there was an astonishing diversity of *computational artifacts*—artifacts that participate one way or another in information processing and automatic computation.

For instance, there was the Turing machine, a mathematical construction, embodying the *idea* of computing, invented in 1936 by the English mathematician and logician Alan Turing, at the time of King's College, Cambridge. Turing proposed that any process that "naturally" or "intuitively" we think of as computing can be realized by a version of the Turing machine.[13] This proposition was dubbed generously the "Turing thesis" by American logician Alonzo Church (1903–1995)—"generously" because Church had, about the same time, arrived at a similar proposition using a different kind of argument. And by the end of the 1960s, it was firmly ensconced as a central theoretical element of the computational paradigm. The Turing machine was undoubtedly a computational artifact but not of the kind most people think of when they use the term "artifact."

Of a more practical nature was yet another concept—the model of the stored-program digital computer conceived in the mid-1940s by a tiny group of pioneers and given voice by the Hungarian American mathematical genius John von Neumann; and thus often called (unjustly according to some), the "von Neumann computer model."[14] Here again is an invention of the human mind, thus an artifact.

The Turing machine and the stored-program computer model, both *abstract* entities, are just two of the varieties of computational artifacts that were represented in the computational paradigm as it was at the end of the '6os—and as reflected in *Curriculum 68*. In fact, one can place the varieties of computational artifacts into three broad classes.[15]

Material artifacts: These are governed by the laws of physics and chemistry. They occupy physical space, and their actions consume physical time. Physical computers and their constituent electronic circuits, input and output devices, and communication networks are the prime examples of material computational artifacts (known, more informally, as hardware).

Abstract artifacts: These exist as symbol structures, devoid of any materiality and thus not subject to physico-chemical laws. The Turing machine and the stored-program computer model are prime examples of this class, but there are others, such as programming languages and algorithms.

Liminal artifacts: These are inherently abstract yet their manipulation causes changes in material artifacts such as the transmission of electrical signals across communication paths, changing the physical states of circuits, activating mechanical devices, and so on. They also require a material infrastructure that enables their work to be done. Elsewhere, I have called such artifacts "liminal" because they straddle—are between and betwixt—the purely abstract and the purely material.[16] Computer programs ("software"), microprograms ("firmware"), and computer architectures are principal examples of liminal computational artifacts.

P.4 THE LIFE AND TIMES OF A PARADIGM

As Thomas Kuhn pointed out a scientific paradigm is a dynamic thing.[17] There are "holes" in its constituent bodies of knowledge that have to be plugged. Controversies must be resolved. Problems suggested by the overall paradigm must be solved. Solutions may have to be revised, even dismissed. The paradigm evolves and grows.

An artificial science is driven by the invention of artifacts. New artifacts stimulated by new social, cultural, technological, and economic needs come into being; and older artifacts, no longer of practical relevance, recede into the background or perhaps disappear altogether from scientific consciousness. The new artifacts give rise to new kinds of problems never recognized before. New questions are asked. The new problems engender new ways of thinking. Circumscribing the new artifacts, entirely new bodies of knowledge and philosophies crystallize: They form "subparadigms." Older bodies of knowledge may be discarded or forgotten, but they never disappear altogether, for one never knows how or when they might become relevant again. Most significantly, however, the fundamental structure of the paradigm does not change.

Kuhn controversially and somewhat disparagingly called this evolutionary growth of a paradigm "normal science" because he wanted to distinguish it from radical challenges to an existing paradigm, so radical that they may cause the existing paradigm to be *overthrown* and replaced by a new one. Thus his terms "extraordinary science" and "paradigm shift"; thus his description of a scientific revolution. In the Kuhnian framework, normal science and paradigm shifts were the twin pillars of how a science progresses: the one incremental, constructive and *evolutionary*; the other iconoclastic, destructive or reconstructive and *revolutionary*.

P.5 THE SECOND AGE OF COMPUTER SCIENCE

Yet the history of an actual science is rarely as well structured as philosophers of science like to portray. Computer science may well have attained paradigmatic status by the end of the '60s. And there was, no doubt, Kuhnian-style normal science at work thereafter, adding bricks to the edifice created in the preceding decades, and enriching the paradigm as a whole. But computer science was still a precociously *young* science at the time. The computational paradigm was anything but stable. New, unanticipated forces, some technological, some economic, some social, some intellectual, impinged on it. The outcome was that entirely new kinds of problem domains arose over the next two decades that both significantly altered existing subparadigms and created new ones. In one or two situations, the core element of the computational paradigm—the von Neumann model of computing—would even be challenged.

Indeed, by the beginning of the 1990s, though there was not a paradigm *shift* in the Kuhnian sense—a revolution—the structure of the computational paradigm looked very different in many important respects from how it was at the end of the 1960s. So much so that it seems entirely reasonable to call the two decades from 1970 to 1990 the *second age of computer science*. My aim in this book is to narrate the story of this second age.

So what were the core themes of this new age?

P.6 THE QUEST FOR SCIENTIFICITY IN REAL-WORLD COMPUTING

First, this was a time when many computer scientists became seriously reflexive about their discipline—not so much whether computer science was a genuine science but about the nature of the science itself. Herbert Simon's notion of the sciences of the artificial mentioned in Section P.2 obviously pointed computer scientists in a certain direction. But outside the AI community—Simon being a seminal figure in AI—it is not clear whether *The Sciences of the Artificial* had much of an *immediate* impact on other computer scientists.

The intellectual origins of computer science lay in logic and mathematics. As the 1960s ended the presence of these venerable disciplines was still very much visible in the shape of two branches of computer science. One was numerical analysis, concerned with the approximate numerical solutions of continuous mathematical problems—which was what began the quest for automatic computation in the first place and that particularly flourished in its first age of computer science.[18] The other was automata theory concerned with the study of highly abstract, mathematical models of computation.

For some computer scientists, automata theory *was* the "science" in "computer science." This suggested, indeed assumed, that computer science was a *mathematical* science. Research papers, monographs, and textbooks devoted to automata theory were full of axioms, definitions, theorems, corollaries, and proofs in the traditions of mathematics and mathematical logic. The problem was that not all computer scientists were convinced that automata theory had any direct influence on, or relevance in, the design, implementation, and understanding of *real* computational artifacts. As Maurice Wilkes, one of the pioneers of automatic computing, put it in an informal talk delivered in 1997, one did not need to know automata theory to build physical computers, algorithms, or software. This can be contrasted, he remarked, to the relationship between Maxwell's equations in electromagnetic theory and radio engineering. If one wanted to design a radio antenna, Wilkes said, one had better know something about Maxwell's equations. Automata theory had no such linkages with much of the rest of computer science (including numerical analysis).[19]

Moreover, there were substantial parts of the computational paradigm that at the end of the '60s did not seem "scientific" at all. For example, a substantial part of computer architecture, the branch of computer science concerned with the functional structure and design of computers, was essentially a heuristic, craft-like discipline having little theoretical content at all.

So also in the realm of programming: Even though there was recognition that this entailed a whole new kind of artifact (liminal in the sense I have used here), deserving a name of its own—"software," coined in the mid-1960s—there was a perceptible sense of inadequacy about the way programs and software were being designed and implemented. In 1968, the term "software engineering" was invented to indicate that software had evolved to a level of complexity that demanded the same kind of intellectual respect that was accorded to the artifacts civil, mechanical, and other contemporary branches of engineering dealt with. Yet it was far from clear what shape this respect should take.

Programming languages—abstract artifacts—had reached a more scientifically respectable state by the beginning of the second age. In particular, the '60s decade was rich in its advancement of a theory of the syntax of programming languages, concerned with the rules governing grammatically correct forms of programs expressed in programming languages. But *semantic* considerations—what the elements of a programming language meant and how to characterize their meanings, how language implementers could unambiguously and correctly translate programs in a given language into machine-executable code, how programmers could objectively interpret the meanings of language elements in the course of writing programs—were a different matter. Theories of the semantics of programming had indeed emerged in the late '60s, but very little of these theories had been adopted by the designers and implementers of "real-world" languages and industrial-grade software. Instead, these designers and implementers resorted to natural language and informality in defining the semantics of their programming languages.

Collectively, these concerns became audible as sounds of discontent and dissent. They represented a desire for a *science of real-world computing* that would play the same role in the design, implementation, and understanding of real, utilitarian computational artifacts as, say, the engineering sciences of strength of materials and fluid mechanics play, respectively, in structural and aeronautical engineering.[20]

As we will see, this quest for *scientificity*—and how this quest elicited responses—constituted one of the most significant markers of the second age of computer science.

P.7 COMPUTING AS A HUMAN EXPERIENCE

This was also a time when some computer scientists developed a heightened consciousness about computing *as a human experience*. This too did not arise de novo. The makers of the first age of computer science had clearly recognized that although automatic computation with minimal human intervention was their fundamental mission, humans were not thereby absolved of all responsibilities. Having built physical computers they could not, in the manner of Pontius Pilate, wash their collective hands and look the other way. After all, at the very least humans had to communicate their intentions to machines. The very first book on programming, authored by Maurice Wilkes, David Wheeler, and Stanley Gill of Cambridge University, was called, significantly, *Preparation of Programmes for an Electronic Digital Computer* (1951). (In Britain, the English spelling "programme" still prevailed to signify this artifact rather than the American spelling that would eventually be generally adopted.)[21] Computer programs, reflecting the users' computational intentions, had to be *prepared* by human beings. The invention and design of the first programming languages, reaching back to German engineer Konrad Zuse's *Plankalkül* in 1945[22] through David Wheeler's invention of an assembly language for the Cambridge EDSAC in 1949[23] to the development by the American John Backus and his collaborators in IBM of FORTRAN[24] in 1957 were clear manifestations of the computer pioneers' awareness of incorporating humans into the computing experience. The remarkable transatlantic collaboration leading to the design of the Algol programming language between 1958 and 1963[25] offered further evidence of this awareness.

But with the emergence and development of the computational paradigm in the late '50s and especially through the '60s, two quite different and contrasting manifestations of the human–computer connection emerged. In both cases it was not the incorporation of humans into the computing milieu but rather the opposite: incorporation of computing into the human experience.

P.8 PROGRAMMING AS AN INTELLECTUAL ACTIVITY

Along one front, in the late 1960s, the Dutch mathematician-turned-"humble programmer" (as he would style himself[26]) Edsger Dijkstra articulated a vision of programming as, fundamentally, a human activity.[27] Dijkstra was among the very first advocates of a new *mentality*: treating programming not only, or not necessarily, as a means for communicating tasks to computers but as an *intellectual activity in its own right* regardless of the presence or absence of physical computers.

This mentality can be viewed as a provocative *challenge* to one of the key onto-logical assumptions underscoring the prevailing computational paradigm in the first age: that it should concern itself with computing as a *mechanical* information process—an assumption that, of course, was rooted in Turing's vision of computing. We will see that although this new human-centered view of programming was rooted in the 1960s it engendered an intellectual *movement* in the next two decades called "structured programming." Human-centeredness formed the second marker of a new age of computer science.

P.9 TOWARD A UNIFIED SCIENCE OF INTELLIGENCE

Neither Dijkstra nor the other makers of the structured programming movement extended their idea of programming as a human activity into the realm of human *cognition* itself. Envisioning programming as an intellectual activity did not imply that the mind itself was a computational object—a *natural* computational object, not an artifact. The idea of the mind as a machine has its own incredibly long history reaching back to antiquity.[28] The modern notion of the mind as a computational ma-chine, however, originated in the 1940s, coincident with the time that digital com-puting was "in the air." For instance, influenced by *analog* computing machines such as the differential analyzer (originally built by American engineer Vannevar Bush in 1931),[29] the British psychologist Kenneth Craik suggested in 1943 that thinking was a kind of computational process.[30] This daring hypothesis was given much more concrete form in 1958 by Allen Newell, Cliff Shaw, and Herbert Simon. In a paper published in an influential psychology journal they suggested just how the newly emergent computational paradigm offered a promising, indeed powerful, model of human problem solving.[31] They backed this model by a working computer program called Logic Theorist, implemented in 1956, which could prove theorems in mathe-matical logic.[32] In effect, this paper was seminal in the emergence in the 1960s of AI as a subparadigm of the computational paradigm that would connect computers with human cognition; and it would persuade psychologists, linguists, anthropologists, and philosophers to join with AI researchers to create an interdisciplinary science of mind they called *cognitive science*.

AI's agenda in this regard was not *exactly* that of cognitive science: The latter's al-legiance was to human (and to a lesser extent, other animal) cognition; computation was a means to that end. AI's agenda, at its most ambitious, was to construct a uni-fied science of intelligence that would embrace *both* artifacts and humans. And, in the 1970s and 1980s, this project evolved in a number of surprising directions.

One was the consolidation of the idea that symbols constituted the fundamental "stuff" of thought in both machines and mind. So a unified science of intelligence would be centered on *symbol processing*.

Along a second path lay what we may call *computational epistemology*. Epistemology—the study of the nature of knowledge—was of course one of the great pillars of the philosophy of mind, having origins reaching back to both Western and Eastern antiquity. But what AI ushered in were new concepts about knowledge, in particular, how knowledge is represented in cognitive systems, both natural and artificial, and how knowledge could be processed in the course of thinking. This new computational epistemology was another major marker of the second age of computer science.

But even as symbolism was being consolidated, even as a symbol-based epistemology was being vigorously explored, the very idea of symbols as the stuff of intelligence and cognition would be challenged. Stimulated by the heady excitement of parallel processing (see the next section) and by the findings of neurobiology, neurologically inspired models of cognition would begin to seriously challenge the symbolist subparadigm governing AI. This was yet another core signifier of the second age of computer science.

P.10   LIBERATING COMPUTING FROM THE SHACKLES OF SEQUENTIALITY

The evolution of electronic computing from its very inception in the early 1940s had always been shaped by the *technological imperative*: the force of the technology with which material computational artifacts—memories, processors, peripheral devices, communication paths—were built. But as the '60s gave way to the '70s this imperative assumed a particularly dramatic form: the phenomenal progress in integrated-circuit (IC) technology resulting in the fabrication of IC chips (the units of production) of sharply increasing density of components and (thus) computing power, and the concomitant decrease in unit cost of circuit components. By the early 1970s the term "large-scale integration" was part of the vocabulary of hardware designers; it had also infiltrated the consciousness of computer scientists. By the end of the decade the new operative term on everyone's lips was "very large-scale integration."

The impact of this rapid evolution on IC technology—some would claim it was a revolution—was manifold. For computer scientists it made practical what had been more a promise in the 1960s: liberating computing from the shackles of sequentiality. Thus the second age of computer science was marked by the exuberant reality of *parallel computing*—inside the computer, between computers, within algorithms, within programs, in programming languages. This prospect had certainly emerged in the first age, but it was in the second age that computer scientists seriously came to *think* parallelism.

P.11   SUBVERTING THE VON NEUMANN PRINCIPLE

A very different phenomenon arose in the '70s and '80s: an act of serious rebellion, in fact. There were some people who wished to subvert the very core of the prevailing computational paradigm—the so-called von Neumann principle. They wished to challenge the von Neumann-style *architecture of computing* as had been adopted since the birth of modern computing. These rebels wanted to be revolutionaries: They desired to bring about a veritable *paradigm shift* in computer science. A part of this desire originated in the annoyance with certain "bottlenecks" inherent in von Neumann architecture, in part in the impatience with sequential computing, in part in the excitement engendered by the architecture of the brain, in part in the perceived imperfections of conventional programming languages, and in part in the advent of very large-scale integration technology.

This subversive mentality and its consequences formed the fifth marker of a distinctive second age of computer science.

P.12 CAVEATS FOR THE READER

So the principal themes that characterize what I am calling the *second age of computer science*, circa 1970–1990 are (a) the quest for scientificity in real-world computing; (b) computing as a human experience; (c) the search for a unified theory of intelligence encompassing both machine and mind; (d) liberation from the shackles of sequentiality; and (e) subverting the von Neumann core of the computational paradigm. These themes appear and reappear in various guises in the seven chapters that follow.

But some caveats are in order.

One. This book is intended to continue the story of computer science beyond where my earlier book *It Began with Babbage* (2014) ended. That book had chronicled the genesis of computer science and had ended circa 1969 by which time (as noted earlier) an academic discipline of computer science and the central features of the computational paradigm had been established. *Curriculum 68* was a manifesto of what I call here the first age of computer science.

The present story covers roughly the '70s and '80s, the two decades forming what I am calling the second age of computer science. But such words as "circa" and "roughly" are important here, especially concerning the beginning of this second age. As the reader will see, this beginning often overlaps the waning years of the first age. Concepts and ideas from the latter appear like the painter's pentimento in this story. Likewise, ending the story at about 1990 does not imply that this age ended abruptly. Rather, circa 1990 was when a pair of remarkable new computational artifacts made their appearances, ones that would have vast social and commercial consequences but also give new shape to computer science itself. These artifacts were the Internet (whose own history began in the 1960s, in fact), an American invention, and the "World Wide Web" invented by the Englishman Tim Berners-Lee. The third age of computer science would begin. Thus it is more appropriate to say that circa 1990 was a marker of the *waning* of the second age.

Two. The chapter sequence in this book is not a narrative that maps onto the sequential passage of historical time. Rather, the seven chapters are seven overlapping "spaces," like adjacent geographic regions, *within* each of which creative events occur in roughly sequential time, but *between* which events are or may be concurrent. The history I have narrated here is thus really seven histories unfolding concurrently over the same historical time period. We may well call this an instance of *parallel history*. But they are not mutually independent. Like all complex and interesting cases of parallel processing, there are interactions, linkages, and cross-connections among these parallel histories. So the reader will find frequent cross references between chapters.

P.13 WHAT KIND OF HISTORY IS THIS?

Historians tend to identify themselves and their writings (*historiography*) with different *genres* of history. So what genre does this book fall under?

As the reader will see, this is a history of creativity, imagination, the origin, and evolution of ideas and knowledge in the realm of computer science; it is also a history of how computer scientists perceived, identified, invented, and/or discovered computational themes and problems, and how they reasoned and critically responded to such problems. Thus this book is primarily a blend of *intellectual history* and *cognitive history*. The former is of older vintage, the latter is of very recent origins.

Intellectual history itself has a history. Previously (under the name of *history of ideas*) it was concerned with tracing how ideas and concepts evolve, grow, and spawn other ideas over historical time.[33] More recently intellectual historians have turned their attention to the study and interpretation of historical texts and the contexts and (especially) the language in which they are produced.[34] It so happens that both the "old" and the "new" versions of intellectual history have places in this book, though more notably as history of ideas.

Cognitive history constructs the history of some creative realms (such as science, art, technology, or religion) by delving into the cognitive attributes of the creative individuals involved. Cognitive-historical explanations of acts of creativity are grounded in such cognitive issues as the goals, needs, or desires wherein ideas originate, the creative being's system of beliefs, knowledge, modes and style of reasoning and critical thinking.[35]

To repeat, the present work lies at the juncture of intellectual history and cognitive history.

But, like any other human endeavor, computer science has its *social* and *cultural* dimensions. For instance, the fruits of a science must be *communicated* or disseminated. Journals are created, societies are formed, conferences are held. Networks of relations are formed. Computer science is no exception to this. Most of the major computing societies—in the United States, Canada, the United Kingdom, Australia, Germany, and India—were established in the first age of computer science including, most notably, IFIP, a kind of United Nations of computing. The second age saw the expansion of this communicative–disseminative aspect of computer science in the form of new societies, journals, and conferences, and the formation of "special-interest groups," "technical committees," and "working groups."

Another sociocultural aspect pertains to gender. In *It Began with Babbage* I noted that the genesis of computer science was almost exclusively the province of men. There were a very small number of women who contributed importantly to the early development of electronic computing, most famously Grace Murray Hopper and Jean Sammet, but the others have sadly and unjustifiably remained largely unknown until very recent times.[36] In the second age women still remained very much a minority. Yet a small but highly visible coterie made their presence felt through the pages of computer science journals, in universities and college academic departments, and at conferences.

Yet another sociocultural feature of the first age was that the people who created computer science were, without exception, from the West.[37] By the end of the second age this demographic would change markedly, some might even say dramatically. Publications would bear names of very definitely non-Western character. The computer science community became increasingly and noticeably multiethnic and multicultural.

Finally, as we will see, computer science as a scientific discipline is practiced both in academia and in the corporate world. And as is well known there are distinct cultural differences between these two societies. Thus the following interesting questions arise: How was the development of computer science *distributed* across these two cultures? Did the two cultures impose their imprint on the nature of their respective activities in computer science? And if so, in what manner?

Some of these sociocultural elements of the second age will appear in this narrative—but only implicitly, insofar as they relate to the intellectual and cognitive story being told. The *sociocultural history* of computer science addressing explicitly such issues as gender, race, ethnicity, and the "two cultures" needs to be written. But this is not that book.

NOTES

1. Or some variant or other-language equivalent thereof. For example, "computing science" or "computational science" in English-speaking lands, *informatik* or *informatique* in European countries.

2. For an account of this history see S. Dasgupta, 2014. *It Began with Babbage: The Genesis of Computer Science*. New York: Oxford University Press.

3. See N. Roe (ed.), 2005. *Romanticism*. Oxford: Oxford University Press.

4. T. S. Kuhn, [1962] 2012. *The Structure of Scientific Revolutions* (4th ed.). Chicago: University of Chicago Press.

5. ACM Curriculum Committee, 1968. "Curriculum 68," *Communications of the ACM*, 11, 3, pp. 151–197.

6. ACM Curriculum Committee, 1968, p. 156.

7. Ibid.

8. Ibid., p. 155.

9. Ibid.

10. P. S. Rosenbloom, 2010. *On Computing: The Fourth Great Scientific Domain*. Cambridge, MA: MIT Press.

11. A. Newell, A. J. Perlis, and H. A. Simon, 1967. "What is Computer Science?," *Science*, 157, pp. 1373–1374.

12. For a detailed examination of Simon's polymathic mind and work see S. Dasgupta, 2003. "Multidisciplinary Creativity: The Case of Herbert A. Simon," *Cognitive Science*, 27, pp. 683–707.

13. A. M. Turing, 1936. "On Computable Numbers with an Application to the *Entscheidungsproblem*," *Proceedings of the London Mathematical Society*, 2, pp. 230–236.

14. The history of the origins of the von Neumann model is complex and somewhat controversial. See Dasgupta, 2014, pp. 108–133.

15. Ibid., p. 4.

16. Ibid.

17. Kuhn, [1962] 2012.

18. Dasgupta, 2014, pp. 10 *et seq*, p. 279.

19. Cited by P. J. Bentley, OUPblog, Oxford University Press, June 18, 2012: www.http//blog.oup.com/2012/06/maurice-wilk. Retrieved February 12, 2014. Wilkes had, in fact, expressed this view in conversations with the present author.

20. S. P. Timoshenko, [1953] 1983. *History of Strength of Materials*. New York: Dover; W. G. Vincenti, 1990. *What Engineers Know and How They Know It*. Baltimore, MD: Johns Hopkins University Press.

21. M. V. Wilkes, D. J. Wheeler, and S. Gill, 1951. *Preparation of Programmes for an Electronic Digital Computer*. Cambridge, MA: Addison-Wesley.

22. F. L. Bauer and H. Wössner, 1972. "The 'Plankalkül' of Konrad Zuse: A Forerunner of Today's Programming Languages," *Communications of the ACM*, 15, pp. 678–685.

23. D. J. Wheeler, 1949. "Automatic Computing with the EDSAC," PhD dissertation, University of Cambridge.

24. J. W. Backus, R. W. Beeber, S. Best, R. Goldberg, L. M. Halbit, H. C. Herrick, R. A. Nelson, D. Sayre, P. B. Sheridan, H. Stern, I. Ziller, R. A. Hughes, and R. Nutt, 1957. "The FORTRAN Automatic Coding System," *Proceedings of the Western Joint Computer Conference*, Los Angeles, CA, pp. 188–197.

25. P. Naur (ed.), et al. (1962–1963). "Revised Report on the Algorithmic Language ALGOL 60," *Numerische Mathematik*, 4, pp. 420–453.

26. E. W. Dijkstra, 1972. "The Humble Programmer," *Communications of the ACM*, 15, 10, pp. 859–866.

27. E. W. Dijkstra, 1965. "Programming Considered as a Human Activity," pp. 213–217 in *Proceedings of the 1965 IFIP Congress*. Amsterdam: North-Holland.

28. M. A. Boden, 2006. *Mind as Machine* (two volumes). Oxford: Clarendon.

29. Dasgupta, 2014, p. 92.

30. K. J. W. Craik, [1943] 1967. *The Nature of Explanation*. Cambridge: Cambridge University Press.

31. A. Newell, C. J. Shaw, and H. A. Simon, 1958. "Elements of a Theory Human Problem Solving," *Psychological Review*, 65, pp. 151–166. See also Dasgupta, 2014, p. 233.

32. A. Newell and H. A. Simon, 1956. "The Logic Theory Machine: A Complex Information Processing System," *IRE Transaction on Informaton Theory*, IT-2, pp. 61–79. See also Dasgupta, 2014, pp. 229–232.

33. For a classic explanation of what the history of ideas is, see A. O. Lovejoy, [1936] 1964. *The Great Chain of Being*. Cambridge, MA: Harvard University Press. For a more recent and out-standing instance of the *practice* of history of ideas see, e.g., R. Nisbet, [1980] 1984. *History of the Idea of Progress*. New Brunswick, NJ: Transaction Publishers.

34. See, e.g., D. La Capra, 1983. *Rethinking Intellectual History: Texts, Contexts, Language*. Ithaca, NY: Cornell University Press.

35. See N. Nersessian, 1995. "Opening the Black Box: Cognitive Science and the History of Science," *Osiris* (2nd series), 10, pp. 194–211, for a discussion of the cognitive history of science. For a broader explanation in the context of creativity in general, see S. Dasgupta, 2016. "From *The Sciences of the Artificial* to Cognitive History," pp. 60–70 in R. Frantz and L. Marsh (eds.), *Minds, Models and Milieux: Commemorating the Centennial of the Birth of Herbert Simon*. Basingstoke,UK: Palgrave Macmillan. For discussions of cognitive history in the context of ancient religious minds, see L. H. Martin and J. Sørensen (ed.), 2011. *Past Minds: Studies in Cognitive Historiography*. London: Equinox. As an instance of the *practice* of cognitive history in creativity studies see S. Dasgupta, 2003. "Multidisciplinary Creativity: The Case of Herbert A. Simon," *Cognitive Science*, 27, pp. 683–707.

36. See, e.g., the memoir by one of the "ENIAC women," women who programmed the ENIAC machine in the mid-1940s, Jean Jennings Bartik, for an absorbing account of her experience in computing in its earliest years. J. J. Bartik, 2013. *Pioneer Programmer* (J. T. Richman and K. D. Todd, eds.). Kirkville, MO: Truman State University Press.

37. See Dasgupta, 2014, "Dramatis Personae," pp. 287–294.

THE SECOND AGE OF COMPUTER SCIENCE

1

ALGOL GENES

1.1 DISSENT IN ALGOL COUNTRY

In 1969 a "Report on the Algorithmic Language ALGOL 68" was published in the journal *Numerische Mathematik*.[1] The authors of the report were also its designers, all academic computer scientists, Adriaan van Wijngaarden and C. H. A. Koster from the Netherlands and Barry Mailloux and John Peck from Canada.

The Algol 68 project was, by then, 4 years old. The International Federation for Information Processing (IFIP) had under its umbrella a number of technical committees devoted to various specialties; each technical committee in turn had, under its jurisdiction, several working groups given to subspecialties. One such committee was the technical committee TC2, on programming; and in 1965 one of its constituent working groups WG2.1 (programming languages) mandated the development of a new international language as a successor to Algol 60. The latter, developed by an international committee of computer scientists between 1958 and 1963, had had considerable theoretical and practical impact in the first age of computer science.[2]

The Dutch mathematician-turned-computer scientist Adriaan van Wijngaarden, one of the codesigners of Algol 60 was entrusted with heading this task. The goal for Algol 68 was that it was to be a successor of Algol 60 and that it would have to be accepted and approved by IFIP as the "official" international programming language.[3]

Prior to its publication in 1969, the language went through a thorough process of review, first within the ranks of WG2.1, then by its umbrella body TC2, and finally by the IFIP General Assembly before being officially recommended for publication.[4]

The words review and recommendation mask the fact that the Algol 68 project manifested some of the features of the legislative process with its attendant politics. Thus, at a meeting of WG2.1 in Munich in December 1968—described by one of the Algol 68 codesigners John Peck as "dramatic"[5]—where the Algol 68 report was to be approved by the working group, the designers presented their language proposal much as a lawmaker presents a bill to a legislative body; and just as the latter debates over the bill, oftentimes acrimoniously, before putting the bill to a vote, so also the Algol 68 proposal was debated over by members of WG2.1 and was finally voted on.[6]

And as in the legislative chamber here too were heard voices of vigorous and powerful dissent. The dissidents dissented because of what they saw was the creation of a language of unnecessary *complexity*, a measure of this being the size of the Algol 68 report—in print it consumed almost 240 pages of dense description in the journal *Numerische Mathematik*. In his 1980 Turing Award lecture, C. A. R. Hoare of the University of Oxford, one of the dissenting members of WG2.1, recounted his experience as one of a group of four entrusted with the "scrutiny" of the language being drafted. Each successive draft document became "longer and thicker" than the last, "full of errors" corrected hastily, and as obscure as the last.[7]

Programming, Hoare would write, is a complex enterprise: It entails the resolution of multiple, interacting, often conflicting, goals, and the programming language as the programmer's fundamental tool should not exacerbate the programmer's task; it should be a means of solution rather than a part of the problem.[8]

And so the dissidents—there were 8 of them out of a total of about 30, including C. A. R. Hoare and Edsger Dijkstra of the Technological University of Eindhoven— chose to write a "minority report."[9] When, in spring 1969, the technical committee TC2 met to discuss the Algol 68 Report it was accompanied by this minority report.[10]

The Algol 68 story does not end here, for even after approval by TC2 and then its sanctification by the IFIP General Assembly, even after the publication of the Report in 1969, the language would further evolve. The design team actually expanded to eight, with additional members from Belgium, the Netherlands, and the United Kingdom.[11] Finally, as had happened with its ancestor Algol 60, a "Revised Report" was published in 1975 in the journal *Acta Informatica*[12] and in book form a year later.[13]

Dissent, discontent, and dissatisfaction are, as is well known, prime movers of creativity.[14] The reason why the Algol 68 project was conceived was that the experience of designing, implementing, and using Algol 60 had yielded a desire to improve on it. Algol 68 was a direct descendant of Algol 60, and its creation reflected the ambition of its four initial inventors, spearheaded by philosopher-in-chief Adriaan van Wijngaarden of the Mathematics Centrum, Amsterdam (but with the advice and consent of the majority of the 30-strong working group WG2.1) to create a highly sophisticated and worthy successor to Algol 60.

And as in natural evolution, artificial evolution retains recognizable ancestral features but is tempered with striking differences. A reader of the Algol 68 Report or of an Algol 68 program and familiar with Algol 60 will have little difficulty in recognizing the genetical imprint of the parent on the child. At the same time the reader will be struck by a startling sense of unfamiliarity with many features of the descendant.

But to comprehend more clearly how and why a successor to a celebrity language that had only been fully defined and "officially" lauded in 1962–1963, was deemed necessary only 2 years later, we need to delve briefly into the Algol 60 story and sense the flavor of this language.

The development of Algol 60 must count as one of the seminal events in the first age of computer science. It was, in fact, a multinational event. Designed by a team of 13 mathematical computer scientists—from Germany, the United States, the United Kingdom, France, Denmark, and the Netherlands—Algol 60 was conceived, designed, and described as a universal machine-independent language. That is, unlike its predecessor FORTRAN, the language paid no homage nor owed allegiance to any particular commercial computer or family of computers. Rather, it belonged to a world of abstract symbolic objects of the kind mathematicians, logicians, and linguists were familiar with. This deliberate distancing from the domain of physical computers ("hardware") was taken to be a virtue of the language. Any Algol 60 program, for example [P21], is a sentence in the language containing syntactic entities such as declarations and statements. Thus to the reader, an Algol 60 program was an abstract artifact. To relate such a text to a real physical machine—to make it into a (liminal) executable program—was the task of the compiler writer.

```
[P21]      xyz: begin
                real x;
                procedure abracadabra;
                    begin
                        x := 0;
                        goto out
                    end;
           ttt:   begin
                    integer y;

                    . . . . . . .
                    abcracadabra
                  end
        out: end
```

The development of Algol 60 was an *experiment*. Not an experiment in the natural scientist's laboratory-centered sense but rather that of the creative artist or novelist. The latter, for example, conceives a new mode of storytelling and constructs a novel that expresses this conception. Readers, as well as the writer, then judge the efficacy of the novel. The whole enterprise, from composing the novel to its critical assessment, comprises the experiment. From its outcome, the novelist may decide to embark on another story that avoids the errors of the first effort and possibly refining his or her mode of storytelling.

So also, the 13 men (and they *were* all men!) embodied their shared vision and broad goals for an ideal programming language in the design of Algol 60 and exposed it to others' critical gazes in the form of reviews, implementations, and use in the real-world environment. The primary goal, in the words of numerical analyst Heinz Rutihauser of the Swiss Federal Institute of Technology (ETH), Zurich, one of the codesigners and author of an influential expository text on the language, was the creation of "*one universal, machine-independent algorithmic language* to be used by all."[15] But there were more specific subgoals. To paraphrase Rutihauser, the language should adhere closely to mathematical symbolism, must facilitate descriptions of numerical processes in publications, and must be "readily" and automatically transformable into machine code.[16]

One marker of any significantly original invention is that the invented artifact will embody *new knowledge* and that this knowledge would diffuse and become absorbed into the relevant paradigm. It would then be used by other inventors, designers, and engineers.[17] Such new knowledge is the artifact's legacy to the world even if or when the artifact itself has been discarded or becomes obsolete.

Such was the case with Algol 60. In the evocative imagery of Alan Perlis of Carnegie-Mellon University, one of the American members of the Algol 60 design team, reflecting on the language in 1981, there is an "Algol gene" in every language that came after; "It is the mother's milk of us all."[18]

Algol 60's success as an experiment in language design lay in a number of new *concepts* that would enter the consciousness of other language designers. Perhaps the most notable of these concepts were the following.

(a) The use of a *metalanguage* to describe precisely, hierarchically, and formally the syntax of a programming language. The particular metalanguage invented for Algol 60 was called the Backus–Naur Form (BNF), named after its American inventor, John Backus of IBM and the Dane Peter Naur, then of Regenecentralen, the editor of the Algol 60 Report who was instrumental in its successful use in defining the syntax of the language.

(b) The concept of *block structure* (exemplified in program [P21] Section 1.2) as a means for hierarchically organizing the structure of programs and for localizing the *scope* (that is, accessibility) of various entities, such as variable declarations.

(c) The description of *procedures* (the Algol term for subroutines) and *functions* (which in the tradition of mathematics produce just a single value as a result) in a unified manner.

(d) The notion of *type* to denote classes of values upon which operations could be performed.

(e) Important advances in *compiler theory* and *technology* produced in the course of implementing Algol 60. They included the development of efficient parsing algorithms ignore for checking the syntactical correctness of Algol programs and the use of a data structure called the *stack* for managing elegantly and efficiently the dynamic processing of blocks and procedures.

I can give in this general history only a whiff of the flavor of these concepts. Consider first the syntax of an identifier—used to identify or name variables or procedures in a program. A BNF description for a (syntactically) legal Algol 60 identifier is given by the following syntactic rules (also called *productions*):

<identifier> ::= <letter> | <identifier><letter> | <identifier><digit>,
<letter> ::= $a|b|c|$ ... $|y|z$,
<digit> ::= $0|1|2|$ ... $|9$.

The entities in angular brackets are called nonterminal symbols and refer to syntactic categories in the language. The symbols <, >, ::=, and | belong to the metalanguage. The symbols in italics are called terminal symbols, and every legal identifier in Algol 60 is composed only of terminal symbols. The first rule states that the nonterminal symbol identifier is either a "letter" or an identifier followed by a letter or an identifier followed by a "digit." The second rule states that the nonterminal symbol letter is the terminal symbol a or b or ... z. The third rule specifies that the terminal symbol digit is one of the terminal symbols 0, 1, ... 9.

These rules define all and only the possible identifiers in an Algol 60 program. For example, the symbols *abc*, *aaa112235*, and *abracadabra* are all syntactically correct identifiers because each can be parsed according to the three rules. On the other hand, *1ab* is not a legitimate identifier in the language because an identifier must begin with a letter.

The BNF metalanguage not only enabled the syntax (also called grammar) of a programming language to be described, it also *defined* that grammar. Thus a nexus was established between the design of programming languages and the science of linguistics (which is fundamentally concerned with *natural* languages), in particular, the language *typology* constructed by American linguist Noam Chomsky in the late 1950s.[19]

As for block structure, organizing a program in the form of blocks was a particularly elegant innovation of the Algol 60 designers. In the program fragment [P21] shown previously in this section, the **real** variable x is declared in the outer block labeled *xyz* (defined by the outermost **begin** and **end**). The scope of this variable—its "visibility" or accessibility—is the entire program fragment; that is, it can be referenced both by the enclosed **procedure** *abracadabra* and the inner block labeled *ttt*. On the other hand, the scope of the **integer** variable y is confined to the inner block *ttt*.

Finally, a function in Algol 60 "returns" a single value as a result just as a mathematical function F when applied to an argument x [symbolically denoted as $F(x)$] returns a value. A function is described in the same fashion as a **procedure** except that a function is associated with a type signifying the value returned. Thus the declaration

---

[P22]	**integer procedure** *gcd* (*m, n*);
	value *m, n*;
	integer *m, n*;
	begin
	. . . . .
	End

---

specifies that the value returned by *gcd* is of type **integer** (and so [P22] is a function declaration). The assignment statement

$$z := gcd \ (a, b)$$

will compute the function *gcd* according to [P22] and assign this integer value into the variable *z*.

1.3   THE CONSEQUENCES OF BOUNDED RATIONALITY

The polymath scientist Herbert Simon was awarded the Nobel Prize in economics in 1978 for his theory of bounded rationality: This argues that because of our cognitive and computational limitations and our imperfect and incomplete knowledge about the world we can rarely make perfectly rational decisions.[20] We are constrained by bounds on our rationality. We try to circumvent this condition by deploying various heuristic strategies. Thus scientists test the correctness of their theories by empirical observations of, or controlled experiments on, predictions of their theories. Engineers test the validity of their designs by simulating their intended artifacts or building prototypes or implementing the artifacts and observing their real-world behavior.

Like everyone else, the Algol 60 designers were boundedly rational beings. Their visions led to the innovations mentioned (and many others of a more technically abstruse nature). But implementations of the language by way of fully operational compilers and using the language to construct real-world programs inevitably revealed the pitfalls of boundedly rational design decisions. Flaws were revealed.

For example, there were unsettled issues in the design of certain language features. There were ambiguities and inconsistencies in the language definition. And there were features in the language that proved to be difficult to implement.

Because of these complications, few compilers (if any) actually implemented the full language as presented in the official IFIP-approved defining document titled "Revised Report on the Algorithmic Language ALGOL 60" (1962–1963).[21] Rather, different implementers imposed their own restrictions on the language, creating distinct *dialects* of Algol 60.[22]

Another limitation was that, despite its designers' overarching goal, Algol 60 (developed, we recall, by people who came from a mathematical background) was heavily oriented toward numerical computation. But by the end of the 1960s nonnumerical computation had emerged as a major domain of computing. For example, one of the most intensely studied problems was that of sorting data files; another was the problem of searching data files for specific items. The vast literature on just these two nonnumerical computational domains became evident when Donald Knuth's encyclopedic volume on sorting and searching was published in 1973.[23] Finally, systems programming, most specifically compiler writing and operating systems development, were fundamentally nonnumerical in nature.

All these factors established the context for the proposal to design Algol 68 as a successor to Algol 60.

1.4   SCIENCE, COMPLEXITY, AND THE SEARCH FOR UNIVERSALS

As we have noted (in Section 1.1) Algol 68 had its detractors even while the language was a work-in-progress. It was accused of being unnecessarily large. The original Algol 68 defining document published in 1969[24] was found by some to be incomprehensible, thus unreadable; a damning indictment for any text, especially a text describing a language. Even the "Revised Report" published in 1975 after considerable review and change to the 1969 "Report"—and the addition of four new members to the design team[25]—did little to mollify the critics. In 1969 the defining document was 239 pages in length; the revised document of 1975 was 5 pages shorter. Indeed, the "Revised Report" was long enough to be published as a respectably sized book in 1976.[26]

The sizes of the two reports were, of course, symptoms of something else: the *complexity* of the language. But what were the reasons for this alleged complexity? For this we must look to the arch aspiration of its designers: their search for a universal programming language.

Universality is, of course, the grandest aspiration of the natural scientist. The great laws, principles, and theories of physics, chemistry, biology, and cosmology achieved renown because of their universal scope. Mathematicians also seek to discover universal principles, either in the form of a small set of axioms and definitions that govern an entire branch of mathematics or as grand, sweeping theorems. Of course, in general, universals in science and mathematics are universal to a greater or lesser extent. Pythagoras' theorem is universal across the set of all right-angle triangles; Kurt Gödel's incompleteness theorems sweep over the whole of arithmetic. Some universals are more universal than others.

But one thing seems certain: Universals in science and mathematics aspire to bring unity to diversity, and thus reduce complexity and enhance our ability to understand their domains.

The situation in the sciences of the artificial—the scientific principles governing artifacts, especially utilitarian artifacts (see Prologue)—should, one would expect, be similar. Computer science is a science of the artificial, so one would expect that computer scientists also aspire to discover or invent universal principles. The overarching goal of the Algol 60 designers was to create a universal programming language for numerical computation. And when the IFIP Working Group WG2.1 mandated a successor to Algol 60 they aspired for a language that would be more universal in scope than Algol 60.

Unfortunately, the notion of universality in the context of utilitarian artifacts—and programming languages, though abstract, are intended to be utilitarian—often implies something quite opposite to the notion in natural science and mathematics. Universal claims for artifacts imply that they are all things to all users. Complexity enters the discourse.

Here lies the source of Algol 68's complexity.

1.5   ALGOL 68 AND THE QUEST FOR UNIVERSALS

One of the most prominent embodiments of the quest for universalism in Algol 68 was its expression orientation. In contrast, Algol 60 was a statement-oriented language: Statements are complete syntactic units that specify a piece of computation that causes changes in the state of the computational process by altering the values of variables or in the flow of control through the computation. Expressions are pieces of computation that produce new values but otherwise does not cause a state change. For example, in the program fragment

---

[P23]   **begin**

. . . . .

$y := (a + b)/2;$

again: **if** $x < y$ **then**   **begin**

$x := x + 1;$

$z := z - 1;$

goto *again*

end

. . . . .

end

---

we see arithmetic expressions $(a + b)/2$, $x + 1$, and $z - 1$, each of which computes a numeric value (of type **integer** or **real** as the case may be) that is assigned to variables y, x, and z, respectively, by way of assignment statements. In addition, there is an **if . . . then** statement whose condition part, $x < y$, is a **Boolean**-valued expression.

And there is the **goto** statement that causes an unconditional change in the flow of control. The two blocks are compound statements. Thus, in Algol 60, statements and expressions are clearly distinct.

The Algol 68 term for expression is formula. And in this language almost every syntactic entity, including statement-like entities, can yield a single value and is thus an expression (or formula). For example, the Algol 68 conditional statement may be used as an expression as in

$$z := \text{if } a > b \text{ then } x \text{ else } y \text{ fi.}$$

The conditional statement (notice the closing bracket **fi**) will return a value (of the variable x or of y depending on whether the expression $a > b$ evaluates to *true* or *false*) that is then assigned to z.

Still more surprising, this same conditional statement may appear on the *left-hand* side of the assignment operator :=:

$$\text{if } a > b \text{ then } x \text{ else } y \text{ fi } := z,$$

in which case its value is the variable identifier x or y to which the value of z will be assigned.

Still more counterintuitive, a sequential statement (called a serial clause in Algol 68) can be an expression. Such a statement might appear as in the following:

[P24]   if $c := (x-y)/2; c > 0$
                                then . . . . .
                                    else . . . . .
                    fi.

Here, the value of the sequential statement in the conditional part is the value of the last expression that will be of type **bool**. So either the **then** part or the **else** part is selected if the value returned is *true* or *false*, respectively.

This expression orientedness of Algol 68—making statements into expressions—is one instance of Algol 68's quest for universals, its designers' aspiration for principles of language design that were akin to scientific laws. But there are others, for example, the facilities provided to construct an infinite number of what in Algol 68 were called modes.

Mode in Algol 68 corresponded approximately to, but extended, the type concept in Algol 60. The term type referred to a class of values on which operations could be performed. Algol 60 recognized just three types, **real** (the class of real number values), **integer** (the class of integers), and **Boolean** (the logical values *true* and *false*). Simple variables were then declared by specifying their types, for example,

real x; **integer** y; **Boolean** z.

Variables that were arrays of values of one of these types could be defined. Thus

$$\text{real array } a\,[\,1:n\,]$$

declares a row of n elements $a[1], \ldots, a[n]$, each of which is of type **real**. The **array** is the only data structure definable in Algol 60, but there is no such thing as an **array** *type*.

The designers of Algol 68 eschewed the term type. They aspired for something grand, something universal, a language that afforded an infinite number of modes. Thus Algol 68 included a small set of "primitive" modes (including **real**, **int**, **bool**, and **char**—this latter, the class of character symbols) out of which as many new modes could be constructed as the programmer desired. The programmer could declare new modes and name them and declare variables as instances of such named modes. Thus:

$$\text{mode matrix} = [1{:}n, 1{:}n] \text{ real};$$
$$\text{matrix } a; \;\; \text{matrix } b$$

would create variables a and b of mode **matrix**.

In addition to the primitive modes, the language provided the programmer with a number of "built-in" modes that could also be used to create other new modes. They included, in particular, **struct** and **union**. For example, the declarations

$$\text{mode complex} = \text{struct } (\text{real } rl, im);$$
$$\text{complex } x$$

would introduce the mode called **complex** as a structure consisting of two **real** components having the identifiers rl and im. The variable x would be of this mode. Its two fields could be individually selected and their values assigned to other variables, as, for example,

$$w := rl \textbf{ of } x;$$
$$z := im \textbf{ of } x.$$

The declaration

$$\text{mode realorint} = \text{union } (\text{real}, \text{int})$$

would create a mode **realorint** as the **union** of modes **real** and **int**. The declaration

$$\text{realorint } a$$

would specify a variable a of this mode that could hold values that were either **real** or **int**.

1.6   ALGOL 68 AND ITS EPISTEMIC COMPLEXITY: THE PRICE OF UNIVERSALISM

As we have noted (in Section 1.1), Algol 68 evoked serious dissent within the IFIP WG2.1 community. The dissidents objected vigorously to the complexity of the language. One marker of this complexity was the length of the original "Report" of 1969. But the document length was merely a symptom of much deeper issues.

The reader may have noticed that two key terms in Algol 60 were replaced in Algol 68: Instead of expression, a common-enough term in mathematical discussions, there was formula, a term used in the natural sciences, especially in chemistry; instead of type there was mode. In fact, as C. H. Lindsey of the University of Manchester (one of the design team's later members) and S. G. van der Meulin, authors of the first textbook on Algol 68 (somewhat ruefully) admitted, an astonishingly large number of new technical terms were needed to be learned in order for someone to master the language.[27] It was as if, though the language was deemed an immediate descendant of Algol 60, Adriaan van Wijngaarden (who was the only person who served on both design teams) and his colleagues sought to distance themselves as far as possible from the Algol 60 milieu. One way this seemed to be effected was the invention not only of new concepts and their names as part of the language itself (e.g., **mode**, **struct**, **union**) but an entirely new *metavocabulary* to describe and explain the language constructs.

New linguistic concepts and new metalinguistic concepts represent new knowledge. The complexity of Algol 68 not only lay in its large number of components that interacted in nontrivial ways[28]—which we may call systemic complexity—it also lay in the amount and variety of new knowledge embedded in the concepts associated with the language. Elsewhere, I have called the knowledge embedded in an artifact as constituting its *epistemic complexity*.[29] Of course, as we noted, the production of new knowledge in the design or invention of a new artifact is a marker of the creativity of the artificer. But such epistemic complexity can also be a hindrance to comprehensibility.

To get a sense of the epistemic complexity of Algol 68, consider as an example the simplest piece of text in the language, a declaration:

real *x*.

The meaning of this declaration—its semantics—is given operationally by its *elaboration*. This is one of the new technical terms that is part of the metavocabulary, and it refers to a process whereby an external object (another technical metavocabulary term to mean some part of a program text) is inspected and thereby causes certain actions to occur.[30]

The preceding declaration, then, is elaborated as follows.

(a) A location for a **real** value is reserved somewhere in memory (of a computer or of a human being).

(b) This **real** value is referred to by a name, say x^*, which is its address.

(c) The identifier x is a representation of x^* in the program text. Identifier x is said to possess the name (address) x^*. ("To possess" is yet another new technical term meaning "an external object is possessing an internal object"; the latter, another element of the metavocabulary, is an instance of a value stored and manipulated inside the computer.[31])

(d) The relevant variable in Algol 68 is not the name x^* but the value referred to by x^*.

(e) The name x^* (being an address) is also a value, so it too must have a mode. The mode of x^* is called **ref real** ("reference to **real**").

To summarize, in elaborating the declaration **real** x, the identifier is made to possess a **ref real** (the name x^* of a **real**) that refers to a **real** value located somewhere in some computer (or human) memory.[32] The reader may also notice that the word value (in the singular) is used in a special way in Algol 68. Perhaps the best way to understand its connotation is to visualize a value as a box associated with a particular **mode** (e.g., **real**) that can hold *instances* of that **mode** (e.g., the **real** number 3.1415).

1.7   IMPLEMENTING ALGOL 68: A FIELD EXPERIMENT IN COMPUTER SCIENCE

Even before Algol 68 was approved by the IFIP, implementation studies had begun. In 1967, one of the designers, Barry Maillioux of the University of Alberta had completed a PhD dissertation titled "On the Implementation of ALGOL 68" at the Mathematics Centrum, University of Amsterdam, under Adriaan van Wijngaarden's tutelage.[33] Another member of the design team, C. H. A. Koster, from the University of Amsterdam, had examined the problem of parsing Algol 68 programs.[34]

Implementation of computational artifacts serves two purposes. In the case of programming languages, for instance, from the practitioner's perspective, it is what makes the language into a usable tool. From the scientist's or theorist's perspective, implementation is the beginning of a field experiment to observe, confirm or falsify, and evaluate the theoretical visions of the language designers.

And so an IFIP-sponsored working conference on Algol 68 implementation was held in Munich in July 1970, primarily to "stimulate the production of ALGOL 68 compilers and thus to put the new language to the test of use," and "to reveal the current status of implementation efforts in many countries and to provide for the interplay of ideas."[35]

The papers presented at this conference—there were individuals or groups from Belgium, Canada, Czechoslovakia (as it was then), France, West Germany (as it was then), Holland, Norway, the United Kingdom, the USSR (as it then was), and the United States—were concerned with particular aspects of compiling. Undoubtedly, the "stars" of the conference were a British team from the Royal Radar Establishment (RRE) who described a complete implementation of a variant of the language they named Algol 68-R. This group stood apart from the other participants for two reasons: One, they were not academic researchers; and two, they had a working compiler already in productive use. The implementers of Algol 68-R, Ian Currie, Susan Bond, and John Morison, all members of RRE's Mathematics Division, along with the superintendent of the Division, Philip Woodward, were scientific civil servants; they were computing practitioners responsible for providing computing support in a government research establishment for the whole organization. And their goal was severely practical: that Algol 68-R would be RRE's main language, to be used primarily for scientific programming but also for such business data processing needs as payroll and inventory control. Thus all software support by the organization's computing service would be "directed towards its use."[36]

For many years the computational *lingua franca* in RRE had been Algol 60.[37] Algol 68-R was to replace Algol 60 as the organization's new *lingua franca* for all its computing activities—scientific and business. The goal of the Algol 68-R project was to produce a robust, "industrial-quality" programming tool.

From the perspective of computer science, however, Algol 68-R offered another kind of significance. As the RRE team realized upon arriving at the conference, they had produced the world's first Algol 68 compiler. Thus Algol 68-R was seminal in empirically demonstrating the feasibility of Algol 68 as a programming language robust enough for real-world use. It was just one experiment, no doubt, but it was an experiment involving stringent demands imposed by a practical computing milieu. From a scientific perspective the Algol 68 effort constituted a stringent experiment. Moreover, as a replacement of Algol 60 it proved to be successful. Its implementers stated that, although comparisons of the efficiency of object code was always a tricky issue, the compiler compared "favourably" with Algol 60 compilers, at least for programs that used features common to both Algol 60 and Algol 68.[38]

Compilability is one face of implementation as a field experiment. The other is usability and user acceptance. Philip Woodward reported that he had "little difficulty" in teaching Algol 68 to potential users. Moreover, he claimed, users became "keener and keener" on the new language once they started using it for real applications.[39]

It is not difficult to imagine the impact of Algol 68-R as an empirical validation of the principles of the language on the latter's designers, especially as prior to the Munich conference they had been apparently unaware that such an implementation already existed.[40] And like any stringent experiment the Algol 68-R experience revealed "various imperfections" of the language, as did other implementations that followed. These became the basis of revising Algol 68 so as to "make the language neater for the user and easier to implement."[41] The ontogeny of Algol 68 would continue for several years. As we noted before, the design team expanded to eight members and eventually the "Revised Report" was published, first in a journal and then as a book-length monograph.[42]

1.8   NICKLAUS WIRTH'S SEARCH FOR SIMPLICITY

At the time that members of IFIP WG 2.1 were debating over a successor to Algol 60, a proposal was put forth by Nicklaus Wirth, then at Stanford University, and Hoare, at the time at Queens University, Belfast. It was referred to simply as Algol X. It was to be an extension of Algol 60 rather than a radically new descendant.[43] But, as we have seen, supporters of the new language strategy won the day, resulting, eventually, in Algol 68.

Wirth in fact had resigned from WG2.1 and went his own way to design and implement a language he called Algol W, based on his and Hoare's original proposal. Algol W was thus one of the Algol dialects that emerged in the mid–late 1960s. Wirth implemented the language on an IBM System/360 computer, and this language was subsequently used in several universities in the early 1970s to teach programming.[44] To write a compiler for Algol W, Wirth also invented and implemented another Algol dialect he named PL360, which was an Algol-like, high-level *assembly language* for the IBM System/360 computers.[45]

To understand the tenor of Wirth's thinking, we may want to relate it to a paper Edsger Dijkstra had presented in 1965, in which he had drawn the reader's attention to programming as a human experience.[46] We can link this notion with Herbert Simon's concept of bounded rationality (this chapter, Section 1.3). Given the limits to our ability to make correct decisions, given that programming is a human activity, an obvious way to fuse these concerns is to try and *avoid* complexity if one can: to resort to *simplicity*.

This might seem counterhistorical to some, as the evolution of technological artifacts is often linked with, both an *increase* in complexity and a progressively attendant dehumanization.[47] Certainly in the realm of computer languages, a technological (though abstract) class of artifacts, if one examines their short evolution from about 1949 (the first assembly language invented by David Wheeler of Cambridge University) to 1969 (Algol 68 and PL/1, IBM's putative successor to FORTRAN) the growth in complexity was there for all to see.

This trend was precisely what the dissidents within IFIP WG 2.1 were protesting against.

In his autobiographical Turing Award lecture of 1984, Nicklaus Wirth recalled his dissatisfaction even with Algol W soon after it had been implemented.[48] Algol W, he had decided, was not only unsuitable for the all-important domain of *systems programming* (that is, programming compilers, operating systems, input–output routines, and other software artifacts that manage computing resources and facilitate applications programming) but also, the size of the Algol W compiler had grown beyond one's cognitive comfort zone. It was these considerations that led Wirth, now ensconced in ETH, Zurich, to embark on the design of a language that, he wrote later, would include only those features he deemed essential and exclude all those features whose implementation might prove cumbersome.[49] His was a search for simplicity.

This was not, of course, unique to Wirth. We find its presence in Dijkstra, as we have seen. As we see in Section 1.9 and later, its pursuit became a major theme in *design methodology* in the 1970s. But what Wirth did was to translate his search for simplicity into a concrete form: the programming language *Pascal*.

Pascal, named in honor of the French *savant* Blaise Pascal (1623–1662)—in 1966 Wirth had published with Helmut Weber a paper describing another dialect of Algol they called EULER, after the Swiss mathematician Leonhard Euler (1707–1783)[50]—was formally announced in 1971 in the very first issue of the journal *Acta Informatica*.[51] The same year the design of a Pascal compiler (written in Pascal!) for the Control Data Corporation (CDC) 6000 computer was published.[52] Also in 1971 Wirth published a paper on developing programs using a technique he called "stepwise refinement," in which he used Pascal notation to represent repetitive computation, but without mentioning Pascal at all.[53] However, its most influential introduction to the programming public, at least to teachers and students of programming, was his slight but elegant book *Systematic Programming: An Introduction* (1973).[54] Herein too, Wirth was coy about mentioning Pascal. In the preface to his book, he mentioned only that the notation used in the book was based on Algol 60.[55] In a later chapter he quietly elaborated on this: The notation used, he wrote, was "a close approximation to Algol 60," though "not identical" to any previous language.[56] A footnote here cited his *Acta Informatica* paper. Only at the very end of the book, in an appendix, do we find an extremely concise but systematic description of Pascal's syntax.[57]

The language was formally and fully presented in a book-length monograph in 1974 (second edition in 1975), coauthored by Wirth and his student Kathleen Jensen.[58]

The originality of Pascal lay in part in its realization of Wirth's search for simplicity; but it was, ultimately, a very obvious descendant of Algol 60. The larger claim of its originality lay in its relationship and contribution to *programming methodology*.

1.9 THE ADVENT OF PROGRAMMING METHODOLOGY

The word methodology is very often, incorrectly, used as a rather grandiose-sounding synonym for method. But methodology really refers to a *body* of methods characteristic of some activity or discipline; or, more abstractly, to the *study* or *theory* of methods.

In 1969, IFIP and its technical committee on programming, TC2, approved the formation of a new working group to deal with the "problem of construction and control or reliable programs." The title given to this working group, suggested by Brian Randell of the University of Newcastle-upon-Tyne, was "Programming Methodology."[59]

We have already noted that in 1965 Edsger Dijkstra had ruminated publicly on programming as a human experience. Like Herbert Simon—though probably unaware of the latter's ideas on bounded rationality (which Simon had mainly addressed to economists and decision theorists)—Dijkstra was concerned with how the human mind could overcome its cognitive limitations in making sense of complex problem situations. In Dijkstra's case this meant the realm of program design and construction.

Historians of computing now recognize that Dijkstra had, in 1965, planted the first seeds of what became *structured programming*, a movement that was a fundamental component of an emerging programming methodology. Dijkstra himself coined this term in 1969 as the title of a "working document" (his words) read at a NATO-sponsored conference.[60]

The evolution of structured programming in the first half of the 1970s and the dispersion of its ideas into other branches of computer science are discussed in the next chapter. Our concern here is how programming methodology became inextricably entwined with the problem of *language design* and of Pascal's place in this entwinement.

This is *not* to claim that language designers had been previously or even contemporaneously indifferent to the issue of program design. Clearly, such seminal texts as Heinz Rutihauser's *Description of ALGOL 60* (1967)[61] and Lindsey and van der Meulin's *Informal Introduction to ALGOL 68* (1971)[62] in presenting their respective languages paid attention to how their respective language's features should be used to program problem solutions. But the advent of programming methodology as a discipline brought the issue of how one *thinks* about the development of computer programs and how such thinking is shaped by and in turn shapes the nature of programming languages to the very forefront of the computer scientist's consciousness. Programming methodology was concerned with the *cognitive process* entailed in program design. And of all the programming languages that had emerged circa 1970, none played a more prominent role in promoting the cause of program methodology than Pascal.

1.10   THE MATTER OF LINGUISTIC RELATIVITY

But then, the question arises: Should programming languages *dictate* the programmer's thought processes? Or should one's thinking about program development have primacy over language issues?

The relationship between *natural* languages and the nature or structure of human thought processes was famously discussed by linguist (and chemical engineer) Benjamin Lee Whorf and his mentor, linguist Edward Sapir of Yale University, in the form of what is called the Sapir–Whorf hypothesis, also called the linguistic relativity principle.[63] The central idea of this principle is that the form and nature of one's language determines, or at least shapes, the structure of our thought processes and our perceptions.

In the domain of programming, the idealist would probably desire that programming languages should be subservient to our thought processes. The realist would more likely insist that some version of linguistic relativity prevails; that the way we think about program development and construction is at the very least *influenced* by our programming language of choice.

1.11   PASCAL, PROGRAMMING METHODOLOGY, AND LINGUISTIC RELATIVITY

In his 1971 paper "Program Development by Stepwise Refinement"[64] Wirth offered a view that was only mildly linguistically relative. He rejected the view that one learns programming by first learning a language, then applying one's "intuition" to "transform ideas into finished programs" by somehow mapping the ideas into language constructs.[65] Rather, he asserted that one should first learn how to design, gradually develop, and construct programs.[66]

In other words, programming methodology should be the focus of the teacher's teaching and the student's learning. Language should hover in the background. Of course, it will have *some* influence. And so the language Wirth used to teach programming methodology in this paper was a "slightly augmented ALGOL 60."[67] That "augmentation" happened to be a new repetitive statement:

$$\textbf{repeat } S \textbf{ until } B,$$

where S is a statement sequence that is repeatedly executed until the Boolean expression B evaluates "true." In fact, this **repeat** statement was one of Pascal's constructs though Pascal was never mentioned in this paper.

Wirth elaborated his philosophy of programming methodology (and by implication, his version of a weak linguistic relativism) in his book *Systematic Programming*.[68] This, as it happened, was the first programming textbook that used Pascal as its *lingua franca*. He is introducing here, he tells the reader, "a notation that is not identical to any of the previous languages"—by which he meant the "big four" of the time, FORTRAN, COBOL, Algol 60, and PL/1. The new notation would "mirror the most fundamental concepts of programming in a natural, perspicacious and concise form." It could be thought as an extension of Algol 60.[69]

Almost apologetically he cited, in a footnote, his 1971 paper on Pascal.[70] There was no further mention of the language in the main text. It is only when readers reached the appendix, describing the syntax of Pascal, that they would comprehend fully that along the way of learning some principles of programming methodology they had also learned a new programming language.

1.12 THINKING MATHEMATICALLY ABOUT PROGRAMS

Pascal emerged from the recondite pages of academic print into the hurly-burly of the real world in 1974 with the publication of a *User Manual and Report*, coauthored by Kathleen Jensen and Nicklaus Wirth.[71] Pascal's syntax was specified in BNF and also in a graphical notation, called syntax graph or syntax diagram, but the semantics was in English.

But, in fact, the language's definition transcended the scope and limitations of such natural language-based semantics and in this a truly original link was established between language design and programming methodology.

To understand this, recall that the IFIP mandate for WG 2.3 was that it would be concerned with the construction and management of *reliable* programs.

For the mathematically inclined person the way to this is obvious: One *proves* that the program is correct. Proofs are the ultimate means of acquiring trust in the correctness of any proposition. But proving a proposition demands that an elaborate and appropriate framework must first be constructed.

Mathematics is, of course, all about proving things. But those "things" are propositions about invented things. The objects of mathematical thought and the properties we discover about these objects belong to a *formal* universe, an invention of the human mind. A mathematical world is formal, not empirical. And the way mathematicians create such a world is by inventing objects (e.g., integers or geometric shapes), defining concepts and properties pertaining to these objects (e.g., defining an acute angle as the angle between two lines meeting at a point as less than 90°), axioms that state principles, relations, or rules about the objects that are taken to be true without further justification (e.g., "the immediate successor to a number is a number"), and certain rules of inference that allow us to deduce a new proposition from others. Such a system is an *axiomatic system*. Given such a system, the mathematician applies the rules of inference in a systematic fashion to build up a chain of reasoning—a proof—that demonstrates some hitherto undiscovered or unproven property about the fundamental objects of interest. The latter is, of course, called a theorem. Once a theorem is proved it then enters and enriches the axiomatic system and can then be used to prove other theorems.

And here lay the power and the glory of the mathematical method. Given an axiomatic system (and providing there are no flaws in the chain of reasoning—after all, constructing proofs is a human cognitive process subject to bounded rationality), once mathematicians agree that a proof is without flaws, the theorem is guaranteed to be true relative to the axiomatic system. The theorem is true "forever." Mathematical truths are immortal in a way empirical truths (that is, "facts" about the empirical world) are not.

Thus, returning to the mandate for IFIP WG 2.3, if we want to construct a reliable program, the program must be situated in a formal world; an axiomatic system must be constructed that would allow us to prove the correctness or reliability of the program.

This is where C. A. R. Hoare enters the conversation.

1.13 HOARE'S MANIFESTO

C. A. R. Hoare was one of the WG 2.1 dissidents and a signatory of the minority report that accompanied the Algol 68 proposal (see Section 1.2). As we have noted, his ideas about language design developed jointly with Wirth[72] found their way into Wirth's Algol W.

Hoare must count as a weak linguistic relativist: In his Turing Award lecture of 1980 he tells us how his construction in 1962 of "Quicksort," an efficient sorting algorithm,[73] was influenced by his discovery of Algol 60's facility for implementing recursive processes (e.g., functions that execute by calling themselves).[74]

Interestingly, Hoare's formal training was in neither mathematics nor science nor engineering. His undergraduate degree from Oxford University (in 1956) was in their famed *Literae Humanioris* ("Greats")—classics and philosophy. Clearly, his background in classics did not impede his developing a mathematical mentality in the course of a professional life, first as an industrial programmer and then as an academic computer scientist.

In 1969 (he was then at Queen's University Belfast) Hoare proposed "An Axiomatic Basis for Computer Programming."[75] The opening paragraph began with the uncompromising assertion that programming was "an exact science," by which he meant that the program text contained all that was necessary to understand the program's properties and deduce the consequences of its execution.[76]

This was written in 1969. That it became an invariant in his computational thinking—a kind of mantra—is evident from his "inaugural" lecture as professor of computing at Oxford (actually delivered in 1986, nine years after his appointment to the chair) titled "The Mathematics of Programming" in which he laid out his manifesto on programming:[77]

> *Computers are mathematical machines.* That is, their behavior can be mathematically defined and every detail is logically deducible from the definition.
>
> *Programs are mathematical expressions.* They describe precisely and in detail the behavior of the computer on which they are executed.
>
> *A programming language is a mathematical theory.* That is, it is a formal system that helps the programmer in both developing a program and proving that the program satisfies the specification of its requirements.
>
> *Programming is a mathematical activity*, the practice of which requires application of the traditional methods of mathematical understanding and proof.

1.14 HOARE'S LOGIC MEETS WIRTH'S LANGUAGE

So what kind of axioms and rules of inference might be appropriate in programming? And how should such an axiomatic system become connected to a programming language? Hoare was not the first to probe these questions. In 1967, Robert Floyd, then of Carnegie Institute of Technology (later Carnegie-Mellon University) published a paper on a formal basis for programming.[78] In his 1969 paper Hoare acknowledged his debt to Floyd, and, sometimes, the logic Hoare presented in 1969 is called "Floyd–Hoare logic"—though more commonly it is known as Hoare logic. The most obvious difference between the two approaches was that Floyd dealt with programs expressed in flowchart form whereas Hoare addressed programs as texts expressed in Algol-like notation.

Concerning axioms, consider the primitive data type integer. Any elementary college textbook on algebra will present certain axioms that apply to integers. For instance, given integer variables x, y, z we have the following axioms:

$$x + y = y + x,$$
$$(x + y) + z = x + (y + z).$$

The first axiom stipulates that addition of two integers is "commutative"; the second axiom stipulates the "associative" nature of addition. So in the case of a program involving integer data these axioms will be assumed to hold.

The simplest complete action in a statement-oriented language (such as Algol 60 or Pascal) is the assignment statement:

$$x := e,$$

where x is a variable identifier and e is an expression. Hoare postulated an *axiom of assignment* that characterized the behavior of assignment statements. As for composite statements such as the sequential or conditionals, their behavior would be specified by rules of inference. Proving a program's correctness, then, would entail establishing a condition that was true before the program begins to execute—called its precondition—and the final or goal condition that is expected to prevail after the program terminates—called its postcondition—and then deploying the axioms and rules of inference to deduce that if the precondition **P** is true then the execution of the program will lead to the postcondition **Q**. Denoting a program by Π, the (meta)notation for this situation would be expressed as

$$\{P\}\ \Pi\ \{Q\}.$$

Later, this notational scheme came to be known as a Hoare formula, and it is the formal equivalent of a theorem in mathematics. Just as we prove a theorem so also, given a Hoare formula for a particular program, we must prove that this formula is correct. In which case the program will be deemed correct.

From Floyd to Hoare to Wirth: This then seemed to be the chain of influence. In 1973, the same year Wirth's *Systematic Programming* appeared, Hoare and Wirth published an axiomatic system defining the semantics of Pascal.[79] Thus did Hoare's logic meet Wirth's language. Pascal became the first programming language for which an axiomatic system was constructed to define its semantics. In addition to the axiom of assignment, axioms were presented for the Pascal data types and its predefined operations. Rules of inference (also called proof rules) for the various Pascal statements were defined.

To give the reader a glimpse of the nature of these axioms and rules of inference, we need to introduce, first a *metanotation*:

(1) Truth-valued propositions or assertions are denoted by boldfaced uppercase letters such as **P, Q, R**, etc.

(2) The symbol $P[x/e]$, where x is a variable identifier and e is an expression, stands for the proposition **P** but with all occurrences of x in **P** replaced by e. (For example, if **P** is $x = 10$ and e is the expression $y + z$ then **P** $[x/y + z]$ is $y + z = 10$.

(3) $\{P\}\ \sigma\ \{Q\}$ is a Hoare formula in which **P** is the precondition and **Q** is the postcondition of the statement σ. This is read as "if **P** is true before σ begins then **Q** is true after σ terminates."

(4) **F1, F2, . . . ,Fn** => **F**, where the F*i*'s and **F** are Hoare formulas, is a rule of inference that states that if it can be shown that **F1, F2, . . . ,Fn** are all true then it follows that **F** is true. **F1, F2, . . . ,Fn** are called the antecedents of the rule of inference and **F** is its *consequence*.

Here, then, is a tiny selection of the axiomatic semantics of Pascal as laid out by Hoare and Wirth.

Axiom of assignment:
$$\{P\ [x/e]\}\ x := e\ \{P\}.$$
Rule of sequential composition:
$$\{P\}\ \sigma_1\ \{Q\}, \{Q\}\ \sigma_2\ \{R\}\ =>\ \{P\}\ \sigma_1; \sigma_2\ \{R\}.$$
*Rule of the **if then else** statement*:
$$\{P\ \&\ \beta\ \}\ \sigma_1\ \{Q\ \}, \{P\ \&\ not\ \beta\ \}\ \sigma_2\ \{Q\}\ =>\ \{P\}\ \text{if}\ \beta\ \text{then}\ \sigma_1\ \text{else}\ \sigma_2\ \{Q\ \}.$$
*Rule of the **while** statement*:
$$\{P\ \&\ \beta\ \}\ \sigma\ \{P\ \}\ =>\ \{P\ \}\ \text{while}\ \beta\ \text{do}\ \sigma\ \{\ P\ \&\ not\ \beta\}.$$

How these rules are actually used is explicated in Chapter 5. Hoare and Wirth did not define the semantics of the complete Pascal language. A notable exception was the **goto** statement. In 1979, a rule of inference for the latter was offered by Suad Alagic and Michael Arbib of the University of Southern California.[80]

1.15   TAKING THE LOW ROAD FROM ALGOL 60

Pascal was regarded by some as a pedagogic tool, to teach programming. Wirth himself used it to teach programming methodology as his book *Systematic Programming: An Introduction* tells us. His later book, *Algorithms + Data Structures = Programs* (1976), was a more advanced elaboration of this same objective.[81] But, of course, Wirth's goal in creating Pascal was also to design and implement an abstract, machine-independent, high-level language (HLL) suited for systems programming. And the two most ubiquitous classes of systems programs in the 1970s were compilers and operating systems.

The complexity of a compiler rests fundamentally on the complexity of the programming language being implemented, the complexity of the architecture of the machine for which the language is to be implemented, and the distance—often referred to as the semantic gap[82]—between the language and the machine's architecture. Thus, ever since the advent of high-level programming languages in the late 1950s, various strategies have been explored to solve the problem of narrowing this semantic gap, some rooted in the realm of languages, some in the realm of computer architecture. And the situation was not helped by the skeptical attitude harbored by many systems programmers, accustomed to using assembly languages toward the use of HLLs for writing operating systems and compilers.

There was another related problem that exercised language designers and implementers: This was the desire for *portability*; how to reduce the effort of transporting a language implemented on one computer to another.

Martin Richards of Cambridge University addressed all three problems—the semantic gap, portability, and keeping the programming ear close to the hardware ground—when he designed and implemented the Basic Combined Programming Language (BCPL) between 1967 and 1969. BCPL was intended, quite specifically, to be a "low-level" systems programming language for writing compilers.[83] Thus, for example, it was a *typeless* language; more precisely, it recognized only one data type, the *bit string*.

BCPL belonged to an altogether different evolutionary linguistic pathway originating in Algol culture, which we may think of as the "low (-level) road": Its immediate ancestor was a language called Combined Programming Language (CPL), designed in the early 1960s by a combined group (thus, perhaps the name) from Cambridge and London Universities, led by Christopher Strachey (Cambridge) and David Barron (London);[84] its immediate descendant was a language called B (like BCPL a typeless language), developed at Bell Laboratories by Ken Thompson and Dennis Ritchie circa 1969, which in turn was the progenitor of C, enormously successful as a practical and widely used language, designed and implemented by Dennis Ritchie between 1969 and 1973. C was intended originally as a low-level systems programming language for writing operating systems.[85] But such was the evolutionary divergence despite Algol 60's influence on CPL that C bears virtually no resemblance to even Pascal.

1.16   VIRTUAL MACHINES, REAL LANGUAGES

But what interests us most about BCPL was how it resolved the problems of portability and semantic gap. The solution was the creation of an idealized *virtual machine* whose architecture was oriented to the programming language itself. The semantic gap between the language and this virtual machine's language (called O-code) was thus quite narrow, making it much easier to translate a BCPL program into O-code. This O-code virtual machine had nothing to do with the machine (or assembly) language for any real computer, and so the translation of a BCPL program into O-code was a one-time affair. The task of translating (either by interpreting or compiling) an O-code program into the machine language of the target computer was thus the only machine-specific task and would vary from one machine to another. The language and its virtual machine could then be *ported* easily from one computer to another.[86]

And despite the considerable difference between BCPL and Pascal, this same virtual machine idea was adopted in implementing Pascal. In the latter case the virtual machine language was called P-code, and Wirth's student at ETH Zurich, Urs Ammann, defined the P-code machine and wrote a compiler to produce P-code.[87]

NOTES

1. A. van Wijngaarden, B. J. Mailloux, J. E. L. Peck, and C. H. A. Koster, 1969. "Report on the Algorithmic Language ALGOL 68," *Numerische Mathematik*, 14, pp. 79–218.

2. The story of Algol 60 is briefly told in S. Dasgupta, 2014. *It Began with Babbage: The Genesis of Computer Science*. New York: Oxford University Press, pp. 215–216.

3. J. E. L. Peck, 1978. "The Algol 68 Story," *ACS Bulletin*, November, pp. 4–6: www.pdfebooks8.org/a/algol-68-pdf-S113. Retrieved Feb. 2, 2015.

4. J. E. L. Peck, 1970. "Preface," pp. v–vi in J. E. L. Peck (ed.), *ALGOL 68 Implementation*. Amsterdam: North-Holland.

5. Peck, 1978.

6. Ibid.

7. C. A. R. Hoare, 1981. "The Emperor's Old Clothes," *Communications of the ACM*, 24, 2, pp. 75–83.

8. Ibid.

9. The full dissident list was E. W. Dijkstra (The Netherlands), F. G. Duncan (UK), J. V. Garside (UK), C. A. R. Hoare (UK), B. Randall (UK), G. Seegmüller (West Germany, at the time), W. M. Turski (Poland), and M. Woodger (UK). Only Woodger among them was a coauthor of the Algol 60 Revised Report.

10. Peck, 1978.

11. The new members were Michel Sintzoff (Belgium), L. G. L. T. Meertens (The Netherlands) and Charles Lindsay nad Richard Frisker (UK).

12. A. van Wijngaarden, B. J. Mailloux, J. E. L. Peck, C. H. A. Koster, M. Sintsoff, C. H. Lindsey, L. G. L. T. Meertens, and R. G. Frisker, 1975. "Revised Report on the Algorithmic Language ALGOL 68," *Acta Informatica*, 5, pp. 1–234.

13. A. van Wijngaarden et al., 1976. *Revised Report on the Algorithmic Language ALGOL 68*. Berlin: Springer-Verlag.

14. S. Dasgupta, 1996. *Technology and Creativity*. New York: Oxford University Press.

15. H. Rutihauser, 1967. *Description of ALGOL 60*. Berlin: Springer-Verlag: p. 5.

16. Ibid., p. 6.

17. For an insightful discussion of the generation and diffusion of technological knowledge, see N. Rothenberg and W. G. Vincenti, 1978. *The Britannia Bridge: The Generation and Diffusion of Technological Knowledge*. Cambridge, MA: MIT Press.

18. A. J. Perlis, 1981. "Transcripts and Presentation," pp. 139–147 in R. L. Wexelblat (ed.), *History of Programming Languages*. New York: Academic Press, p. 147.

19. N. Chomsky, 1957. *Syntactic Structures*. The Hague: Mouton.

20. Simon has written about bounded rationality in many articles and books. For a concise account, see H. A. Simon, 1987. "Bounded Rationality," pp. 266–268 in J. Eatwell, M. Milgate, and P. Newman (eds.), *The New Palgrave: A Dictionary of Economics, Volume 1*. London: Macmillan.

21. P. Naur et al. (eds.), 1962–1963. "Revised Report on the Algorithmic Language ALGOL 60," *Numerische Mathematik*, 4, pp. 420–453.

22. This strategy prompted IFIP Working Group WG2.1 to issue, in 1964, an "official" set of restrictions on the language and naming the resulting version the *IFIP Subset Algol 60*. See Rutihauser, 1967, pp. 8–9, 303–308.

23. D. E. Knuth, 1973. *The Art of Computer Programming. Volume 3. Sorting and Searching*. Reading, MA: Addison-Wesley.

24. van Wijnggarden et al., 1969.

25. van Wijngaarden et al., 1975.

26. van Wijngaarden et al., 1976.

27. C. H. Lindsey and S. G. van der Meulin, 1971. *Informal Introduction to ALGOL 68*. Amsterdam: North-Holland, Appendix 3, Glossary.

28. H. A. Simon, 1962. "The Architecture of Complexity," *Proceedings of the American Philosophical Society*, 6, pp. 468–482. Reprinted in H. A. Simon, 1996. *The Sciences of the Artificial*. Cambridge, MA: MIT Press.

29. S. Dasgupta, 1997. "Technology and Complexity," *Philosophica*, 59, 1, pp. 113–140.

30. Lindsey and van der Meulin, 1971, pp. 347, 353.

31. Ibid., p. 345.

32. Ibid., pp. 31–32.

33. B. J. Maillioux, 1967. "On the Implementation of ALGOL 68," PhD dissertation, University of Amsterdam.

34. C. H. A. Koster, 1969. "Syntax Directed Parsing of ALGOL 68 Programs," *Proceedings of the Informal Conference on ALGOL 68 Implementation*. University of British Columbia, Vancouver, British Columbia.

35. J. E. L. Peck 1970. "Preface," pp. v–vi in J. E. L. Peck (ed.), *ALGOL 68 Implementation*. Amsterdam: North-Holland, p. v

36. I. F. Currie, S. G. Bond, and J. D. Morison, 1970. "ALGOL 68-R," pp. 21–34 in Peck 1970, p. 21.

37. P. M. Woodward, 1970. "Introduction," p. 28 in Currie, Bond, and Morison, 1970.

38. Currie, Bond, and Morison, 1970, p. 23.

39. Woodward, 1970.

40. Peck, 1978.

41. Ibid.

42. Van Wijngaarden et al., 1975.

43. N. Wirth and C. A. R. Hoare, 1966. "A Contribution to the Development of ALGOL," *Communications of the ACM*, 9, pp. 413–432.

44. My own first encounter with Algol culture was, in fact, via Algol W, as a graduate student in the University of Alberta, circa 1972.

45. N. Wirth, 1968. "PL360, A Programming Language for the 360 Computers," *Journal of the ACM*, 15, 1, pp. 34–74.

46. E. W. Dijkstra, 1965. "Programming Considered as a Human Activity," pp. 213–217 in *Proceedings of the 1965 IFIP Congress*. Amsterdam: North-Holland.

47. The classic text on these aspects of the history of technology is L. Mumford, 1967. *Technics and Civilization*. New York: Harcourt Brace Jovanovich.

48. N. Wirth, [1984] 1987. "From Programming Language Design to Computer Construction," pp. 179–190 in Anon., 1987. *ACM Turing Award Lectures: The First Twenty Years 1966-1985*. New York: ACM Press / Reading, MA: Addison-Wesley, pp. 182–183.

49. Wirth, 1987, p. 183.

50. N. Wirth and H. Weber, 1966. "EULER: A Generalization of ALGOL and Its Formal Definition." Part I, *Communications of the ACM*, 9 (1), pp. 13–25; Part II, *Communications of the ACM*, 9(2), pp. 89–99.

51. N. Wirth, 1971a. "The Programming Language PASCAL," *Acta Informatica*, 1, 1, pp. 35–63.

52. N. Wirth, 1971b. "The Design of a PASCAL Compiler," *Software: Practice & Experience*, 1, pp. 309–333.

53. N. Wirth, 1971c. "Program Development by Stepwise Refinement," *Communications of the ACM*, 14, 4, pp. 221–227.

54. N. Wirth, 1973. *Systematic Programming: An Introduction.* Englewood-Cliffs, NJ: Prentice-Hall.

55. Wirth, 1973, p. xii.

56. Ibid., p. 31.

57. Ibid., pp. 155–162.

58. K. Jensen and N. Wirth, 1975. *PASCAL User Manual and Report* (2nd ed.). New York: Springer-Verlag.

59. M. Woodger, 1978. "A History of IFIP WG 2.3 (Programming Methodology)," pp. 1–5 in D. Gries (ed.), 1978. *Programming Methodology. A Collection of Articles by Members of IFIP WG 2.3.* New York: Springer-Verlag.

60. The paper was not published until 1976 but by then, led by Dijkstra himself, there was already a rich literature on structured programming. See E. W. Dijkstra, [1969] 1979. "Structured Programming," pp. 43–50 in E. N. Yourdon (ed.), 1979. *Classics in Software Engineering.* New York: Yourdon Press.

61. Rutihauser, 1967.

62. Lindsey and van der Meulin, 1971.

63. B. L. Whorf, 1964. *Language, Thought and Reality.* Cambridge, MA: MIT Press.

64. Wirth, 1971c.

65. Ibid., p. 322.

66. Ibid.

67. Ibid.

68. Wirth, 1973.

69. Ibid., p. 31.

70. Wirth, 1971a.

71. Jensen and Wirth, 1975.

72. Wirth and Hoare, 1966.

73. C. A. R. Hoare, 1962. "Quicksort." *Computer Journal*, 5, 1, pp. 10–15.

74. Hoare, 1981.

75. C. A. R. Hoare, 1969. "An Axiomatic Basis for Computer Programming," *Communications of the ACM*, 12, 10, pp. 576–580, 583. Reprinted, pp. 89–100 in Gries, 1978. All page references are to the reprinted version.

76. Hoare, 1969, p. 89.

77. C. A. R. Hoare, 1986. *The Mathematics of Programming.* Inaugural Lecture, University of Oxford. Oxford: Clarendon Press.

78. R. W. Floyd, 1967. "Assigning Meaning to Programs," pp. 19–32 in Anon., *Mathematical Aspects of Computer Science, Volume XIX.* Providence, RI: American Mathematical Society.

79. C. A. R. Hoare and N. Wirth, 1973, "An Axiomatic Definition of the Programming Language PASCAL," *Acta Informatica*, 2, pp. 335–355.

80. M. A. Arbib and S. Alagic, 1979. "Proof Rules for Gotos," *Acta Informatica*, 11, pp. 139–148.

81. N. Wirth, 1976. *Algorithms + Data Structures = Programs.* Englewood Cliffs, NJ: Prentice-Hall.

82. G. J. Myers, 1982. *Advances in Computer Architecture* (2nd ed.). New York: Wiley: pp. 17.

83. M. Richards, 1969. "BCPL—A Tool for Compiler Writing and Systems Programming," *Proceedings, AFIPS Spring Joint Computer Conference*, 34, pp. 557–566.

84. D. W. Barron, J. N. Buxton, D. F. Hartley, E. Nixon, and C. Strachey, 1963. "The Main Features of CPL." *Computer Journal*, 6, 2, pp. 134–143.

85. The classic text on C is B. W. Kernighan and D. M. Ritchie, 1978. *The C Programming Language*. Englewood Cliffs, NJ: Prentice-Hall. See also D. M. Ritchie, 1993. "The Development of the C Language," *ACM SIGPLAN Notices*, 28, 3, pp. 201–208.

86. M. Richards and C. Whitby-Strevens, 1981. *BCPL, The Language and Its Compiler*. Cambridge: Cambridge University Press.

87. U. Ammann, 1977. "On Code Generation in a PASCAL Compiler," *Software—Practice and Experience*, 7, 3, pp. 391–423.

2

ABSTRACTIONS ALL THE WAY

Creative people are driven by certain inner forces, inner needs that are part cognitive, part affective. One such force is intellectual curiosity: the need to know or understand.[1] Another compelling drive is dissatisfaction with the status quo.[2] We saw this as the force that impelled Nicklaus Wirth into creating Pascal (Chapter 1, Section 1.7). But few in the emerging computer science community of the first age of computer science epitomized this characteristic more fiercely than Edsger W. Dijkstra. In his case his discontent was with the direction programming had taken in the 1960s. And the strength of his dissatisfaction was never more evident than in a letter to the editor of the *Communications of the ACM* in 1968.[3]

The practice of communicating new scientific results by their discoverers in the form of compact letters to the editors of scientific journals was, of course, well established in the natural sciences. The British journal *Nature* (London) had established this tradition right from its inaugural issue in 1869. But in an upstart discipline, as computer science still was, this practice as a means of scientific communication was quite unusual. (In one of his celebrated handwritten "EWD notes," Dijkstra, reflecting retrospectively, explained that his short paper was published as a letter to bypass the usual publication pipeline and that the editor who made this decision was Nicklaus Wirth.[4])

Dijkstra had long been concerned with the question of program quality and how one may acquire confidence in the reliability or correctness of a program. But, as the title of the letter—"Goto Statement Considered Harmful"—tells us, the object of his discontent lay in the use of the **goto** statement—the unconditional branch available in one notation or another in most programming languages, including Algol-like ones. Dijkstra claimed that the quality of the *programmers* decreased as a function of the frequency of the **goto** statements in their programs. And so he proposed that the **goto** should be banished from all high-level programming languages.

In vilifying the hapless **goto** Dijkstra appealed to one of the themes he had explored in 1965.[5] This was his theme of the "limited capacity" brain grappling with the complex structure of programs. Now he localized one of the sources of this complexity. He pointed out that our intellectual powers are adapted to the mastery of "static relations," whereas we are poorly adapted to grasping processes that unfold over time. Hence our problems with programs. On the one hand, a program is a *text*—a *static* symbol structure. On the other, they evoke computational *processes*, a *dynamic* phenomenon. We write and read a program as we write and read any text, a sentence (or statement) at a time in linear fashion.

To understand what we read, we would desire that the computational process evoked by a program has a direct correspondence to the ordering of the text. According to Dijkstra it becomes the programmer's obligation to reduce this "conceptual gap" between the static program text and the dynamic computational process; to make transparent the correspondence between the program text laid out in *space* and the computational process as it unfolds over *time*. For example, if statement $S1$ evokes process $P1$ and statement $S2$ evokes process $P2$, we would hope that if $S2$ follows $S1$ in space then $P2$ succeeds $P1$ in time, and vice versa.

This is where the **goto** statement might upset the applecart: the execution of a **goto** can cause this space–time, static–dynamic, text–process ordering to be seriously disrupted. As a simple example, consider the following:

$$L1: x \leftarrow E1;$$
$$\text{If } B1 \text{ then goto } L2;$$
$$Y \leftarrow E2;$$
$$\text{If } B2 \text{ then goto } L1;$$
$$L2: \ldots.$$

Here, if the Boolean expressions $B1$ and $B2$ both evaluate to false, the computational process will adhere to the textual ordering of the statements. If either $B1$ or $B2$ (or both) evaluates to true then the process evoked by the execution of the statements will violate the textual ordering. One can then envision the effect of the presence of multiple "unbridled" uses of the **goto** in a program. It would be "an invitation to make a mess of one's program."[6]

Much ink was spilt over Dijkstra's admonition, both in support and protest, so much so that the issue came to be known as the **goto** controversy. As it happened, the idea of minimizing the use of the **goto** did not originate with Dijkstra, as Donald Knuth pointed out in a masterly analysis of the controversy in 1974.[7] The idea of minimizing, if not entirely eliminating, the **goto** from programs was very much in the air in the mid-to-late 1960s. A notable instance was the work, in 1968, by William McKeeman, James Horning, and David Wortman at Stanford University on the design and implementation of a program (called a compiler–compiler) for producing the front end of a compiler for XPL, a PL/1 dialect;[8] this program used the **goto** only once.[9] The reason why Dijkstra's name came to be so intimately entwined with this controversy—with the precept of "**goto**-less programming"—was in part because prior expressions of this idea were less well known, in part because he made the "case against the **goto**" uncompromisingly explicit, in part because by implication he deemed programmers who freely used the **goto** somewhat inferior, and in part because very soon after the publication of his letter his "Notes on Structured Programming" appeared in public. His advocacy to dispense with the **goto** became linked, rightly or wrongly, with the principles he promulgated in this monograph.

2.2 "NOTES ON STRUCTURED PROGRAMMING"

The term structured programming (SP for convenience) was due to Dijkstra. It was first used in the title of his "Notes" paper of 1969 as one of his widely circulated "EWD notes," then as a paper presented at a NATO-sponsored conference, also in 1969, and, most effectively, as the lead monograph in the book *Structured Programming* (1972), coauthored with C. A. R. Hoare and the Norweigian O.-J. Dahl of the University of Oslo.[10]

Dijkstra never actually *defined* the term; Nor did it appear in the text of "Notes." Perhaps the postmodernist spirit of the times and the postmodernist notion of deconstruction[11] gave license to the numerous deconstructions and interpretations of the concept by different computer scientists.[12]

One way or another, the impact of "Notes" was immediate.[13] Donald Knuth would write, in 1974, "It is impossible to read the recent book *Structured Programming* . . . without having it change your life."[14] Indeed, Knuth used the word revolution—meaning an intellectual or scientific revolution, a paradigm shift in Thomas Kuhn's sense[15]—to characterize the impact of Dijkstra's "Notes" in particular and the book *Structured Programming* in general.

Like many significant ideas that constitute intellectual history, no single, precise, pithy definition suffices for the concept of SP as enunciated in Dijkstra's monograph. Rather, it is best to trace the train of thinking he laid out in the early sections of "Notes" to obtain an understanding of the term according to the Dijkstravian gospel.

2.3 THE GOSPEL ACCORDING TO DIJKSTRA: A POPPERIAN APHORISM

To begin with, SP was concerned with programs as objects: the program text itself. This was the object—the artifact—with which SP dealt. But this was not "just" static text: It was text that evoked computation; thus it had a dynamic face to it. But the physical aspect of the program, the fact that it entails processes in material computers, thus making it into what I have called a liminal artifact (see Prologue, Section P.4), was of no concern to Dijkstra. Thus efficiency concerns vis-à-vis execution on physical machines was not to be the programmer's "primary consideration."[16]

The problem that underlay the invention of SP was a concern that Dijkstra had voiced in 1965:[17] how to demonstrate that a given program—as text and as computation—is reliable or correct? Programmers must not only produce a correct program; they must also *demonstrate* its correctness.[18]

Dijkstra's conception of a program as an abstract, dynamic, textual artifact excluded it from the empirical domain: Programs were not real-world objects. Thus the conventional wisdom of empirically *testing* the behavior of a program on sample input (test) data was not an option. But even if one assumes that one could *simulate* the execution of program texts there was another rather profound reason why Dijkstra rejected the very idea of program testing. He pointed out that program testing entailed selecting only a small subset of the potentially unbounded set of possible inputs to a program. And that demonstrating that the program produces correct outputs for each such sample inputs does not *prove* that the program is free of errors (bugs). It would demonstrate only that the program works for a subset of the data that it may have to deal with. And, within the scope of that subset, it might reveal the presence of bugs.

This led to Dijkstra's much-quoted aphorism:

"Program testing can be used to show the presence of bugs but never to show their absence."[19]

If one reads Dijkstra's published works, especially his books, one realizes that he is not particularly given to citing other people's works. His "Notes," for example, does not have a single citation. So we do not know whether he was familiar with, or had been influenced by, the Austrian-British philosopher of science Sir Karl Popper's principle of *falsifiability*, which had been presented in his books *Conjectures and Refutations: The Growth of Scientific Knowledge* (1965) and *The Logic of Scientific Discovery* (1968), both of which (especially *Logic*) were already well known very soon after their publications.[20]

Popper's falsifiability principle stated that no amount of empirical evidence in support of a theory about an empirical phenomenon suffices to *prove* that the theory is true because there is a potentially unknown number of instances of the phenomenon. Thus the proposition "All swans are white" can never be proved to be true, no matter how many instances of white swans are observed, for there is always the possibility of a black swan being observed in the future. On the other hand, observation of a single black swan is logically sufficient to falsify the proposition "All swans are white." Popper argued that propositions are "scientific" only if they are in principle falsifiable.

Dijkstra's aphorism was thus entirely in accord with Popper's falsifiability proposition. Testing programs was to Dijkstra of the same logical category as observing empirical phenomena was to Popper. Neither could prove anything. But just as a single counterexample sufficed to falsify a scientific proposition so also the observation of a bug sufficed to refute any claims of correctness of a program.

The rejection of program testing was one of the fundamental tenets of SP, one that categorically disrupted a basic conventional wisdom. So what then was the programmer's alternative option?

2.4 THE GOSPEL ACCORDING TO DIJKSTRA: THE WAYS OF MATHEMATICS

Dijkstra's hopes lay in the ways of mathematics. As he would recall in 2001, by 1970 he had spent a decade arguing that programming should be a mathematical enterprise.[21]

This was, of course, the same path that led Robert Floyd and then Tony Hoare to construct an axiomatic basis for assigning meaning to computer programs (see Chapter 1, Sections 1.12, 1.13). We see here a confluence of ideas held by like-minded computer scientists. But Dijkstra was hesitant about treating a *given* program with its precondition and postcondition (Chapter 1, Section 1.13) as akin to a mathematical theorem to be proved. In fact, he rejected the very idea. Programs can be very large and complex in structure, and the task of constructing a systematic, logical proof of its correctness, even for small programs, would be simply unmanageable. Rather, one should begin with a precondition and a postcondition and *develop* a program such that, given the precondition, a program is constructed to lead to the satisfaction of the postcondition when (and if) the program terminates. In other words, to arrange "the programming task so as to make it better amenable to mathematical treatment."[22]

This leads us to an understanding of what Dijkstra meant by the word structured. In fact, it really referred to two things: structured as applied to *programs* and structured as applied to the process of *program development*.

2.5 THE GOSPEL ACCORDING TO DIJKSTRA: WELL-STRUCTURED PROGRAMS

Dijkstra returned to an issue he had raised in his **goto** letter: What could be done to reduce the conceptual gap between the program text represented in text space and the associated computation that unfolded over time?[23] This was certainly one of the key issues guiding his gospel, for it was the key to the intellectual management and understanding of programs.

His answer was that each of the significant statement types in the program text should satisfy the single-entry, single-exit (SNSX for convenience) principle: A statement is "entered" at precisely one point and is "exited from" precisely one point.

The assignment being, by definition, atomic and indivisible will obviously satisfy the SNSX principle. But so must the other major statement types. For example, using Pascal-like notation, each of the following statement types must satisfy SNSX:

Sequential statement:	$S_1; S_2; S_3: \ldots ; Sn.$
Alternative statements:	**if** B **then** S.
	If B **then** S_1 **else** S_2
	case E **of** i_1: S_1; i_2: S_2; $\ldots$; in: Sn.
Repetition statements:	**while** B **do** S
	repeat S **until** B.

In which case, constructing a program using these statements, each satisfying the SNSX property, will also satisfy this property.

Dijkstra called such a composition a well-structured program. It is a way of preserving a correspondence between a program text "spread out in space" and the computation it evokes "evolving in time." Clearly constructing such well-structured programs demands a certain discipline.

2.6   ABSTRACTION AND HIERARCHY IN DESIGN

Enforcing a discipline on the form a program should take—ensuring its well-structuredness—was thus one crucial element of the SP ideology. But, Dijkstra insisted, there was a second, equally crucial, element: imposing a discipline on the program development or design process.

The core of this discipline is to draw on the twin principles of abstraction and hierarchy. The "role of abstraction" in programming, he wrote, "permeates the whole subject."[24] Dijkstra, in fact, had been probing the constructive roles of abstraction and hierarchy in program design for some years, certainly reaching back to (once again) his 1965 IFIP Congress paper.[25] More definitively, he had shown in 1968 how the principles of abstraction and hierarchy were applied by him and his collaborators at the Technological University of Eindhoven to design a multiprogramming operating system.[26]

Abstraction can be manifested in a number of ways, but the particular instance Dijkstra alluded to was what he called "stepwise composition."[27] Nicklaus Wirth used the almost identical term, "stepwise refinement.[28]

In "Notes," Dijkstra illustrated, with an example, the combined principle of abstraction and hierarchy—stepwise composition—as he had enunciated in 1965.[29]

Step 1. Identify the parts or subtasks of the overall programming task.
Step 2. Construct complete functional specifications for the individual parts.
   That is, specify precisely what each individual part must do.
Step 3. Demonstrate that the overall goal task is achieved provided one has parts
   that satisfy the respective specifications.
Step 4. Construct recursively the individual parts to meet the specifications
   ensuring that the parts are mutually independent in their actions.

The recursive feature appears because the individual parts implemented in Step 4 may be of such complexity that each may itself have to go through Steps 1–4. This recursive refinement of a more abstract to a more concrete representation will continue until the individual parts identified and specified in Steps 1–3 are simple enough to be realized in Step 4 using the programming language at hand.

So here, abstraction refers to the fact that the functional specification of a part constructed in Step 2 only reveals *what* that part does, whereas the refinement in

Step 4 reveals *how* that part is implemented to satisfy the specification. The specification in Step 2 is an abstraction of the implementation in Step 4, and the latter in turn is a refinement of the former. The two—specification and refinement—form a *hierarchy*.

Nicklaus Wirth's 1971 paper on this same paradigm referred to Dijkstra's circulated EWD 249 (1968) version of "Notes," so it would seem that Wirth was influenced by the latter. He speaks of decomposing, in successive steps, tasks or instructions into more detailed instructions, the decomposition ending when all the instructions are expressed in a programming language.[30]

But, of course, it was not only the instructions that had to be refined. As refinement of the task proceeded, the data objects on which the tasks operated also had to be "refined, decomposed and structured, and so it seemed natural to *refine program and data specifications in parallel*."[31]

2.7 WIRTH'S *SYSTEMATIC PROGRAMMING*: STRUCTURED PROGRAMMING WITHOUT TEARS

Nicklaus Wirth's book *Systematic Programming: An Introduction* (1973) was not only the first *textbook* to introduce Pascal and its axiomatic semantics (see Chapter 1, Sections 1.7, 1.10, 1.13); it was also the first written expressly to *teach* SP. Indeed, this book is really an elegant primer that introduces and unites, with minimal fuss and few tears, Pascal, the notion of axiomatic semantics, proving program correctness (which we consider in Chapter 5), and the twin components of SP—the nature of well-structured programs and the method of stepwise development of programs. *Systematic Programming* was where the intellectual products of the minds of Dijkstra, Hoare and Wirth fuse in a single document—even though neither Hoare's 1969 paper on "An Axiomatic Basis for Computer Programming"[32] nor Dijkstra's "Notes on Structured Programming" are mentioned anywhere in the book.

In discussing stepwise development, Wirth pointed out that this process could be characterized as a "top-down" strategy for problem solving.[33] On the other hand, he noted, it was also possible to envision to begin with the programming language and to group certain typically occurring sequences of instructions into "primitive procedures" or "action clusters"[34]—this latter term due to the Danish scientist Peter Naur, the editor of the Algol 60 reports (see Chapter 1, Section 1.2).[35] Wirth then suggested that such primitive procedures could be the building blocks to construct higher-level procedures. This approach—"from the depths of primitive, atomic machine instructions to the problem at the surface"—constituted a "bottom-up" approach.[36]

Wirth then observed that in practice neither a strictly top-down method nor a strictly bottom-up approach can be realistically applied but rather some hybrid, though with the top-down approach being probably the dominant one.[37]

2.8   THE HAPLESS **GOTO**: PROBING A CONTROVERSY

The impact on the programming community of the SP idea as a disciplined approach to programming along the lines proposed by Dijkstra and then articulated by Wirth was immediate. Moreover, this effect was not confined to academic computer scientists or theorists. The December 1973 issue of the widely read and respected computing "trade" magazine *Datamation* (now long gone) carried a series of short articles on the topic by authors from the industrial sector.[38] These authors' common concern was to explore how SP could be used efficaciously to build industrial-grade software.

That same year an elegant textbook written by Richard Conway and David Gries of Cornell University showed how the structured programming style could be used to write programs in PL/1 and PL/C (the latter a Cornell implementation of PL/1).[39]

This flurry of literary activity did not mean that a single or unified interpretation of the SP idea prevailed among the authors. In a letter to the editor of *Communications of the ACM* published in November 1974, David Gries identified at least a dozen different features different writers had identified as characteristic of SP.[40] But, undoubtedly, the single largest bone of contention surrounded the **goto** statement that Dijkstra had so vilified in 1968.[41] "Notes on Structured Programming" completely eschewed the **goto**, in keeping with Dijkstra's conception of a well-structured program. And though Pascal had the **goto** as one of its constructs, Wirth's expositions of "program development by stepwise refinement" in both his 1971 paper (by that title) and his book *Systematic Programming* (1973) ignored the **goto** altogether (save for a single and dismissive mention of the statement in the latter[42]).

But good computer scientists, like all good scientists—indeed, like all good scholars of any ilk—are trained to be critical and skeptical. They are not willing to adopt precepts simply because some guru ordained them so. And so they were not quite willing to un-questioningly accept the elimination of the **goto** from their armory of programming tools. Thus the so-called **goto** controversy, which, as we have noted, began in the late 1960s, prevailed well into the 1970s. But if this controversy was to be treated as a *scientific* controversy it would have to be probed objectively, bolstered with empirical evidence, theoretically sound reasoning, or both.

The source of the controversy was not whether well-structured programs (which according to Dijkstra's gospel were **goto**-less) could be written to specify arbitrary computations. This had been answered theoretically: In 1966, the Italian theorists Corrado Böhm and Guiseppe Jacopini of the International Computational Centre, Rome, showed that an arbitrary flowchart program (a "flow diagram") could always be transformed into an equivalent program consisting only of sequential composition, iteration, and repetition forms.[43] Five years later Edward Ashcroft and Zohar Manna of Stanford University explored how flowchart programs that included unconditional branches (**goto**s) could be transformed into equivalent repetition-based (**while**) programs.[44] Thus there was a *formal* (that is, axiomatic) foundation for eschewing the **goto**.

The crux of the controversy was, in fact, something else. Two related issues, in fact. First, whether there were situations in which the absence of the **goto** actually *obscured* the transparency or clarity of programs; and second, whether avoidance of the **goto** led to *inefficiencies* in the program—for example, the necessity of additional statements or variables. These were not of mere theoretical interest. They were practical issues of concern to programmers and, more seriously, to language designers. Thus, in 1971, a team at Carnegie-Mellon University led by William Wulf designed and implemented a program language they called BLISS[45]—which, like Algol 68 was expression oriented—that did not have the **goto** at all. From their experience of programming with this language,[46] in 1972 Wulf and his team came to the "inescapable conclusion" that "the purported inconvenience of programming without a **goto** is a myth."[47] As for efficiency, the proper means of achieving this, Wulf remarked, should lie with an optimizing compiler. There were others, such as Martin Hopkins, who maintained that although composing well-structured programs à la Dijkstra was a laudable enterprise and should be encouraged, there were situations in which elegance and simplicity needed to be sacrificed to the altar of efficiency; in such situations the **goto** made complete sense.[48]

Donald Knuth belonged to this latter group. Although he was not the first to probe into the **goto** controversy, he was, undoubtedly, its most thorough, most intensive, and, because of his formidable international reputation, most influential contributor—to both sides of the issue. First, briefly, in 1971, in collaboration with Robert Floyd, his colleague at Stanford University, he argued the case for avoiding the **goto**.[49] Three years later at much greater length, he argued the cases for how and when one might effectively deploy the **goto**.[50]

He followed a methodology of investigation that was, by that time, widely established in programming and language research: the methodology adopted by Dijkstra and Wirth.[51] It was a version of what physicists called the *thought experiment*: Postulate problem situations, construct programs on paper using an appropriate notation to solve them, and analyze their resulting structure and behavior. (Such thought experiments in computer science were sometimes denigrated as playing with "toy problems" or "toy programs." But like all experiments, in the imagination or in the laboratory, these were expressly designed to probe a phenomenon in a highly controlled, oftentimes idealized, and humanly understandable setting.)

Consider, as an instance of this kind of thought experiment, a problem situation with which Knuth began: This involved an array of m elements A and an element x. The problem was to search A to find the position where x was located (if at all). If x was not present in A, it would have to be inserted in A as an additional element. Another array B of m elements would record in $B[i]$ the number of times the value in $A[i]$ had been searched.

Knuth presented two solutions to this problem in an Algol-like notation, one with the **goto**, the other without. And he pointed out that from the perspective of efficiency there was not much to choose between them. On the other hand, he remarked, the **goto**-less solution was "slightly less readable."[52] So, he concluded, not much is gained by avoiding the **goto**. But then "readability" or "understandability" or "clarity" is a subjective matter—of one's familiarity with the notation, of one's taste, of one's aesthetic preference.

Knuth then offered a third solution, and making some assumptions about how the compiled program would execute on a "typical" computer [thus transcending Dijkstra's dismissal of the physical computer from consideration (see Section 2.3)], he estimated that the third solution would save some 20% or 30% (depending on the assumptions) on running time.

This gives just a hint of the nature of Knuth's thought experiments with and without the use of **goto**s in programs. Many other problem situations and their solutions followed. And at the end of a long paper he concluded that, when writing in an Algol-like language, there are situations in which the use of the **goto** does no harm. He also pointed out that new language constructs had been invented in lieu of the **goto** that could serve the same logical function as the **goto** without inviting one to "make a mess of one's program" as Dijkstra had warned (see Section 2.1).[53] An example was the systems programming language BLISS mentioned earlier this section.

But are programs that have **goto**'s *well structured*? As far as Dijkstra (and Wirth) were concerned, well-structuredness entails the hierarchical preservation of the SNSX property of its components. Knuth's solution with the **goto** violated SNSX. So the program was surely *not* well structured. What then did Knuth mean by "structured programming"? His answer was categorical. It should have nothing to do with eliminating or avoiding **goto** statements.[54] Rather, citing Hoare, it is the systematic application of abstraction in the design and documentation of programs.[55]

So for Hoare—and, presumably for Knuth—it was the principle of hierarchical abstraction, the second part of Dijkstra's characterization (Section 2.6), what Wirth called stepwise development, that constitutes the core of SP.

2.9   PLAYING HIDE AND SEEK WITH INFORMATION

The common ground amid the disagreements and debate over what SP "really is" is that it entailed the use of hierarchical abstraction in the program design process. In top-down design, for example (stepwise development in Wirth's language), each stage of the process entails constructing a specification of the parts or modules or of a complete program—let us generically call any such part or whole a *software artifact*—and then implementing the specification in terms of more "primitive" artifacts. Each of the latter, in turn, would have to be defined in terms of specifications and *they* would be implemented, and so the process would unfold in a recursive fashion until modules are simple enough to be described in the relevant programming language.

The task of program specification thus emerged as a distinct and nontrivial component of the software development process. In the case of large system programs—for example, operating systems and compilers—software designers came to realize that the specifications would serve two complementary roles. On the one hand it informed the *user* of that software artifact what it does so that the user can deploy it accurately and correctly. On the other hand it informed the *implementer* of that software artifact the requirements the implemented software would have to satisfy. So the task of a specification was to serve as an interface (i) between the software artifact and its user and (ii) between the specification writer and the implementer.

Contemporaneous to the emergence of SP, Canadian software theorist David Parnas (then of Carnegie-Mellon University) was unhappy about the state of the art of specification techniques. In 1972 he published two papers that together sought to alleviate this dissatisfaction.[56]

Parnas posed for himself a problem of clarification: What form *should* a program module specification take? His answer was a concept he called *information hiding*. And this concept took the form of a set of connected imperatives:[57]

(1) The specification of a software artifact must afford the implementer of that artifact *all and only* the information about the artifact's intended use as necessary for the implementer to correctly create the artifact.

(2) The specification must provide all and only the information necessary for a user to deploy correctly the software artifact.

(3) The specification must be stated sufficiently *formally* that it can be tested automatically (i.e., computationally) for internal consistency and completeness.

(4) Yet the specification must be expressed in a language that is "natural" for both user and implementer.

In other words, ideally, the specification of a software artifact should reveal all and only the information necessary for the user to use it; and, on the other hand, to commit the implementer to only a necessary and sufficient set of requirements the software artifact must meet. All other information about the artifact is *hidden* from the user or from the implementer. This has also been called the "least commitment" strategy: It affords the implementer the greatest flexibility to choose and make decisions as to the best possible means of implementing the specification. Information hiding, then, was another principle of abstraction. And the principle of information hiding—the set of four prescriptive heuristics previously mentioned—found its way into the mainstream of programming methodology (see Chapter 1, Section 1.8) and in this regard became highly influential. Thereafter, no discussion of programming methodology, in particular, the topic of specifications, could sensibly ignore the principle of information hiding.

A small example of a specification adhering to Parnas's precepts is shown here.

---

module   TABLE

1.   TABLE = {<*name, dob*> | *name* is of type **charstring**, *dob* is of type **digitstring**}

2.   Initially TABLE = { }

3.   Let *namein* (TABLE) denote the set of *names* in TABLE

4.   For a given *name* ∈ *namein* (TABLE), let TABLE(*name*) denote a *dob* such that <*name, dob*> ∈ TABLE

5. **operation** INSERT (*name, dob*):
    if <*name, dob*> $\notin$ TABLE **then return** TABLE $\cup$ {<*NAME, DOB*>}
             else return ERROR

6. **operation** DELETE (*name*):
    if *name* $\in$ *namein* (TABLE) **then return** TABLE—{<*name*, TABLE (*name*)>}
             else return ERROR

7. **operation** LOOKUP (*name*):
    if *name* $\in$ *namein* (TABLE) **then return** TABLE (*name*)
             else return ERROR

end module

This specifies a **module** that is to implement a data structure called TABLE consisting of pairs of ordered values, denoted <*name, dob*> each of a specified type. Initially, TABLE is empty. Three operations, INSERT, DELETE, and LOOKUP, are defined on TABLE. The notation used here combines mathematical (specifically set-theoretical) and programming symbols. The specification hides such details as to what kind of data structure is to be used to implement TABLE and the procedures that will implement the three operations.

2.10 THE EMERGENCE OF ABSTRACT DATA TYPES

In 1976 Nicklaus Wirth published a textbook titled *Algorithms + Data Structures = Programs*,[58] an advanced and much more extensive presentation of his philosophy of programming than was outlined in *Systematic Programming*. The title of the book reflects this philosophy: There are no programs without data structures. Thus programming methodology and programming language design needed to be as concerned with the theory of data types as with algorithms.

It is always rather wonderful to see how different streams of intellectual thought and ideas merge into or influence one another. A case in point is the concept of *type*. Mathematical philosophers and logicians were of course long familiar with this concept. In his classic work *Introduction to Mathematical Philosophy* (1919), the British philosopher and logician Bertrand Russell explained that as "a rough-and-ready indication," such entities as individuals, relations of classes to individuals, and relations between classes are instances of types.[59]

Programming theorists and language designers appropriated the logician's concept of type to distinguish clearly between different kinds of data objects: A data object of type **integer** is distinct from a data object of type **boolean**; a data object of type **array of integer** is distinct from one of type **integer**. Most programming languages recognize explicitly the distinction between data types, as we have seen in the cases of Algol 60, Pascal, and Algol 68 (the latter under the name **mode**) in the previous chapter.

But how do computer scientists define type? C. A. R. Hoare in "Notes on Data Structuring" (1972), the second of the monographs in the book *Structured Programming*, examined in great detail the characteristics of data types. He defined a type—specifically data type—as an encapsulation of a class of values along with a set of primitive operations defined on it such that only these operations can be applied to values of that type.[60]

As Hoare pointed out, the meaning of the operations may be formally specified by a set of axioms. Consider, for example, an "ennumeration" type. Such a type consists of an explicit enumeration of all the possible values of that type. In programming notation its general form may be

type $T = (k_1, k_2, \ldots, k_n)$.

For example:

type *suit* = (*club, diamond, heart, spade*),

which simply ennumerates the possible values of the type *suit*.

Given an enumeration type T and variables of type T, the following axioms apply[61]:

(1) $T.min$ is a T (that is, the smallest element of T is also of type T).
(2) If x is a T and $x \neq T.max$ then $succ(x)$ is a T.
(3) The only elements of T are specified in (1) and (2).
(4) $succ(x) = succ(y)$ implies $x = y$.
(5) $succ(x) \neq T.min$.

Consider now, the entity TABLE described in Section 2.9: It comprises of a class of values (ordered pairs of character and digit strings) along with the operations INSERT, DELETE, and LOOKUP defined on them. So this specification of a module that adheres to the information hiding rules prescribed by David Parnas is *also* a description of a data type, but in abstract terms: that is, a description that says nothing about how the entity will be implemented.

The idea of "packaging" a complex type of data along with its attendant procedures that operate on instances of that data type was first proposed and implemented circa 1967 by Ole-Johan Dahl and Kristen Nygaard of the Norwegian Computing Center and the University of Oslo in their simulation programming language SIMULA 67.[62] Their term for this packaging was *class*. One form of a **class** declaration would be

class name of class (parameter list)
            declarations of variables
            declarations of procedures
end name of class

In the mid-1970s, influenced in great part by the SIMULA class concept, the notion of and the term *abstract data type* emerged in the computer science literature to signify objects of the kind described in Section 2.9 as TABLE. It isn't entirely clear who coined the term but the concept may well have been in the air in the first half of the '70s. Barbara Liskov and Stephen Zilles of the Massachusetts Institute of Technology (MIT) published a paper in 1974 with "abstract data types" in the title.[63] But it is possible that both the concept and the term were due to John Guttag whose doctoral research on "The Specification and Application to Programming of Abstract Data Types" was carried out between 1972 and 1975 at the University of Toronto.[64] As he put it, an abstract data type refers to "a class of objects" that is specified in "representation-independent" fashion.[65] In a later paper, Guttag and his colleagues at the University of Southern California, Ellis Horowitz and David Musser, offered a definition that varied slightly from this: They defined a "data type specification" (or, alternatively, abstract data type) as the abstract (representation-independent) specifications of operations on a given data type.[66] Indeed, the data type is *defined by* the abstract specification of the operations.

The abstract data type idea enabled software developers to think of building data structures in the same way they thought of building algorithmic structures. If SP entailed the systematic and disciplined use of abstraction (and its alter ego, refinement) in the specification and design of algorithmic procedures and systems, abstract data types facilitated the design of data structures: first the specification of a particular data type, then an implementation that satisfied the specification.

For the reader to get a sense of the *perspectival* change induced by introduction of the abstract data type concept, consider as an example the computational artifact called the *stack*. This is a data structure that was widely used, especially since the first implementations of Algol 60, to manage memory allocation during the execution of block-structured programs.[67]

Following Donald Knuth in Volume 1 (1968) of his *The Art of Computer Programming*, a stack may be characterized as follows: (a) There exists a data structure called a "linear list" that is an ordered set of elements comprising a "first" element, followed by a "second" element, and so on, and ending with a "last" element.[68] (b) A stack is then a linear list wherein all insertions, deletions of, and accesses to, elements are made at one end of the list.[69] By convention this end is called the "top" of the stack; insertion is called "push" and deletion "pop."

Viewed as an abstract data type, a stack would be characterized *only* by the nature of the operations defined for it. Its "identity" is established by these operations. How a stack actually "looks" is hidden away—the information hiding principle at work; so also the *algorithmic* realization of the operations is abstracted away.

The issue of great interest to abstract data type theorists was how to describe abstract data types: What kind of a *specification language* would be appropriate for this task? The broadly accepted answer that had emerged by the beginning of the 1980s was the deployment of an algebraic technique.[70] Thus, viewed as an abstract data type, a stack could be specified algebraically along the lines described by Michael Melliar-Smith of (what was then) the Stanford Research Institute in 1979:[71]

type stack [object]
>   **operations**
>>   CREATE ( ) → stack;
>>   PUSH (object, stack) → stack;
>>   POP (stack) → stack;
>>   TOP (stack) → object ∪ undefined;
>>   NEW (stack) → Boolean
>   **axioms**
>>   For all *s*: stack, *o* : object
>>   [1] NEW (CREATE ( )) = true
>>   [2] NEW (PUSH (*o*, *s*)) = false
>>   [3] POP (PUSH (*o*, *s*)) = *s*
>>   [4] TOP (CREATE ( )) = undefined
>>   [5] TOP (PUSH (*o*, *s*)) = *o*

end stack

The operation CREATE creates a new stack. PUSH and POP preserve the stack. TOP is a function that returns either an object if the stack is nonempty or an "undefined value" otherwise. NEW is a Boolean function (or predicate) that returns either "true" or "false" depending on whether the stack has just been CREATEd. Note that the syntax of the operations is specified using algebraic notation. Thus, for example, PUSH (*o*, *s*) maps from the domain that is the Cartesian product of the types object and stack to the range that is the type stack.

The semantics of the operations and functions—their behavior—is defined by axioms that show the effects of the operations or their interactions. The task of the specification designer is to ensure that the axioms define the behavior of the operations as completely, unambiguously, and economically as possible. Notice that the semantics of PUSH and POP is defined collectively by axioms [2] and [3]. In fact, the operations and their governing axioms become, in effect, a calculus (just as Boolean algebra is a calculus) that allows one to infer the effects of combinations of the operations. For example, one can *infer* that the expression

$$\text{TOP (POP (PUSH } (b, \text{PUSH } (a, s)))) = a,$$

where *a* and *b* are objects and *s* is of type stack. We can infer this by first applying axiom [3] to the subexpression POP (PUSH (*b*, PUSH (*a*, *s*))) that yields the subexpression PUSH (*a*, *s*) as a value that becomes the argument to the function TOP. Now applying axiom [5] to the resulting subexpression TOP (PUSH (*a*, *s*)) returns the value *a*.

But there is a sense in which *data* are a world of their own. And there is a sense in which "ordinary people" desire to, need *to*, interact with the data world just for the sake of extracting interesting or important information that may be hidden there. And when the data world is really huge, masses of data stored, what in the 1960s were called "data banks," such ordinary people—users of data banks—were supremely uninterested with the issues programmers and software engineers obsessed over; issues such as data types for instance. Nor were they interested in how data were organized and stored in computer memory—literally, physical data banks. Indeed, they would much prefer that how data were organized and stored in physical memory was opaque to their gaze. They too sought abstraction: a view of the data world at a level of abstraction appropriate to their needs.

In the 1960s this was broadly called information storage and retrieval;[72] but by the mid-1970s, this term had come to connote a specific kind of activity—organization and automatic retrieval of *documents* as might be held in an archive, for instance. And the organization and retrieval of information from large banks of data became the purview of *database theory and management*.

Database researchers were also interested in the problem of data abstraction but in a different sense than how programming theorists understood this term. The database researcher's compelling question was this:

What is the "proper" abstraction level at which to *view* the data world that one may access, retrieve, and alter its contents in deliberate ignorance of its storage and representation within a physical computer memory system?

Perhaps the most consequential answer to this question was offered by Edgar Codd of the IBM Research Laboratory in San Jose, California. In a seminal paper published in June 1970—one that launched a veritable research industry as well a technology—Codd proposed that one should view the data world in terms of *relationships* between its denizens. Codd called this a *relational model of data*.[73]

He used the term relation in its mathematical sense. In algebra, given a set of *sets* $\{S_1, S_2, \ldots, S_n\}$ R is a relation defined over these sets if it is the case that there is a set φ of n elements such that the first element of φ is from S_1, the second element is from $S_2, \ldots$, and the nth element is from S_n. R is then said to be an n-ary relation or to be of *degree n*.

For example, if $n = 2$, we have a *binary* relation R_2 defined on two k-element sets S_1, S_2 (where $x_i \in S_1$, $y_i \in S_2$):

$$R_2 = \{ <x_1, y_1>, <x_2, y_2>, \ldots, <x_k, y_k> \}$$

An example of a *ternary* relation ($n = 3$) is the following, depicting (say) a warehouse inventory of parts:

R3:	Vendor	Part #	Quantity
	A	1	20
	A	2	15
	B	2	10
	C	3	20
	D	1	10
	D	5	15

Here, the columns named vendor, part #, and quantity constitute *domains*—the first, second, and third respectively—of this relation **R3**.

The contents of a data bank may, of course, change with time: The data world is a dynamic, "open-ended" world. In mathematical terms this means that an *n*-ary relation is time-varying because it may be subject to addition or deletion of existing *n*-tuple entities or even constituents of an *n*-tuple entity. Thus, for the preceding example, new rows (3-tuples) may be added corresponding to a new <vendor, part #, quantity> entity; or, if vendor *A* (say) is dropped by the warehouse's list of vendors, the first two rows of the array may be deleted; or, if the quantity of a particular part # for a particular vendor is increased or decreased over time, then the quantity field in the relevant row will be changed.

This algebraic model supports in turn formalization of actual *operations* that can be performed on what are clearly *data structures*—usually in array form as in the preceding example. Because relations are defined on sets, typical *set operations* can be applied to these kinds of relations: for example, *permutation* (interchanging the columns of a relation), *projection* (selecting certain columns and disregarding the others, and removing from the resulting structure any duplicates in the rows), and *join* (the union of two relations). So in each case, a relation is input to an operation that produces as output a new relation.

In the case of the relation **R3**, part # and vendor columns may be permuted. A projection that selects part # and ignores vendor yields the following relation:

R4:	Part #	Quantity
	1	30
	2	25
	3	20
	5	15

As for the join operation, given the following pair of binary relations,

R5:	Vendor	Part #	Quantity
	A	1	20
	A	2	15

R6:	Vendor	Part #	Quantity
	C	3	10

their join would yield this:

R7:	Vendor	Part #	Quantity
	A	1	20
	A	2	15
	C	3	10

Now, mathematically, the concept of *function* is a special case of the relation concept. Given two sets X, Y, a function F is a relation that ascribes to each element of X (called the *domain* of the function) a *single* element of Y (the *co-domain* or *range* of the function). In mathematical notation,

$$F : X \rightarrow Y,$$

That is, for an element $x \in X$, there is some $y \in Y$ such that $y = F(x)$. And because two or more functions can be *composed*, Codd described how two or more relations can be composed. Finally, Codd also presented a procedure called *normalization* that can transform a complex relation not inherently depictable as an array of n-tuples—for example, a hierarchical tree structured relation—into a "normal" or "canonical" array form.

In other words, Edgar Codd's relational view of data gave rise to a *calculus* that could manipulate and transform a relation-based data structure into another relation-based data structure. Indeed, one can even envision the relational model of data as a very general kind of abstract data type comprising relations as data objects along with a set of operations defined on them.

2.12   MIGRATING ABSTRACTION INTO LANGUAGE

In the second half of the 1970s a new generation of experimental programming languages came into being. They constituted a new generation in that some fundamentally novel concepts and ideas formed the core elements of these languages. Many of

them were descendants of Pascal, and they included Nicklaus Wirth's own new creation, Modula (circa 1977),[74] Per Brinch Hansen's Concurrent Pascal (circa 1975),[75]
Alphard (circa 1976), designed at Carnegie-Mellon University by William Wulf, Mary
Shaw, and their associates,[76] and CLU (circa 1977), built at MIT by Barbara Liskov and
her collaborators.[77]

A common theme uniting these languages was the concern for migrating the
theoretical principles of abstraction—the fruits of the SP movement, the information hiding principle, the abstract data type concept, modularization, and the
specification–implementation coupling—into the language domain: to provide linguistic mechanisms that supported abstraction principles. This concern for abstraction was further reinforced by another concern that had been very much in computer
scientists' consciousness since the mid-1960s but that became an imperative only in
the mid-1970s: *parallel programming.*

We defer the discussion of this latter—which effected something like a paradigm
shift (in Thomas Kuhn's sense) in programming theory and methodology—to a later
chapter. What interests us here is the shape of abstraction mechanisms in actual programming languages. And, arguably, the one language that dominated the second half
of the 1970s and into the 1980s was *Ada,* and so it is to this language we turn as our
exemplar of this new generation of languages.

2.13 CASTING ABSTRACTIONS IN STEEL(MAN)

When the US Department of Defense (DoD) speaks, the research and development
community sits up and listens. For there are potential, lucrative contracts to be had.
So it was in the computing community in the mid-1970s.

This was the time when the DoD was concerned with the multibillion dollars cost of
the software used in its embedded computer systems—computers and software that
were integrated with noncomputational military systems. Something like 450 programming languages were being used in these embedded systems, most of them not very
widely used.[78] To alleviate the cost of software development and maintenance across
its range of embedded systems wherein such a massive number of, often incompatible,
languages and dialects were used, it was decided, in 1974, that a single (or "common")
DoD-wide programming language should be used. All contractors offering software for
the agency's embedded systems would have to write in this common language.

Toward this end, the DoD established, in 1975, a working group, called the "Higher
Order Language Working Group" (HOLWG) to consider the development of such a
common language.

Thus began a language project whose scope and scale made pigmies of the Algol
60 and Algol 68 projects. It entailed the development and publication of a succession
of *requirements specifications* between 1975 and 1978 called STRAWMAN (1975),
WOODENMAN (1975), TINMAN (1976), IRONMAN (1977), and, finally STEELMAN

(1978), each of which was reviewed extensively, leading to the next revision. In the course of these developments, HOLWG evaluated over 20 existing languages, some of which were being used already in embedded DoD systems, some employed in related applications in Europe, some new research languages that satisfied specific requirements, and, finally, languages widely known and used outside the DoD such as COBOL, FORTRAN, Pascal, and PL/1.[79] A large roster of evaluators from universities, corporations, and government agencies from the United States and Europe was tasked with the evaluation of the existing stock of languages against the evolving requirements; none was deemed satisfactory. However, some languages, specifically, Pascal, Algol 68, and Pl/1, were thought to be promising as starting points for a new language.[80]

An interesting aspect of this stage of the project was the establishment of a *design competition* of the sort well known in civil engineering and architecture. Designers were invited to submit their proposals against what was then the "current" set of requirements, IRONMAN. An additional mandate was that the proposals should be derivatives of Algol 68, Pascal, or PL/1. From some 15 proposals, 4—*all* based on Pascal—were selected as the shortlist, and their design teams from CII Honeywell Bull (France), and Intermetrics, SofTech, and SRI International (all from the United States), were invited to begin their respective, parallel design efforts in the summer of 1977. After further sifting of the four proposals, two—from CII Honeywell Bull and Intermetrics—were selected as the finalists in the spring of 1978. Soon after, the final requirements document STEELMAN was published. In May 1979, CII Honeywell Bull was selected as the "winner" and awarded the contract to develop the complete design of their language against the STEELMAN requirements. This came to be called Ada,[81] in honor of the remarkable 19th-century English mathematician Ada, the Countess of Lovelace, Lord Byron's daughter and close associate of Charles Babbage in his Analytical Engine enterprise.[82] The design team was led by Jean Ichbiah of CII Honeywell Bull (France) and was international in cast with 10 other members. In the course of language design, another 37 people were involved in one way or another, and a veritable plethora of others who served as critics. To repeat, this was a multinational, industrial-scale design project. The first reference manual for the language was published in 1981.[83]

Ada's major homage to abstraction was the concept of the *package*. As the word makes clear, the idea was to provide a facility to bundle together—encapsulate—declarations and operations that are related to one another and that form a *unit*. The package concept first, enabled the separation of specification and implementation, second, enabled the enforcement of information hiding, and, third, offered a facility for modularization.

In its most general form, a package consists of a *specification* and a *body*:

```
package PACKAGE_NAME is
        <visible part>
        <private part>
end PACKAGE_NAME
```

```
package body PACKAGE_NAME is
      . . .
end PACKAGE_NAME
```

The package specification is a set of declarations of data objects with or without operations, and the corresponding package body describes the implementation of the specification. So a package specification can be used to define an abstract data type. As for information hiding à la Parnas, this was achieved not only by separating specification from body but also, within the specification, separating what is "visible" (to the outside world) to what is "private" to the package.

As an illustration, consider the much-beloved exemplar of the stack; in particular, the construction of a *stack-of-integers* in Ada. For simplicity, let us consider the four operations PUSH, POP, NEW, and TOP specified earlier in this chapter (Section 2.10).

A description of the package specification may be along the following lines:

```
package STACK_TYPE is
      procedure PUSH  (x : in INTEGER);
      procedure POP ( y : out INTEGER):
      function NEW return BOOLEAN;
      function TOP return INTEGER;
      STACK_OVERFLOW, STACK_UNDERFLOW: exception
end STACK_TYPE
```

So the specification defines the entity STACK_TYPE as a bundle of operations and functions that are defined only in terms of their input and output features. The two **exceptions** are "alarm" signals associated with this specification.

A possible implementation of STACK_TYPE is given by the following package body:

```
package body STACK_TYPE is
      STACK : array (1 .. 1000) of INTEGER;
      STACK_POINTER : INTEGER range 0 .. 1000 := 0;
      procedure PUSH ( x : in INTEGER) is
            begin
                  if STACK_POINTER = 1000 then
                        raise STACK_OVERFLOW
                  end if;
                  STACK_POINTER := STACK_POINTER + 1:
                  STACK ( STACK_POINTER) := x;
            end PUSH;
```

```
      procedure POP (y : out INTEGER) is
         begin
            if STACK_POINTER = 0 then
                  raise STACK_UNDERFLOW
            end if;
            y := STACK (STACK_POINTER);
            STACK_POINTER := STACK_POINTER-1;
         end POP;


      function NEW return BOOLEAN is
         begin
               return POINTER = 0;
         end NEW;


      function TOP return INTEGER is
         z : INTEGER;
         begin
               POP (z);
               PUSH (z);
         end TOP
```

The package body, then, reveals that a data object of the structured type **array** is used to implement the stack and that the semantics of the operations and functions are defined operationally. Given this package, another module can utilize this package by way of a **use** statement:

use STACK_TYPE

If one chooses to implement STACK_TYPE in a different manner (e.g., using a different data structure than an array) the package body only will be affected, but not the package specification.

2.15   THE RISE OF CLASS CONSCIOUSNESS

In a lecture delivered in 1966 at a NATO Summer School on programming, C. A. R. Hoare pointed out that if one wanted to solve some problem on a computer dealing with some aspect of the world one would have to create a computational model of this aspect and represent its constituent objects.[84] Hoare's name for these representations was *record*. In general, a record may have a number of components or (in Hoare's terminology, "attributes"), each of which would be represented by a *field* within the record. For example, a record representing an employee would have fields designating their attributes: name, date of birth, year of first employment, hourly rate, and so forth.

Records represent individual objects; of course, in common discourse, objects sharing common properties can be grouped together into what we often refer to as *classes*. So also, Hoare proposed, records can be grouped into *record classes*—for example, records representing a number of individual employees in an organization may constitute the record class employee. So the records of a particular class would all have identical structures.[85]

Hoare's idea of representing objects and object classes as records and record classes would find their way via his theory of data structuring[86] into the SP movement, and into such programming languages as Pascal (as the structured data type **record**), Algol 68 (as the type named **struct**), and PL/1 (as the type STRUCT). But what is especially surprising is that his hierarchical notion of records and record classes (the latter an abstraction of the former) initiated the birth and evolution of a computational *style* in which computation was perceived as a system of objects wherein objects interacted by sending messages to one another. This came to be called the *object-oriented style*.

Like structured programming, object-oriented programming became a movement as distinctive and consequential as the former. The two were contemporaneous and in a way interpenetrating because the idea of hierarchical abstraction was central to both. But the object-oriented style properly evolved into a movement in the 1970s and the early 1980s. In this sense object-oriented programming was a "younger contemporary" of and constituted a later generation in programming style than SP.

At the same NATO Summer School where Hoare presented his idea of record classes, the Norweigian computer scientist Ole-Johan Dahl described a language called SIMULA designed by Dahl and Kristen Nygaard of the Norwegian Computing Center in Oslo. SIMULA was yet another descendant of Algol 60 but intended for computer simulation applications. At the time of the NATO Summer School Dahl and Nygaard were both implementing the language—later it would be called Simula I—and contemplating an improved version that would be "a general-purpose programming language but with simulation capabilities."[87]

The new language was called Simula 67, and it would establish itself as one of the two or three major simulation programming languages in the 1970s.[88] Appropriating Hoare's record class concept and records as instances of record classes, Dahl and Nygaard created the **class** construct as the core constituent of Simula 67.

We have encountered the **class** construct before: Recall that it was the source of the idea of the abstract data type à la John Guttag and others (see Section 2.10). But the Simula 67 **class** promised much more than Hoare's record class, which, after all, was a means of structuring data. As Nygaard and Dahl described in 1981 in their detailed history of the SIMULA languages, the central concept in Simula 67 was something called object class.[89] This was a generalization of the Algol 60 block concept (Chapter 1, Section 1.1) in the following sense: In Algol 60, a block is a piece of program text bracketed by the **begin** and **end** reserved words, whereas a block *instance* is the dynamic manifestation of the block within a computation process. Thus an Algol 60 block "may be identified with the *class* of its potential activations."[90] The problem with the Algol 60 block, for Dahl and Nygaard in the context of SIMULA, was that the "lifetime" of a block instance was limited to the duration of the block activation during program execution. This characteristic was central to the principles of abstraction, hierarchy, and information hiding in Algol 60 but not what the SIMULA designers desired. They needed SIMULA programs to specify blocks whose instances "outlived" its natural activation. This led to the Simula 67 concept of a class and a class instance as a generalization of the Algol 60 block–block instance feature.

A procedure that is capable of giving rise to block instances that survive its call will be known as a class; and the instance will be known as objects of that class.[91]

This was how Dahl and Hoare, in "Hierarchical Program Structures," the third monograph in *Structured Programming*, explained the class–object relationship. And as an example, they presented the case of representing a histogram. A histogram is a method for depicting the frequency of numeric values of a random variable distributed over disjoint uniform intervals. For instance, a histogram may show the distribution of heights in an adult population over the range 4–7 ft (say) in intervals of 6 in. Thus the intervals would be 4 ft, 4 feet 6 in., 5 ft, 5 ft 6 in., . . . , 6 ft 6 in., 7 ft. Another histogram may show for the same population the distribution of weights, and a third by age distribution. As these examples suggest, the histogram can be declared as a class. Such a declaration specifies the data structure representing intervals along with any procedures that may operate on the data structure.

The histogram **class** would then have the following skeletal form:[92]

```
class histogram (x, n);
        array x; integer n;
        begin
                declaration of variables
                procedure tabulate (. . .)
                . . . .
                end tabulate;
                procedure frequency ( . . . )
                . . . .
                end frequency;
                . . . .
        end histogram.
```

The procedures tabulate and frequency, which perform some computations over the contents of x, are said to be "attributes" of the histogram **class**. And instances of this **class** can then be created as objects from elsewhere in the program by way of statements such as

height :- **new** histogram (. . .);
weight :- **new** histogram (. . .);
age :-   **new** histogram (. . .).

So three (suitably parameterized) objects, height, weight, and age, of the **class** histogram are created. Each will *inherit* the attributes of the class so that one may activate the procedures tabulate and frequency for any one of these objects.

2.15   OBJECTS OF ONE'S OWN

Among those who were influenced by SIMULA, as a "promise of an entirely new way to structure computations" was Alan Kay.[93] He remarked on this influence in 1993 while reconstructing the origins of Smalltalk, the first object-oriented programming language he and his colleagues created between 1972 and 1980. (Smalltalk evolved through several versions through the '70s, named, most notably, Smalltalk-72, Smalltalk-74, Smalltalk-76, and Smalltalk-80.) He first encountered SIMULA—Simula I as the version that then existed—in 1966 as a graduate student at the University of Utah.

The idea that caught his particular attention was a form of what in Simula 67 became the class entity "that acted like masters [which would] . . . create instances, each of which was an independent entity."[94] But Kay was not interested in improving SIMULA: evolution was not in his mind but something more radical, something akin to a revolution: "an entirely new way to structure computations."[95]

All insights have a context. Kay's reflections on the possibility of "a new way to structure computations" arose in the context of the first ideas about *human–computer symbiosis* that were starting to be explored in the late 1960s, when the first interactive computing systems with graphical interfaces were being conceived and built at a number of centers across the United States, including the University of Utah. One of his mentors there was David C. Evans who, in 1968, induced Ivan Sutherland to join the computer science department in Utah from MIT. In 1963 Sutherland had invented Sketchpad, a graphical human–machine communication system, one of the very first of its kind, that could be used "to draw electrical, mechanical, scientific, mathematical, and animated drawings" directly on a computer screen using a "light pen."[96]

In fact, remarkably, at least 3 years before Hoare's record class and, following that, the publication of Dahl and Nygaard's class–object concept, Ivan Sutherland had conceived the idea of a "master" pictorial structure and its instances that inherited the master's attributes, and he incorporated it into Sketchpad. For example, as Sutherland would explain, in drawing an electrical circuit, its transistor components would be created as instances of a "master transistor drawing." So if this master drawing is altered, the change would automatically appear in all its instances in the drawing.[97]

David Evans introduced Alan Kay to Sketchpad well before Sutherland moved to Utah. The impact, as he would recall, was immediate and electric.[98]

The master structure–instance structure concept, he realized, was one of Sketchpad's "big ideas"[99]—which, he discovered soon after, was replicated in SIMULA. But there was another "big idea" embedded in Sketchpad: the invention of interactive computer graphics. What he had more or less stumbled on was the first intimation of *personal computing*: an individual user with a machine of one's own, a solipsistic person–machine universe. This was a dramatic departure from the tradition of centralized mainframe computing then prevalent. So rather than thousands of "*institutional* mainframes" there would be "millions of *personal* machines" that for their users would be outside institutional control.[100]

There were other influences on Kay. Robert Barton, on the faculty at Utah, had been the principal designer of the Burroughs Corporation B5000 computer, the first of the highly innovative family of Algol-oriented computers Burroughs would build in the 1960s and 1970s.[101] Barton had articulated, in a lecture, his principle of "recursive design": "to make the parts have the same power as the whole."[102] This had prompted Kay to think "of the whole as the entire computer" comprising a multitude—"thousands"— of little computers "each simulating a useful structure."[103]

And there was also the South African–born Seymour Papert, codirector of MIT's Artificial Intelligence Laboratory, and a protégé of the preeminent Swiss development psychologist Jean Piaget, who had applied Piagetian theories of learning to the building of an interactive system called LOGO, which teaches small children how to do geometry.[104] Kay's encounter with LOGO led him to realize that the "destiny" of personal computing lay in "something more profound" than just a personal computing "vehicle": rather, a "personal dynamic *medium*."[105]

Thus, through a route that meandered its way through a number of machines, languages, and programming ideas, Alan Kay arrived at a generalized view of an object-based personal computing system: the idea of multiple machines sending to and accepting requests from one another.[106]

In effect, an object à la Dahl and Nygaard (or Sutherland) would be a computer of sorts of its own. As Kay put it in 1993, using a term that would later become commonplace in computing vocabulary, each object would become a "*server* offering *services*" to other objects. Objects would thus enter into a "server–servee" relationship.[107]

But it was not until after Kay joined the newly formed Xerox Palo Alto Research Center (PARC) in 1970, having completed his PhD dissertation at Utah, that object-oriented thinking began to evolve into a mature shape. The vehicle of this maturation was the object-oriented programming language Smalltalk, designed and implemented in successive versions between 1972 and 1980 by Kay, Dan Ingalls, and Adele Goldberg.[108] The first version of Smalltalk (Smalltalk-72) emerged about the same time as Kay's Xerox PARC colleagues, led by Charles Thacker, were designing and implementing Alto, an experimental personal computer.[109] Smalltalk was implemented on Alto as one of its interactive programming environments.[110]

So what did the object-oriented style in its Smalltalk incarnation actually entail? We may paraphrase a concise account by Adele Goldberg, one of Kay's principal collaborators on Smalltalk design:[111]

(1) An object-oriented program comprises a society of objects interacting by sending and receiving messages to one another.

(2) An object is an abstraction of a computational agent. It has its own private memory and so can hold information. It can manipulate its private information. It can perform operations.

(3) An object's set of operations constitute its interface with other objects. However, the interpretations of these operations are hidden from other objects. Thus the principle of information hiding is exercised.

(4) Objects sharing common attributes—the same kind of private memory or data types and a common set of operations—constitute a class. Objects of a class are its instances. The objects of the same class inherit its attributes.

(5) A class may also have subclasses that inherit attributes of the (parent) class. Thus an inheritance hierarchy of classes, subclasses, and objects may prevail.

(6) Programming in the object-oriented style entails creating new classes, creating new objects as instances of a class, and specifying sequences of operations on these objects.

(7) An object performs an operation in response to a *message* sent to it by another object. Message-passing is the sole means of passing control among objects in an object-oriented computation.

(8) Associated with a message is a procedure or method describing how the receiving object should respond.

2.16   A CULTURAL REVOLUTION?

From Hoare to Dahl and Nygaard to Kay: Such was the evolutionary pathway to the object-oriented style. Some people characterized this as a shift from one programming paradigm to another: from the "procedural" or "imperative" paradigm or style to the object-based style. But more generally, this was a transformation entailing a marked difference in what we might well call their protagonists' computational *cultures*. Hoare, Dahl, and Dijkstra were embedded in the mainframe culture, which originated with the first, one-of-a-kind experimental machines of the 1940s, passed through the commercial computers of the 1950s that users shared in a one-user-at-a-time mode, to the multiprogrammed, time-shared machines of the 1960s (and into the 1970s).[112] Even the advent of the low-cost minicomputers in the early 1970s that enabled small organizations such as an individual academic department or laboratory to disassociate itself (up to a point) from centralized installations did not seriously challenge the mainframe culture.

Alan Kay, as we have seen, was nurtured in an embryonic personal computing culture (exemplified by Ivan Sutherland's Sketchpad system and Papert's LOGO environment); and, in turn, he returned the compliment. Along with others, including his Xerox PARC colleagues who created Alto, he helped develop the embryo into a mature personal computing culture: a culture in which individuals possessed their own computing environment and that encouraged a very different way for users to interact with their dedicated environment. This culture enabled a kind of human–machine symbiosis that was hitherto unknown. The mainframe culture by its very nature represented a disassociation of the human user from the computing environment. Programming was no doubt a human activity à la Dijkstra, but there was a clear demarcation between the human being and the machine. That demarcation was demolished in this new culture. It is not unreasonable to think of this as a cultural revolution.

NOTES

1. S. Dasgupta, 1996. *Technology and Creativity*. New York: Oxford University Press, pp. 25–27.
2. Ibid.
3. E. W. Dijkstra, 1968. "Goto Statement Considered Harmful" (letter to the editor), *Communications of the ACM*, 11, pp. 147–148.
4. E. W. Dijkstra, 2001. "What Led to 'Notes on Structured Programming.'" EWD1308-0.
5. E. W. Dijkstra, 1965. "Programming Considered as a Human Activity," pp. 213–217, in *Proceedings of the 1965 IFIP Congress*. Amsterdam: North-Holland.
6. Dijkstra, 1968, p. 147.
7. D. E. Knuth, 1974. "Structured Programming with **goto** Statements," *Computing Surveys*, 6, 12, pp. 261–301. Reprinted, pp. 17–90 in D. E. Knuth, 1992. *Literate Programming*. Stanford, CA: Center for the Study of Language and Information. All page references to this article are to the reprint version.
8. W. M. McKeeman, J. J. Horning, and D. B. Wortman, 1970. *A Compiler Generator*. Englewood Cliffs, NJ: Prentice-Hall.
9. Knuth, 1974, p. 23.
10. E. W. Dijkstra, 1972. "Notes on Structured Programming," pp. 1–82 in O.-J. Dahl, E. W. Dijkstra, and C. A. R. Hoare (eds.), 1972. *Structured Programming*. London: Academic Press.
11. The central tenet of deconstruction is the relativism of truth: that truth is always relative to a particular standpoint or intellectual framework influencing the person assessing the truth of a proposition. For a recent concise account of postmodernism and deconstruction theory, see C. Butler, 2002. *Postmodernism: A Very Short Introduction*. Oxford: Oxford University Press, esp. pp. 16–19.
12. See the brief review by D. Gries, 1974, "On Structured Programming" (letter to the editor), *Communications of the ACM*, 17, 11, pp. 655–657. Reprinted, pp. 70–74 in D. Gries (ed.), 1978. *Programming Methodology. A Collection of Articles by Members of WG 2.3*. New York: Springer-Verlag.
13. I still recall, as a graduate student in the early 1970s at the University of Alberta, my excitement on first reading the EWD version of "Notes," which one of my teachers (and later colleague) Barry Mailloux, codesigner of Algol 68 (see Chapter 1) had made available.
14. Knuth, [1974] 1992, p. 18.
15. T. S. Kuhn, [1962] 2012. *The Structure of Scientific Revolutions* (50th anniversary edition). Chicago: University of Chicago Press.
16. Dijkstra, 1972, p. 6.

17. Dijkstra, 1965.

18. Dijkstra, 1972, p. 5.

19. Ibid., p. 6.

20. K. R. Popper, 1965. *Conjectures and Refutations: The Growth of Scientific Knowledge.* New York: Harper & Row; K. R. Popper, 1968. *The Logic of Scientific Discovery.* New York: Harper & Row.

21. E. W. Dijkstra, 2001. "What Led to 'Notes on Structured Programming.'" EWD1308-0.

22. Ibid.

23. Dijkstra, 1972, p. 16.

24. Ibid., p. 11.

25. Dijkstra, 1965.

26. E. W. Dijkstra, 1968. "The Structure of 'THE' Multiprogramming Operating System," *Communications of the ACM*, 11, 5, pp. 341–346.

27. Dijkstra, 1972, p. 26.

28. N. Wirth, 1971. "Program Development by Stepwise Refinement," *Communications of the ACM*, 14, 4, pp. 221–227.

29. Dijkstra, 1965.

30. Wirth, 1971, p. 221.

31. Ibid.

32. C. A. R. Hoare, 1969. "An Axiomatic Basis for Computer Programming," *Communications of the ACM*, 12, 10, pp. 576–580, 583.

33. N. Wirth, 1973. *Systematic Programming: An Introduction.* Englewood Cliffs, NJ: Prentice-Hall.

34. Ibid.

35. P. Naur, 1969. "Programming by Action Clusters," *BIT*, 9, pp. 250–258.

36. Wirth, 1973.

37. Ibid.

38. D. D. McCracken, 1973. "Revolution in Programming: An Overview," *Datamation*, 19, 12, pp, 50–52; J. R. Donaldson, 1973. "Structured Programming," *Datamation*, 19, 12, pp. 52–54; E. F. Miller and G. E. Lindamood, 1973. "Structured Programming: A Top-Down Approach," *Datamation*, 19, 12, pp. 55–57; F. T. Baker and H. D. Mills, 1973. "Chief Programmer Teams," *Datamation,* 19, 12, pp. 58–61.

39. R. W. Conway and D. G. Gries, 1973. *Introduction to Programming: A Structured Approach Using Pl/1 and PL/C.* Cambridge, MA: Winthrop Publishers.

40. D. G. Gries, 1974. "On Structured Programming" (letter to the editor), *Communications of the ACM,* 17, 11, pp. 655–657.

41. Dijkstra, 1968.

42. Wirth, 1973.

43. C. Böhm and G. Jacopini, 1966. "Flow Diagrams, Turing Machines and Languages With Only Two Formation Rules," *Communications of the ACM*, 9, 5, pp. 366–371.

44. E. A. Ashcroft and Z. Manna, 1972. "Formalization of Properties of Parallel Programs," pp. 250–255 in *Proceedings of the 1971 IFIP Congress*, Volume 1. Amsterdam: North-Holland.

45. W. A. Wulf, D. B. Russell, and A. N. Haberman, 1971. "BLISS: A Language for Systems Programming," *Communications of the ACM*, 14, 12, pp. 780–790.

46. W. A. Wulf, 1971. "Programming Without the GOTO," pp. 408–413 in *Proceedings of the 1971 IFIP Congress,* Volume 1. Amsterdam: North-Holland.

47. W. A. Wulf, 1972. "A Case Against the GOTO," pp. 85–100 in *Proceedings of the 25th National ACM Conference*. New York: ACM. Reprinted, pp. 85–100 in E. N. Yourdon (ed.), 1979. *Classics in Software Engineering*. New York: Yourdon Press, p. 93.

48. M. E. Hopkins, 1972. "A Case for the GOTO," pp. 787–790 in *Proceedings of the 25th National ACM Conference*. New York: ACM.

49. D. E. Knuth and R. W. Floyd, 1971. "Notes on Avoiding **go to** Statements," *Information Processing Letters*, 1, pp. 23–31.

50. Knuth, [1974] 1992.

51. Dijkstra, 1972; Wirth, 1971.

52. Knuth [1974] 1992, p. 25.

53. Ibid., p. 70.

54. Ibid., p. 70.

55. Ibid., p. 73.

56. D. L. Parnas, 1972a. "A Technique for Software Module Specification With Examples," *Communications of the ACM*, 15, 5, pp. 330–336. Reprinted, pp. 5–18 in E. Yourdon (ed.), 1982. *Writings of the Revolution*. New York: Yourdon Press. D. L. Parnas, 1972b. "On the Criteria to be Used in Decomposing Systems Into Modules," *Communications of the ACM*, 15, 12, pp. 1053–1058. Reprinted, pp. 141–152 in E. Yourdon (ed.), 1979. *Classics in Software Engineering*. New York: Yourdon Press. All page references are to the reprints.

57. Parnas, 1972a, p. 5.

58. N. Wirth, 1976. *Algorithms + Data Structures = Programs*. Englewood Cliffs, NJ: Prentice-Hall.

59. B. Russell [1919], Reprint n.d., p. 53 in *Introduction to Mathematical Philosophy*. New York: Touchstone/Simon & Schuster.

60. C. A. R. Hoare, 1972. "Notes on Data Structuring," pp. 83–174 in Dahl, Dikstra, and Hoare, 1972, pp. 92–93. Italics added.

61. Hoare, 1972, p. 167.

62. O.-J. Dahl, B. Myhrhaug, and K. Nygaard, 1968. *The SIMULA 67 Common Base Language*. Oslo: Norweigian Computing Centre.

63. B. H. Liskov and S. Zilles, 1974. "Programming with Abstract Data Types." *SIGPLAN Notices*, 9, 4, pp. 50–59.

64. J. V. Guttag, 1975. "The Specification and Application to Programming of Abstract Data Types," PhD dissertation, Department of Computer Science, University of Toronto.

65. J. V. Guttag, 1977. "Abstract Data Types and the Development of Data Structures," *Communications of the ACM*, 20, 6, pp. 396–404.

66. J. V. Guttag, E. Horowitz, and D. R. Musser, 1978. "The Design of Data Type Specifications," pp. 60–79 in R. T. Yeh (ed.), *Current Trends in Programming Methodology*. Volume 2. Englewood Cliffs, NJ: Prentice-Hall, p. 61.

67. See B. Randell and L. J. Russell, 1964. *Algol 60 Implementation*. New York: Academic Press.

68. D. E. Knuth, 1968. *The Art of Computer Programming. Volume 1. Fundamental Algorithms*. Reading, MA: Addison-Wesley, p. 234.

69. Knuth, 1968, p. 235.

70. Guttag, 1977; S. N. Zilles, 1980. "An Introduction to Data Algebra," pp. 248–272 in D. Bjørner (ed.), *Abstract Software Specifications*. Berlin: Springer-Verlag; B. Liskov, 1980. "Modular Program Construction Using Abstractions," pp. 354–389 in Bjørner (ed.), 1980; B. Liskov and V. Berzins, 1979. "An Appraisal of Program Specifications," pp. 276–301 in P. Wegner (ed.), *Research Directions in Software Technology*. Cambridge, MA: MIT Press; P. M.

Melliar-Smith, 1979. "System Specification," pp. 19–65 in T. Anderson and B. Randell (eds.). *Computing Systems Reliability.* Cambridge: Cambridge University Press.

71. Melliar-Smith, 1979, p. 37.

72. G. R. Salton, 1968. *Automatic Information Organization and Retrieval.* New York: McGraw-Hill.

73. E. F. Codd, 1970. "A Relational Model of Data for Large Shared Data Banks," *Communications of the ACM*, 16, 6, pp. 377–387.

74. N. Wirth, 1977. "Modula: A Language for Modula Programming," *Software—Practice and Experience*, 7, 1.

75. P. Brinch Hansen, 1975. "The Programming Language Concurrent Pascal," *IEEE Transactions on Software Engineering*, 1, 2, pp. 195–202.

76. W. A. Wulf, R. L. London, and M. Shaw, 1976. "An Introduction to the Construction and Verification of Alphard Programs," *IEEE Transactions on Software Engineering*, 2, 4, pp. 253–264.

77. B. Liskov, A. Snyder, R. Atkinson, and C. Shaffert, 1977. "Abstraction Mechanisms in CLU," *Communications of the ACM*, 20, 8, pp. 564–576.

78. D. A. Fisher, 1978. "DoD's Common Programming Language Effort," *IEEE Computer*, March, pp. 24–33: p. 26.

79. Ibid., p. 29.

80. Ibid., pp. 28, 33.

81. Ada is a registered trademark of the US DoD.

82. For a discussion of Ada Lovelace and her contribution to Babbage's Analytical Engine project, see S. Dasgupta, 2014. *It Began with Babbage: The Genesis of Computer Science.* New York: Oxford University Press, chapter 2.

83. Anon., 1981. *The Programming Language Ada Reference Manual.* (Proposed Standard Document, United States Department of Defense.) Berlin: Springer-Verlag.

84. C. A. R. Hoare, 1966. "Record Handling." Lectures, The NATO Summer School, Villard-de-Lans, September 12–16, p. 1.

85. Ibid.

86. Hoare, 1972.

87. O.-J. Dahl, 2001. "The Birth of Object Orientation: The Simula Languages." http:// www. olejoahndahl.info/old/birth-of-oo.pdf. Retrieved Jan. 21, 2016.

88. K. Nygaard and O.-J. Dahl, 1981. "The Development of the SIMULA Languages," pp. 439–480 in R. L. Wexelblat (ed.), *History of Programming Languages.* New York: Academic Press.

89. Ibid., p. 461.

90. O.-J. Dahl and C. A. R. Hoare, 1972. "Hierarchical Program Structures," pp. 175–220 in Dahl, Dijkstra, and Hoare, p. 178.

91. Ibid., p. 179.

92. Ibid., p. 182.

93. A. Kay, 1993. "The Early History of Smalltalk," *ACM SIGPLAN Notices*, 28, 3, pp. 1–54, p. 6.

94. Ibid., p. 5.

95. Ibid., p. 6.

96. I. C. Sutherland, [1963] 2003. "Sketchpad: A Man-Machine Graphical Communication System." Tech. Report 574. Computer Laboratory, University of Cambridge: http://www.cl.cam.ac.uk/TechReport/. Jan. 27, 2016. The quote is from the abstract.

97. Ibid., p. 27.

98. Kay, 1993, p. 5.

99. Ibid.

100. Ibid., p. 7.

101. R. S. Barton, 1961. "A New Approach to the Functional Design of a Digital Computer," pp. 393–396, *Proceedings of the Western Joint Computer Conference*. Los Angeles, CA: n.p.; E. I. Organick, 1973. *Computer Systems Organization: The B5700/6700 Series*. New York: Academic Press.

102. Kay, 1993, p. 6.

103. Ibid.

104. S. Papert, 1980. *Mindstorms: Children, Computers and Powerful Ideas*. New York: Basic Books.

105. Kay, 1993, p. 10.

106. Ibid., p. 11.

107. Ibid.

108. A. Kay and A. Goldberg, 1977. "Personal Dynamic Media," *Computer*, 10, 3, pp. 31–41; D. H. H. Ingalls, 1978. "The Smalltalk-76 Programming System: Design and Implementation," pp. 9–16 in *Proceedings of the 5th Symposium on the Principles of Programming Languages*. New York: Association for Computing Machinery; A. Goldberg, 1984. "The Influence of an Object-Oriented Language on the Programming Environment," pp. 141–174 in D. R. Barstow, H. E. Shrobe, and E. Sandwall (eds.). *Interactive Programming Environments*. New York: McGraw-Hill.

109. C. P. Thacker, E. M. McCreight, B. W. Lampson, R. F. Sproull, and D. R. Boggs, 1979. "Alto: A Personal Computer." Xerox Corporation. Reprinted, pp. 549–580 in D. P. Siewiorel, C. G. Bell, and A. Newell, 1982. *Computer Structures: Principles and Examples*. New York: McGraw-Hill.

110. Ibid., p. 570.

111. Goldberg, 1984, pp. 142–143.

112. S. Dasgupta, 2014. *It Began with Babbage: The Genesis of Computer Science*. New York: Oxford University Press, chapters 8, 15.

3

IN THE NAME OF ARCHITECTURE

3.1   BETWEEN A SOFT AND A HARD PLACE

When Caxton Foster of the University of Massachusetts published his book *Computer Architecture* in 1970, this term was only just being recognized, reluctantly, by the computing community.[1] This despite an influential paper published in 1964 by a group of IBM engineers on the "Architecture of the IBM System/360."[2] For instance, ACM's "Curriculum 68" made no mention of the term in its elaborate description of the entire scope of computing as an academic discipline.[3]

Rather, in the late 1960s and well into the '70s terms such as computer organization, computer structures, logical organization, computer systems organization, or, most blandly, computer design were preferred to describe computers in an abstract sort of way, independent of the physical (hardware) details.[4] Thus a widely referenced paper by Michael Flynn of Stanford University, published in 1974, was titled "Trends and Problems in Computer Organization."[5] And Maurice Wilkes, even in the third edition of his *Time-Sharing Computer Systems* (1975) declined to use the term computer architecture.[6]

Yet, computer architecture as both an abstract way of looking at, understanding, and designing computers, and as a field of computer science emerged in the first years of the '70s. The Institute of Electrical and Electronics Engineers (IEEE) founded a Technical Committee on Computer Architecture (TCCA) in 1970 to join the ranks of other specialist IEEE TCs. The Association for Computing Machinery (ACM) followed suit in 1971 by establishing, alongside other special-interest groups, the Special Interest Group on Computer Architecture (SIGARCH). And in 1974, the first of what came to be the annual International Symposium on Computer Architecture (ISCA) was held in Gainesville, Florida. By the end of the decade a series of significant textbooks and articles bearing the term computer architecture(s) had appeared.[7]

The reason for naming an aspect of the computer its "architecture" and the reason for naming an academic and research discipline "computer architecture" can be traced back to the mid-1940s and the paradigm-shaping unpublished reports by John von Neumann of the Institute of Advanced Study, Princeton, and his collaborators, Arthur Burks and Herman Goldstine.[8] One of the crucial points made by von Neumann et al. was to make a distinction between what they called the *logical design* of a computer and its *physical implementation*. The former was an abstract thing, the latter a material one.

In the computing realm, then, concerns about abstraction were not the prerogative of only the programmer or software designer (see Chapter 2). From von Neumann on, computer designers found it *necessary* to make a distinction between the computer—a real one, not a mathematical device such as the Turing machine—as an abstract artifact and the computer as a material one. In the case of the computer, abstraction meant that many elements of the material artifact—hardware, the nature of the physical components, and their physical technology—were to be suppressed, hidden, or abstracted away as irrelevant in some significant contexts. And the necessity for abstraction arose in three distinct contexts.

First, the same issue of the intellectual management of complexity that was concerning Dijkstra and others in the programming realm and that led to the development of programming methodology in the early 1970s (see Chapter 2) prevailed in the realm of computer design. Ultimately, a physical computer is a digital system made of circuits and other electromagnetic and mechanical elements—just as a living body is a system of different kinds of cells. In the latter case evolution by natural selection produced certain organizations and differentiations among cells, leading to functionally distinct entities such as tissues, tissues organized and differentiated into organs, and organs organized and differentiated into anatomical–physiological systems. Although this organization and differentiation were biological phenomena, they were also of *cognitive* significance because biologists and physicians could study, understand, diagnose, and treat the living body by focusing on the subsystems and their interactions in a disciplined, hierarchic manner.

So also, by a process of reasoning that was both logical and heuristic, computer designers, from the very first days of electronic digital computers, came to organize and differentiate functionally and logically distinct subsystems and their interactions. They came to realize that what made a digital system into a computer *was* such organization and differentiation. They came to realize also that a single organizational design could be implemented using different physical components. And they came to realize that a particular organization of functionally distinct subsystems effectively defined a *class* of computers. Revealing the physical components and the attendant technology resulted in a particular instance of that class. We are reminded of Parnas's information hiding principle and the class–object relationship that led to object-oriented thinking (Chapter 2, Sections 2.9, 2.13, 2.14).

These considerations led to another context. Imagine a person whose business is to design computers as abstract artifacts. In that case a material artifact—a physical computer—can be implemented to satisfy this abstract design. If we call this abstract design an architecture then it serves the same role as a specification does in the software realm. An architecture thus functions as a specification that the hardware builder must implement. Any computer design methodology that takes a top-down approach necessitates the design of the architecture—a specification—*before* decisions about implementations can begin.[9]

Finally, there is yet another context that necessitates the separation of the computer-as-abstract artifact from the computer-as-physical artifact. And that is that the abstract artifact can serve as an *interface* between the physical computer and its environment. That environment, of course, will be computational in nature, comprising programs and information systems. A computer's architecture thus serves as a mediator between what programmers and other users desire of their computer and what the physical computer can actually deliver.

Herein lies the ultimate responsibility of the computer architect: to create an interface that mediates between software and hardware, between two realities; between the Scylla of software and the Charybdis of hardware. The architect's task is made even more vexing because both Scylla and Charybdis change over time. As we will see, the history of computer architecture in the second age of computer science is a history of how computer architects navigated between the two.

These then were the issues that produced, beginning with von Neumann in the 1940s, through the first age of computer science (1950s and 1960s), an awareness of both an aspect of the computer called its architecture and the discipline of computer architecture. But, as we will see, it was during the 1970s that computer architecture became a widely accepted name for the field (though there still remained some holdouts[10]) and its study in relation to both software and hardware and as a design discipline became a distinct subparadigm of computer science.[11]

3.2 DEFINING COMPUTER ARCHITECTURE: SETTING THE TONE

But what exactly does computer architecture *mean*? The answer to this was a matter of some disagreement even at the time the concept and the term had entered the mainstream of computing discourse. As previously noted Gene Amdahl, Gerrit Blaauw, and Frederick Brooks of IBM cast the first stone in 1964 by defining what *they* meant by the term: By the "architecture of the IBM System/360" they were referring to the attributes of the machine as visible to the lowest-level (assembly language) programmer.[12] In theory this included that and only that which was needed to be known to write (assembly language) programs for the computer. A computer's architecture à la Amdahl et al., is the interface between the physical machine and its programmers. All else—the means by which this interface is implemented—is suppressed from the programmer's gaze.

In the case of the IBM System/360, this architecture was defined in the official IBM System/360 manual.[13] Interestingly, Gerrit Blaauw and Frederick Brooks also described this architecture in a special issue of the *IBM Systems Journal* on the System/360 in 1964, but did not use the word architecture; their preferred term was "logical structure."[14] The System/360 architecture or logical structure comprised the following components:

Functional structure: showing all the principal functional components—memory organization, input–output processors (channels in IBM language), central processing unit, input–output control units and devices—and their interconnections.

Internal logic of the central processing unit: showing control unit, general registers set, floating-point registers set, main memory, the arithmetic and logic units, and their interconnections.

Data types and data formats: showing the representation of the principal data types as they would be encoded in main memory.

The repertoire of processing unit operations: fixed-point arithmetic, floating-point arithmetic, logical operations, decimal arithmetic.

Instruction formats: five basic types for representing instructions.

Instruction sequencing: outline of the cycle of events resulting in instruction execution.

Interrupt conditions: describes the different types of conditions—pertaining to input–output operations, unusual conditions during program execution, call to the supervisor program, external interrupts from the control panel or other special devices, and hardware malfunction—that the programmer must be aware of.

Program status word: the format of a special element that holds the state of computation at any given time, which can be saved when an interrupt occurs and can be restored after servicing the interrupt.

Input–output: description of input–output commands, command formats, and their processing by means of input–output processors (channels).

Amdahl and his IBM colleagues presented their view of architecture by example. Theirs was what we might call a pragmatic or engineering, view. The scientific approach would entail, at the very least, some attempt to articulate an explicit, generalized model of what constitutes an architectural description. This was precisely what Gordon Bell and Allen Newell of Carnegie-Mellon University attempted 6 years later. At the Spring Joint Computer Conference in 1970, they presented a generalized and quite abstract conception of computer architecture that entailed viewing the computer at two distinct abstraction levels.[15] The year after, they re-presented this model and applied it to characterize a variety of architectures in their enormously valuable book *Computer Structures: Readings and Examples* (1971).[16]

3.3   ARCHITECTURE ACCORDING TO BELL AND NEWELL:
      THE PROCESSOR–MEMORY–SWITCH LEVEL

Their starting point was a rather fundamental and quite abstract view of the matter. They conceived a digital system as a discrete *state-based* entity; that is, a system that is in some discrete state at any given moment of time and that experiences discrete changes of state over time. Precisely what the physical features of these states are were abstracted away. A digital computer is a special instance of a digital system that manifests three characteristic features.

First, the state of the system is realized by information held in memories. Second, the system is composed of (usually) a small number of functional subsystems that are logically interconnected by information paths along which information may flow. Third, associated with each functional subsystem is a set of operations that enable changes of state of either that subsystem or one of its proximate neighbors. So the individual subsystems are themselves discrete state-based components whose behaviors determine the state changes of the computer as a whole.

So one can envision a computer system as a *network* (or its mathematical equivalent, an algebraic graph) in which the nodes of the network (or the vertices of the graph) are state-based subsystems and the links between the nodes (edges connecting the vertices of the graph) are the information paths.

Now there was nothing intrinsically original about this conception. Computer scientists and computing practitioners were long accustomed to depicting the structure of computer systems in terms of block diagrams or flowcharts. Indeed, the "functional structure" of the IBM System/360 as envisioned by Gene Amdahl et al. (Section 3.2) corresponded to this model and was represented graphically as such by them.[17] Bell and Newell's originality lay in their offering a quasi-formal way of depicting a computer's functional structure using their state-based digital model. In particular, they identified seven distinct *component types,* each defined by the nature of the kinds of operations it can perform. Giving special prominence to the component types they called "processor," "memory," and "switch," they named this mode of description the *processor–memory–switch* (PMS) model.

As the name suggests, the processor (P) as an abstract component type is responsible for interpreting and executing a program's instructions; the memory component type (M) stores information; and the switch type (S) establishes or breaks links between other components. The four remaining component types are the link (L), responsible for transferring information between other components either directly or by means of switches; a data operation (D) performs operations on and transforms information; the transducer (T) changes information from one form to another; and control (K) evokes and controls other components. The P component type is taken to be the most complex in that it may itself be internally composed of the other types.

Thus the functional structure part of the IBM System/360 architecture described by Amdahl et al. can be modeled as a PMS description. Bell and Newell parted ways

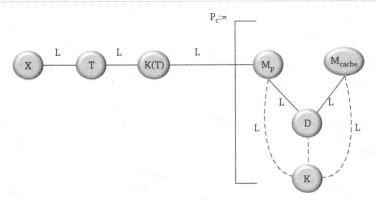

FIGURE 3.1 A simple PMS architecture.

from Amdahl and his colleagues, not only in quasi-formalizing this kind of description but also in postulating *PMS as a distinct architectural level*. In other words, Bell and Newell suggested that a computer's architecture can be defined or described at the PMS level of abstraction. The PMS formalism offered a framework for characterizing a computer's architecture at a particular level of abstraction.

And because a PMS architecture is a network-like structure, it can best be represented in pictorial (graphical) form.

In Figure 3.1, the solid lines denote data links (information paths); the dotted lines represent control links; X denotes the external environment; Pc signifies a central processor; Mp represents a primary (or main) memory; Mcache signifies a cache memory; and Pc is composed of Mp, Mcache, D, K, and the data and control links. The external environment is linked to a transducer (T) that will transform the input–output from and to the environment into a form suitable for the computer or for the environment. K(T) is the controller associated with the transducer and is directly linked to Mp. Although not shown in this diagram, additional attributes of the individual components (such as word size and memory size of Mp) may be specified alongside the component names.

But as we saw, for Amdahl, Blaauw, and Brooks, a computer's architecture entails much more than a functional structure. Bell and Newell, of course, recognized this. A PMS description was only one *aspect* of architecture. There was yet another formalism they had in mind, at another abstraction level.

3.4 ARCHITECTURE ACCORDING TO BELL AND NEWELL: THE INSTRUCTION SET PROCESSOR LEVEL

They named this formalism *instruction set processor* (ISP). That is, the focus was on the processor (P) component type within the PMS formalism. In addition to a PMS architecture for a computer, there was another aspect of the computer that fell within the architectural realm: the internal structure and behavior of the processor. Referring once more to the architecture of the IBM System/360, an ISP description would embrace all but the functional structure part of that description.

In fact, the focus of an ISP description was *behavior*. Figure 3.1 shows the structure of a particular Pc, a central processor; we see that it comprises a network of two memories, a data operation component, a control unit, and interconnection links. But this PMS description tells us nothing of how this network behaves: the pattern of state changes it affords.

The most effective representation of stage changes in a computational artifact is by way of procedural descriptions: algorithms or programs. This was precisely how Bell and Newell thought: The architecture of a processor's behavior would be described in procedural form. An ISP description would be a procedural description expressed in a program-like notation.

It was almost inevitable that the original *notation* Bell and Newell invented to describe ISP architectures should evolve into a computer description *language*. Such a language not only facilitates rigorous definitions of an ISP architecture; if implemented, the language could serve as the front end of a simulator and other *design tools* for computer architecture design and evaluation.

The transformation of ISP into a language called ISPS (instruction set processor specifications) by way of an intermediate version called instruction set processor language (ISPL) and its implementation was primarily the work of Mario Barbacci of Carnegie-Mellon University (along with collaborators) in the second half of the 1970s.[18]

We get a sense of the nature of an ISP architecture (expressed in ISPS) from the following example of a very simple, hypothetical computer that we will call X.[19] We can articulate this architecture in a hierarchic (top-down) fashion.

At the most abstract level it comprises five distinct components:

```
X :=    Begin
               ** Memory.state **
               ** Processor.state **
               ** Instruction.format **
               ** Effective.address.calculation **
               ** Instruction.interpretation **
        End
```

Each asterisked entity specifies the name of the corresponding component. The state of the computer as a whole is defined by the combination of the contents (mathematically, their "Cartesian product") of the *state variables* that are declared in **Memory.state**, **Processor.state**, and **Instruction.format**. For example,

---

    ** Memory.state **
        Mp ＼ Primary.memory[0:255]<0:11>
    ** Processor.state **
        PC ＼ Program.counter<0:7>
        ACC ＼ Accumulator<0:11>
    ** Instruction.format **
        IR ＼ Instruction.register<0:11>,
        Op ＼ Operation<0:2> := IR<0:2>,
        Ibit ＼ Indirect.bit <> := IR<3>,
        Adr ＼ Address<0:7> := IR<4:11>

At the next stage of refinement, the behavior of the computer at the ISP architectural level is specified in the last two major components:

    ** Effective.address.calculation **
        Z ＼ Effective.address<0:7> :=
        Begin
            If Ibit Eql 0 ⇨ Z ← Adr;
            If Ibit Eql 1 ⇨ Z ← Mp[Adr]<4:11>
        End
    ** Instruction.interpretation **
        ICycle ＼ Interpretation.cycle :=
        Begin
            Repeat
                Begin
                    IR ← Mp[PC] NEXT
                    PC ← PC + 1 NEXT
                    IExec ( )
                End
        End

Behavior is a dynamic entity that causes state change over time. In the component called **Effective.address.calculation**, the semicolon separating the two conditional statements signifies concurrency: The two conditional tests are performed in parallel. In **Instruction.interpretation**, the symbol NEXT signifies sequential execution. IExec ( ) is a call to another procedure that effects the execution of instructions. Here is a partial specification of this procedure:

```
** Instruction.cycle **
IExec \ Instruction.execution :=
        Begin
          Decode Op ⇨
            Begin
              0 \ AND := ACC ← ACC And Mp[Z( )],
              1 \ TAD :=  ACC  ← ACC + Mp[Z( )], ! Two's comp add
                2 \ ISZ := Begin
                              Mp[Z] ← Mp[Z( )] NEXT
                              If Mp[Z] Eql 0 ⇨ PC ← PC + 1
                            End,
                3 \ . . .
                4 \ . . .
                5 \ JMP := PC ← Z( ),
                6 \ RET := Begin
                              PC ← Mp[Mp[0]]  NEXT
                              Mp[0] ← Mp[0]–1
                            End
              7 \ . . .
            End
        End
  End
```

Here, the notation Z( ) represents a call on the address evaluation function Z.

There are many other features of ISPS that enable other, more complicated types of behavior of an ISP-level architecture to be expressed. In fact, in 1970, at the same conference where Gordon Bell and Allen Newell unveiled their PMS and ISP formalisms, Bell and his collaborators at Digital Equipment Corporation (DEC) also presented a paper announcing and describing the architecture of the DEC PDP-11 family of minicomputers—at the time, the highest-end family of DEC minicomputers.[20] Up until this time, many computer architectures had been described in the published literature. What was striking about this paper was that the authors defined, not completely, but in some detail, the ISP architecture in ISPL, an earlier version of the ISPS language.

Gordon Bell and his coworkers did not *invent* the concept of computer architecture. As we have seen, the pioneers in this matter were Gene Amdahl and his IBM colleagues in 1964. Nor, in fact, in specifying the DEC PDP-11 in ISP notation, were they the first to attempt to provide a rigorous specification of the architecture of a real, commercially produced computer. The credit for this lay also within the portals of IBM: In the same 1964 issue of the *IBM Systems Journal* in which a set of definitive papers were published on various aspects of the IBM Systems/360, there was another remarkable paper by three members of the IBM Thomas J. Watson Research Center. The paper was titled "A Formal Description of SYSTEM/360" and it was authored by

Adin Falkoff, Kenneth Iverson, and Edward Sussenguth.[21] They used the programming language APL ("A Programming Language") invented by Iverson in 1962[22] to describe the System/360 architecture.

But what Gordon Bell and Allen Newell achieved—augmented by their collaborators, most notably, Mario Barbacci—was to classify and delineate the scope of computer architecture within a quasi-formal general framework defined by the PMS and ISP descriptive systems. This was a significant achievement for the subsequent attempt, through the '70s and '80s, to develop a discipline of computer architecture. This agendum was a striking feature of the second age of computer science.

3.5 BUT *HOW* TO DISCIPLINE COMPUTER ARCHITECTURE?

As pointed out (in Section 3.1), by the beginning of the 1970s, there had emerged a notion of computer architectures-as-artifacts that serve as interfaces between software and hardware. Moreover, as the work by Bell and Newell indicates, these artifacts were, on the one hand, intrinsically abstract but, on the other, their effectiveness was made possible only when they were cast in silicon. For this reason, we can claim that, like programs, architectures constitute a class of *liminal* computational artifacts (Prologue, Section P.3).

There were interesting consequences of the idea of architectures-as-interfaces.

First, the design of computer architectures can begin only with the identification of a set of goals mandated from "above" (put forth from the realm of the computer's intended computational environment): programming languages and their compilers, operating systems and other system software, and the range of applications that would ultimately run on the implemented computer. At the same time, the design can also begin only by identifying the constraints imposed from "below"—the available physical technology. As we noted, the architect's task is to navigate carefully and judicially between the Scylla of software needs and the Charybdis of hardware constraints.

Second, we have also seen that computer architecture, as artifact, as design discipline, and as field of research emerged in a distinctive sort of way by this name only at the beginning of the 1970s. Computer architecture had escaped from its capture as a metaphor (drawn in part from the discipline practiced by architects whose domain is the spatial and functional design of buildings; and in part from its metaphorical use in other domains such as embodied in the expression "architecture of matter"). The metaphor could be dispensed with by computer designers. Instead computer architecture became a technical term, referring to a class of computational artifacts and their design, ensconced in its own little (sub)paradigm.

(Not all, however, elected to adopt this term. To take an example, in writing about the design of the Cambridge CAP computer in 1979, a highly innovative machine that enforced a sophisticated form of memory protection, its principal creators Maurice Wilkes and Roger Needham of Cambridge University devoted a detailed chapter to what they called its hardware design without mentioning the term architecture, even though what they described was, in fact, its architecture.[23])

Thus, in the course of the '70s and '80s the idea of a *discipline* of computer architecture as a genuine science of the artificial (see Prologue, Section P.2) emerged as a serious topic of conversation. And at the center of this conversation was this: *How should one discipline computer architecture?*

3.6 THE TECHNOLOGICAL IMPERATIVE (OR THE CHARYBDIS FACTOR)

In the realm of computing, the word technology (when used without further qualification) normally refers to the electronic, physical, or mechanical means for the *physical implementation* of computers. From the very beginning of electronic computing the computer designer's thoughts about computer design were shaped, indeed dictated, by what we may call the technological imperative: what the physical technology of the time permitted or constrained. Indeed, when Saul Rosen of Purdue University wrote his important historical survey of computers in 1969 (and published in the inaugural issue of ACM's *Computing Surveys*) he organized the development of commercial computers into first, second, and third "generations" according to the dominant technology deployed.[24] And there was concomitantly a temporal ordering of these generations. Thus the first generation (circa 1950–circa 1956) was characterized by vacuum tubes for processors, ultrasonic and electrostatic technologies for memory, and electromagnetic and electromechanical means for secondary storage and input–output devices. The second generation (circa 1956–circa 1964) was dominated by the first uses of discrete transistor circuits for processors and ferrite (magnetic) core memories. The third generation (circa 1964–1969, the year during which Rosen's survey ended) was characterized by the deployment of monolithic integrated-circuit (IC) "chips" for processors.

The dominant evolutionary theme in both processor (or logic) and memory technology from the mid-1950s on was the remarkable increase in the number of "gates" (the basic digital circuit elements) per chip. The trend in the growth of such chip density was encapsulated in 1965 by Gordon Moore, then of Fairchild Semiconductors (and later, cofounder of Intel Corporation), in an empirical "law"—which came to be known as Moore's law—that proclaimed that the number of components on an IC chip had doubled every year between 1959 and 1965, and it would likely continue to double every year over the next decade.[25] In 1979, Moore would note that this rate had decreased slightly to doubling chip density every 2 years.[26]

Depending on the device density of IC chips, it became customary to classify them according to the extent of their integration: "small-scale integration" (SSI), "medium-scale integration" (MSI), "large-scale integration" (LSI), and, by the beginning of the 1980s, "very large-scale integration" (VLSI).

The actual device count delineating these classes varied slightly from one author to another.[27] But what was significant from the computer designer's perspective was the increase in the functional capacity ("functionality") of the chips when viewed as building blocks in implementing computers. For example, in MSI technology, functional building blocks such as small arithmetic-logic units (ALUs) were available. The advent of LSI facilitated the implementation of small (2K words, 8 bits/word) "read-only" memories (ROMs) and even small (8-bit) *microprocessors*—an entire processor on a single chip. The advent of VLSI opened up the possibility of building much larger (16-bit or 32-bit) microprocessors.

At the time computer architectures emerged as a distinct class of abstract computational artifacts and computer architecture began to be recognized as a field of teaching and research in its own right (1970–1972), LSI technology was very much in evidence. Any reflections on the design of computer architectures were dominated by the implications of this technology.

So it was when Caxton Foster wrote an article in 1972 titled "A View of Computer Architecture."[28] Caxton's "view"—of the way computer architecture would go over the next 25 years—was entirely dominated by the prospect of a "complete computer on a chip." Foster used the term microcomputer to refer to such single-chip artifacts, and he speculated on what might a microcomputer "look like" in the future.[29] His main architectural concerns were the word length and the speed (in millions of instructions per second, MIPS) of such computers-on-a-chip, and, of course, the cost of such microcomputers. There were hardly any other architectural matters under discussion.

To repeat, the technological imperative extended its fierce pressure—the Charybdis of our metaphor—on the computer architect. But this imperative not only constrained what was or was not economically feasible, but it also *liberated* the architect's imagination in directions previously unimagined. The disciplining of computer architecture in the LSI–VLSI era turned out to be a history of this liberated, yet disciplined, imagination, as we see in Section 3.7.

3.7 "NOW YOU SEE ME, NOW YOU DON'T"

Computer architecture as a branch of computer science owes much to the people who created the IBM System/360. For not only did they introduce the concept of (and the word) architecture into computational discourse; they also conceived a remarkable idea of an (ISP-level) architecture as an artifact of its own that could be realized as physically different computers with different performance–cost characteristics. There was no actual machine corresponding to the System/360. *That* was the name of an architecture. The reader of the article by Gerrit Blaauw and Frederick Brooks describing the 360 architecture was told, just in passing, that the architecture described in the paper was embodied in six different physical computers called, by IBM, *models:* These models gave rise to six different physical computers called 360/30, 360/40, 360/50,

360/60, 360/62, and 360/70.[30] (Later, the models 360/65, 360/67, 360/85, and 360/91 would be added.) Other than this off-the-cuff mention, all that the reader learns are the details of a single, monolithic architecture.

The means by which the IBM designers implemented the same ISP architecture on a range of physical machines with distinct physical, performance, and cost characteristics was yet another of the 360 project's contribution to computer architecture and design. The technique was called microprogamming, invented in 1951 by Maurice Wilkes of Cambridge University.[31]

To understand this technique, consider the ISPS description of the hypothetical computer earlier in this chapter (Section 3.5). It is not difficult to see that the behavioral sections **Effective.adddress.calculation** and **Instruction.interpretation** describe procedures. It is the responsibility of the computer's *control unit* [the K element of the PMS description (Section 3.4; see Figure 3.1)] to implement these procedures by issuing *control signals* in proper sequence to effect these actions within the computer. Imagine now that instead of a rather complicated control *circuit* one implements these procedures as a *program* stored in a *memory*, and this program is interpreted by an appropriate sequencer, which issues control signals as per the individual "instructions" of the program. This program was called, by Wilkes, a microprogram, its constituent instructions, microinstructions, and the memory holding the microprogram, the control memory. Implementing **Effective.address.calculation** and **Instruction.interpretation** as a microprogram made the task of implementing the control unit of a computer much more orderly and comprehensible.

Until 1964, microprogramming had remained a "laboratory-scale" technique. Several research or one-of-a-kind computers had been built between 1951 and 1964[32]—including one by Wilkes and his colleagues. But it was the IBM System/360 that launched microprogramming as an industrial-quality, industrial-scale technique that effected a transfer of research knowledge into industrial technology. This was, then, the third major contribution of the IBM System/360 project to computing.

There was yet a fourth. Using microprogramming, the 360 engineers implemented microcodes (microprograms) that executed on the physical model and that interpreted not only the System/360 instruction set; but also the instruction set of *other* previous and commercially successful IBM computers, notably the IBM 1401, a widely used second-generation (technology-wise, that is) general-purpose computer. This allowed former users of the 1401 who had "upgraded" to a 360 model to execute its 1401 software on the 360 model. When running 360 software, the 360 model (e.g., 360/40) *was* a System/360 machine; when running 1401 the 360 model *became* a 1401 machine. This process of simulating, in effect, one computer by another through microprogrammed means was termed emulation. The 360/40 (say) was the *host* computer, and the emulated architectures, the System/360 and the 1410, were termed target computers. Depending on which emulator was running on the host, the target was now a 360, now a 1401 (Figure 3.2).

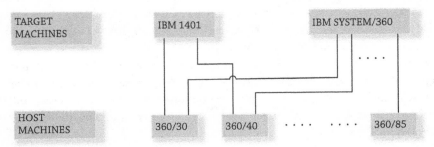

FIGURE 3.2 Emulating different target machines on many "biased" hosts.

3.8 THE DISCOVERY (OR INVENTION) OF MICROARCHITECTURE

The 360 and 1401 emulators were held in a ROM. That is, the microcode was "hard-wired" into the pertinent 360 model (e.g., 360/30) control store. This reflected the technological imperative of the time: Only ROMs were fast enough and economically feasible for the task at hand. Besides which, the user could not alter the microcode.

But even before the end of the 1960s semiconductor memory technology had evolved to the point that the writable control store (WCS) had become both economically and performatively feasible. A new vista was unveiled.

Robert Rosin (then) of the State University of New York (SUNY), Buffalo, was not the first to grasp the new possibilities opened up by the advent of economical and fast WCSs. Others, such as Harold Lawson of Linköping University in Sweden and Helmut Weber [Nicklaus Wirth's collaborator in the design of the programming language EULER (see Chapter 1, Section 1.7)] had been exploring new pathways for emulation to work.[33] But, writing in 1969, Rosin presented a concise, comprehensive, and timely account of the newly emerging vistas afforded by the availability of the WCS.[34]

One prospect was to build a general-purpose emulator to support programming languages. Rosin's idea was to design an "intermediate language architecture" for a given host machine to which compilers for different programming languages would translate their respective source programs. [This is the converse of the idea of a single virtual machine corresponding to a specific programming language such as the O-code machine for BCPL and the P-code machine for Pascal (see Chapter 1, Section 1.16) that would emerge in the late 1970s.] A microprogram stored in WCS would then interpret the code for the intermediate language architecture. Imagine, so the narrative went, that a "user"—a systems programmer, say—of a computer containing, in addition to a ROM-based microprogram that defined the computer's "native" ISP-level architecture, an "empty" WCS. Then the user could bypass the native architecture and emulate any particular target machine architecture he or she desired.

Writing microprograms would no longer be solely the prerogative of the computer designer or hardware engineer. More important, the computer user would no longer be beholden to the computer maker, as far as the ISP-level architecture was concerned. The user would become a "systems microprogrammer" or, from a different perspective, an architect and create an "ISP architecture of one's own" that would efficiently support the particular computing environment of interest to him or her. This user–systems microprogrammer would have access to the computer at a level of abstraction just above the level of the physical hardware itself. Thus a new architectural level was discovered by users. Over time it was referred to as microarchitecture.

3.9   THE UNIVERSAL HOST MACHINE: AN AUDACIOUS ACT OF LIBERATION

The idea of emulation, as we have seen, originated in 1964 with the IBM System/360 family of mainframe computers. IBM continued to use microprogramming and emulation in the successor to the 360, the System/370 family, first introduced in 1970 and evolved through its various models over the next two decades. Another series of mainframes that enthusiastically adopted microprogramming and emulation was the RCA Spectra series in which emulators were implemented not only for RCA's previous computers but also IBM's second-generation 1401 and 1410 machines.[35]

But it was the availability of the WCS that moved emulation into the public domain, so to speak. The very idea of one physical machine with its own inherent identity disguising itself as another captured the imagination of a segment of the computer architecture community. To some, the raison d'être of microprogramming was no longer the orderly implementation of a computer's control unit as envisioned by Maurice Wilkes in 1951. Rather, the new Young Turks took his analogy between programming and microprogramming seriously and "ran with it." They came to believe that microprogramming was another "kind" of programming whose real raison d'être was the emulation of any machine of one's fancy by another. And the ultimate liberation of microprogramming from the clutches of the hardware designer was the audacious and exhilarating idea of a user-microprogrammable computer with no inherent, native, ISP-level architecture of its own. Rather, it would serve as a *universal host machine* (UHM) that the (micro)programmer could (micro)program to emulate any target machine. This genre of microprogramming, to be distinguished from the "traditional" genre, was called dynamic microprogramming. The contrast between the old and the new is depicted in Figures 3.2 and 3.3.

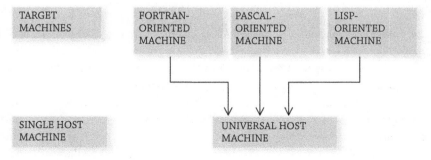

FIGURE 3.3 Emulating different target machines on one "unbiased" host.

It is not entirely clear as to exactly who was the originator of this audacious idea. At any rate the concept of the UHM can certainly be traced back at least to Robert Rosin's influential paper of 1969 wherein he asked the question: "Is there a microprogramming environment which best supports" typical programming environments?[36] In answer to his rhetorical question, he suggested that by shifting the property of universality down to the microprogramming level, designers could offer support for different programming environments by way of emulators, each tailored to meet the requirements of such environments.[37]

Did Rosin derive this idea from elsewhere? It could well have been, for in his paper he described a contemporary commercial machine, Standard Computer Corporation's IC-900 which, he noted, was "designed to support the emulation of a wide variety of machines" including the ISP architectures of some of the leading mainframe computers of the day such as the System/360, Univac 1108, and the General Electric GE 645.[38]

What was astonishing was the rapidity with which the idea of UHMs propagated within both the academic and industrial communities. On the one hand, the theoretical idea of the UHM was further developed in the early 1970s, most notably by Rosin at SUNY Buffalo, Michael Flynn at Stanford University, and their respective coworkers [39] On the other hand, several UHMs were designed and implemented in both industrial[40] and academic sectors—in the United States, Denmark, and Japan, among other places.[41] Indeed, this remarkable interest in UHM (or user-microprogrammable) microarchitectures, prompted the publication, in 1976, by Alan Salisbury, a US Army engineer and scientist, the first book devoted entirely to this topic,[42] followed 2 years later by a more specialized monograph, by Elliot Organick and J. A. Hinds of the University of Utah, devoted to the Burroughs B1700 UHM.[43]

There is, however, a curious footnote to these efforts. The computer at the microprogramming level came to be called, by some, a microprocessor. This usage was confusing because, by the early 1970s, entire processors on a chip were also called microprocessors. Eventually the latter usage prevailed.

3.10 THE "HORIZONTAL" ARCHITECT

In 1970 Samir Husson of IBM published *Microprogramming: Principles and Practics*.[44] If the publication of a textbook is a hallmark of the maturation of a discipline—a characteristic of what it means to attain paradigm status (in Thomas Kuhn's sense[45])—then Husson's book surely announced the emergence of a subparadigm of computer science. For it was the first, and impressively authoritative, tome on the state of the art of microprogramming circa 1970. One may call it as representing the first era of microprogramming, spread over the two decades since Wilkes's invention of the technique in 1951—an era wherein microcode was wired into read-only semiconductor memories and the task of emulation was constrained by the objectives of the computer manufacturer. And, as far as the latter was concerned, "what went on below the surface of the functional processor stayed below the surface." Microprogramming was an esoteric craft practiced by hardware designers.

The advent of UHMs and dynamic microprogramming ushered in not only a new level of architecture—microarchitecture—but also a new breed of architects: Let us call them *microlevel architects*. Their task was to ponder, analyze, conceive, and design the architecture of UHMs with the interest of the user–microprogrammer in mind. Indeed, to some extent the user–microprogrammer and the microlevel architect were joined at the hip. The concerns that had driven the traditional computer architect at the ISP level of abstraction were migrated down to the microarchitectural level, introducing its own set of problems that were absent at the conventional ISP level. The principal microarchitectural concerns were the organization of the control store (or microprogram memory), the design of *microinstruction* formats, ways of addressing and fetching microinstructions from memory, and the design of the "microinstruction execution cycle" (analogous to the ICycle). The microlevel architect was dealing with the computer almost at the level of physical hardware, involving the issue and transmission of digital control signals that would enable hardware units to do their jobs. And these concerns brought forth design and research issues that were foreign to ISP-level architecture.

Most prominent of these were twofold: First, the execution of a microinstruction (MI) is rigorously synchronized with one or more *clock cycles*; second, and most distinctive, an MI holds possibly several, more primitive *microoperations* (think of these as multiple opcodes), such that the execution of the microinstruction induces the *parallel* activation of multiple microoperations, each such activation causing a control signal to be issued to some specific unit in the physical computer. Such microinstructions, encoding for several microoperations, could be quite large in width (one commercial UHM, the Nanodata QM-1, had a 360-bit-wide microinstruction format), for which reason they were named horizontal microinstructions; and the execution of a microinstruction would result in *parallel processing* at a very low level of abstraction. Microprogramming such machines came to be called horizontal microprogramming and was quite unknown at the ISP-level architecture for most conventional mainframe computers in the 1970s.

3.11 FORMALIZING MICROPROGRAMMING: TOWARD A (TINY) SCIENCE OF THE ARTIFICIAL

Samir Husson's book, published in 1970, spoke to the "Principles" of microprogramming. These were heuristic principles: rules and precepts discovered empirically and intuitively that worked (to a greater or lesser degree) to the designers' satisfaction. To use Herbert Simon's term, these principles were *satisficing*: good but not necessarily the best.[46] The design of microprogrammed control units through the 1950s and 1960s, like the design of computer architectures in general, was a heuristic activity; to the extent it was scientific, it was a heuristic science.

The advent of UHMs and dynamic microprogramming seemed to act like a magic wand. Computer scientists of a theoretical disposition suddenly became interested in the topic. If the 1960s witnessed a heuristic science, the 1970s and the early 1980s saw the emergence of a *formal* science of microprogramming—at least some aspects of it. Indeed, no part of computer architecture showed more vividly how a science of the artificial of the kind Herbert Simon had envisioned can emerge out of a well-trodden heuristic domain than we saw in the case of microprogramming.

This metamorphosis is illustrated by comparing the two important survey papers on microprogramming published in the ACM journal *Computing Surveys* in 1969 with a paper published in 1979, exactly a decade later in the same journal. In 1969, Maurice Wilkes, the inventor of microprogramming, reviewed the "Growth of Interest in Microprogramming."[47] The same year Robert Rosin discussed "Contemporary Concepts of Microprogramming and Emulation."[48] Both papers were authoritative and comprehensive. And neither showed the remotest hint of formal theory. Reading Wilkes and Rosin, the reader comes to the inescapable conclusion that here was a field full of heuristic principles and empirical practices.

The present writer was the author of the 1979 paper.[49] Its subject matter was "The Organization of Microprogram Stores." And its concerns were of a kind the mathematically minded would find congenial. The opening segment of the paper announced that its aim was to survey aspects of microprogram store organizations and to review progress in techniques for their *optimization*.[50]

Such optimization techniques are excellent examples of the transformation of heuristic-based scenarios in architecture design into formal problems that could draw on mathematics and logic or allied topics to solve them, thus enabling *provable* claims about the solution. These were exactly the kinds of problems in designing artifacts for which a *science of design*—of the sort Herbert Simon envisioned[51]—emerged in parts of computer science.

3.12   THE CASE OF THE MICROWORD LENGTH-MINIMIZATION PROBLEM

To get a sense of the nature of these optimization problems, consider as an example the following situation.

Suppose the microlevel architect is designing a control store. To do this he begins by identifying the complete set of microoperations (MOs) for the machine. For the sake of exposition suppose this set consists of just eight MOs:

$$M = \{ m1, m2, \ldots, m8 \}.$$

Each MO corresponds to a control signal that is issued when the MO is activated. The activation of the MO and the duration of the control signal are controlled by a clock signal. That control signal enables some sort of activation of a hardware unit in the computer at the microarchitectural level.

MOs are encoded in "fields" of the microword. So if two MOs, say mi, mj are encoded in the same field, then a 1-bit configuration of the field will represent mi, another bit pattern mj. But obviously, if mi, mj share the same field they cannot be activated simultaneously. If, however, mi, mj are located in two different fields of the microword, then they *can* be activated in parallel.

So, given the set M of MOs for a given machine, in designing the microword organization a pair of MOs mi, mj in M will be *potentially parallel*—we'll denote this as mi || mj—if there is no *conflict* between their respective resource requirements. Based on this condition, a set of "potentially parallel" sets of MOs is constructed. Suppose for this example set M, this set of parallel sets is:

$$P1 = \{ m1, m3, m5 \},$$
$$P2 = \{m2, m5\},$$
$$P3 = \{ m2, m4\},$$
$$P4 = \{m1, m3, m8\},$$
$$P5 = \{m1, m5, m7\},$$
$$P6 = \{ m1, m4, m6\},$$
$$P7 = \{ m2, m8\},$$
$$P8 = \{m5, m6\},$$
$$P9 = \{m6, m8\}.$$

So each parallel set identifies MOs that, taken pairwise, are potentially parallel. This suggests that ideally these MOs should *not* be encoded in the same field of the microword organization because the potentially parallel MOs can be activated in parallel from the same microword. For example, in $P1$, we have $m1$ || $m3$, $m1$ || $m5$, and $m3$ || $m5$. So we would *like* to place $m1$, $m3$, and $m5$ in distinct fields of the microword.

Generally, then, the architect's aim should be to determine a microword organization comprising a set of fields of *minimal length* (in terms of the number of bits) that allows each potentially parallel set of MOs to be activated from MIs contained in the control memory's set of microwords. This, then, is the *microword length-minimization problem*. The point is that the smaller the microword length, the more compact the control memory in its breadth dimension.

For example, given the preceding set of potentially parallel sets of MOs, three possible organizations are

(1) <M1, M2>; <M3, M6, M7>; <M4, M5, M8>: The number of fields = 3; number of bits required for encoding the fields (including the "no-op" option, the situation in which no MO from a field is to be activated from a MI) $B = 2 + 2 + 2 = 6$ bits.

(2) <M2, M3, M6, M7>; <M4, M5, M8 >; < M1 >: The number of fields = 3; $B = 3 + 2 + 1 = 6$ bits.

(3) <M2, M6 >; < M4, M5, M8 >; <M3, M7 >; <M1 >: The number of fields = 4; $B = 2 + 2 + 2 + 1 = 7$ bits.

As it happens, the minimal ("optimal") solution for this problem is B = 6 bits, so either solution (1) or (2) is a possible microword organization.

The microword length-minimization problem was first posed by Scott Schwartz of Honeywell for ROMs in 1968.[52] Interestingly, Schwartz's paper was presented at a major conference devoted to the mathematical theory of switching circuits and automata. It was an exceedingly formal paper, replete with theorems and proofs. A later articulation of the problem and its solution, offered in 1970 by A. Grasselli and U. Montanari of the Instituto di Elaborazione dell' Informazione in Pisa, Italy, and by Sunil Das, Dilip Banerji, and A. Chattopadhay of the University of Ottawa in 1973 also drew on the theory of switching circuits.[53] In contrast to the switching-theoretic approach of these papers, Carlo Montangero, also from the Instituto di Elaborazione dell'Informazione, Pisa, produced, in 1974, a solution that drew on algebraic graph theory.[54]

All these papers were concerned with the minimization of ROM microwords. Moreover, none of them addressed the complications introduced by microoperation timing issues. The problem of minimizing WCS microwords, unifying this problem with that of the ROM case and incorporating clock cycles and timing, was addressed in 1975 by the present author, then at the University of Alberta.[55]

Interest in this problem did not wane. Other efforts followed adopting other techniques, including optimization using a linear programming algorithm (in 1976)[56] and the use of a version of the branch-and-bound heuristic algorithm (in 1977).[57] In 1981, the problem of optimizing the word length of ROMs was revisited by Charles Silo and James Pugley of the University of Maryland and Albert Jeng of the Computer Sciences Corporation.[58] These authors drew on characteristics of a particular IC technology to consider the design of "multivalued ROMs."

To reiterate a point made earlier, few aspects of computer architecture were so intensely subjected to mathematical formalization as was the problem of microword length minimization. This was quintessential science of design as envisioned by Herbert Simon.

3.13   THE LINGUISTIC IMPERATIVE (OR THE SCYLLA FACTOR)

We have noted that computer architecture—both as artifact and as discipline—had to navigate between the Charybdis of hardware constraints and the Scylla of software desiderata. One of the historical ironies of the technological imperative leading to the advent of dynamic microprogramming is that this same imperative also strengthened the software resolve. For these phenomena opened up the possibility of users of UHMs designing ISP-level architectures at a higher level of abstraction than conventional ISP architectures. One of the intentions of such higher-level ISP architectures was to support, explicitly and directly, high-level programming languages. Such designs came to be called *high-level language* (HLL) or *language-directed* (LD) *architectures*. And, as an architectural *style*, this had nothing to do intrinsically with dynamic microprogramming or UHMs. The latter were just one avenue to such HLL architectures.

As we have done so many times, we can trace the origins of this idea to the early 1960s. And when it was first conceived, the idea was quite radical. In 1961, Burroughs

Corporation began work on the development of a mainframe computer that would support both the compilation and execution of Algol-like block-structured programs. The outcome, created through the '60s and the early '70s, was a family of machines called the Burroughs B5000/B6000/B7000 series that, to use a cliché, blazed a remarkable new trail in the realm of computer architecture.[59]

The central architectural innovation in the Burroughs machines was the *stack* as the memory organization to manage the dynamics of storage management for block-structured programs, of the kind pioneered by the Algol 60 designers and adopted by the designers of PL/1, Algol 68, and Pascal (see Chapter 1). The stack was, of course, well understood as a data structure, and we have encountered its functional characteristics in our discussion of the development of abstract data types (see Chapter 2, Section 2.10). To understand its role in managing block-structured program execution, consider, as a hypothetical example, the following Algol 60 program fragment named P.

Now, block-structured programs obey what is called the scope rule. The scope of a declared object refers to the region of the program in which the declared object is accessible and can be referred to by its identifier. In the case of Algol 60, the scope of a declared object is the block in which it is declared and all the blocks declared inside this block, unless the same identifier is used for another declared item within an inner block. In the case of program P, the scope of variables y, z declared in block A is the entire program. The scope of x is the entire program *except* inside block C, because the identifier x appears there in another declaration. The scope of variables m, n declared in block B is that block as well as blocks C and D. The scope of variables p, q is confined to block C.

Blocks defined at the same level (in this example, C and D) also obey the scope rule. The scope of p, q does not extend to block D just as that of s, t does not extend to block C.

```
A:   begin comment block at level 0;
        real x, y, z, integer n;
        "executable statements"
B:          begin comment block at level 1;
               integer m, n;
               "executable statements"
C:                 begin comment block at level 2;
                      integer p, q;  real x;
                         "executable statements"
                   end comment block C
D:                 begin comment block at level 2;
                      real s;  boolean t;
                         "executable statements"
                   end comment block D;
            end comment block B;
      end comment block A
```

The other important property of block-structured programs is that at run time blocks are dynamically created and deleted. Its declared objects are thus also dynamically created (on entry to the block during execution) and destroyed (on exit from the block). For program P this means that during execution the variables in block A are the first created and the last to be destroyed; those in B are the next to be created and the second last to be destroyed; and so on. Figure 3.4 shows four states of the stack when the preceding program is executing codes in blocks A, B, C, and D respectively. The shaded parts indicate variables that are not accessible in the relevant stack states.

This combination of the scope rule and the *last-in–first-out* (LIFO) property of block-structured programs suggests that the stack is a natural memory organization for managing storage allocation during program execution. Architecturally, the stack is more than a data structure; it is the computer's *memory* as far as the programming environment is concerned: Space in this memory is allocated on entry into a block and deallocated on exit from the block in a LIFO mode.

If the stack is an *architectural* feature (rather than a *programmed* data structure), the hardware must support it. Thus instructions such as

BLOCK *reserved amount*

UNBLOCK *reserved amount*

would be part of the machine's ISP-level instruction set.

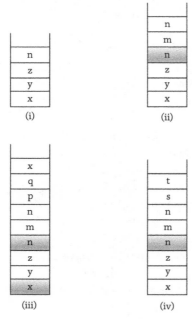

FIGURE 3.4 States of the stack during program execution: (i) in block a; (ii) in block b; (iii) in block c; (iv) in block d.

There is, of course, much more to a *stack machine* architecture than this, for that same memory can be used to execute assignment statements and procedure calls and returns. All these contributed to the architecture of HLL or LD machines.

3.14   THE SEMANTIC GAP AND ITS DISCONTENT

The Burroughs computers are early embodiments of a philosophy of computer architecture that strove to distance itself from the status quo: a philosophy that preached a language-driven, or more generally, a computing-environment-driven style of computer architecture (the Scylla factor in the large, so to speak). But we must note that the Burroughs designers, perhaps the most avant garde among corporate computer makers of the time, were not the sole advocates of this philosophy. In the commercial sector, for example, the UK-based English Electric Company produced their KDF.9 computer, roughly contemporaneously to the early Burroughs project.[60] The KDF.9 achieved a measure of celebrity because one of the very first Algol 60 compilers, called the Whetstone compiler, was designed and implemented by Brian Randell and L. J. Russell to run on the KDF.9.[61]

Also in the United Kingdom, a few years later, beginning in 1968, International Computers Limited (ICL) initiated the design and production of the ICL 2900 series, also intended to support efficiently a range of common programming languages.[62] But the architecture of this series was itself indebted to that of the MU5 computer, designed at the University of Manchester between 1967 and 1974 by a team led, initially, by Tom Kilburn, one of the world's pioneers of electronic computing.[63] The MU5 had many significant architectural features, but the one that catches our attention was the statement made in a grant application made to the UK Science Research Council in 1967:

"An instruction set will be provided which will permit the generation of more efficient object code by the compilers."[64]

Perhaps it is also no coincidence that John Iliffe was employed by ICL at the time the company was developing the 2900 machines. Iliffe had very definite—and advanced—views of what a computer's architecture should be like, which he expressed in a slim monograph titled *Basic Machine Principles* (1972; first edition 1968).[65] Iliffe's discontent with the existing architectural paradigm was manifold. One prominent example was his antipathy to the constraints the so-called von Neumann style of architecture imposed on the structure of a computer's main memory: the "linear store" as he called it, the sequential addressing of memory locations. For, he pointed out, the data and program storage needs of actual computational problems did not cohere with the linearity of stores.[66]

The constraint of the linear store compelled the programmer (or the compiler) to "reduce the problem to linear form"[67]—to map the "natural" structure of a computation (in particular its data structures) onto the linearity of the von Neumann store. And, early in his book, he gave a hint about his basic philosophy: Computer design should proceed "from the outside," "working inwards": "top down," beginning with the requirements imposed by the programmer's needs toward machine hardware. In Iliffe's worldview Scylla was the prime mover.[68] It was not architecture as an artifact alone that concerned Iliffe but also the architecture *design method*.

A shared worldview was thus clearly in evidence at the time, expressed by both avant garde companies like Burroughs Corporation, English Electric, and ICL, and by academic computer scientists such as Iliffe—who though industrially employed seemed to have a distinctly academic bent of mind, as his writing revealed—and Robert Barton of the University of Utah, another early champion of LD architectures.[69]

And as the 1970s progressed, as the design of programming languages, compiler technology, and operating systems evolved to create richer and increasingly complex computing environments, a new term was coined to capture the discontent felt by a segment of the computer architecture community: *semantic gap*—to characterize the chasm between the concepts governing the actual programming environment and the concepts embodied in actual ISP-level architecture. If the ISP-level architecture was intended to be the interface between hardware and computing environment (Section 3.1), the semantic gap was a massive indictment of the failure of traditional ISP architectures; for it represented the schism between the computing environment and conventional architecture (Figure 3.5).

Probably it was the industrial computer scientist Glenford Myers, of the IBM Systems Research Institute who firmly implanted the term semantic gap into the architectural consciousness in 1978 with his book *Advances in Computer Architcture* (second edition 1982).[70] The term was probably "in the air" before Myers, and he himself seemed to attribute it to a 1973 report by Ugo Gagliardi of Harvard University.[71]

The title of Myers's book is also noteworthy. The principles expressed in it, he seemed to be telling his readers, were avant garde principles of architecture design. He was advocating something akin to a paradigm shift in the realm of computer architecture.

The semantic gap, Myers pointed out, was on many fronts but most notably in the context of programming languages and operating systems. Using PL/1 as a "typical" programming language of the 1970s and the IBM System/370 as the exemplar of ISP-level architectures of the decade, Myers laid out in no uncertain terms the schism between language and architecture. For example, the array—"the most frequently used data structure in PL/1 and most programming languages"[72]—is implemented in PL/1, Ada, and other languages in terms of a host of concepts for referring to, representing,

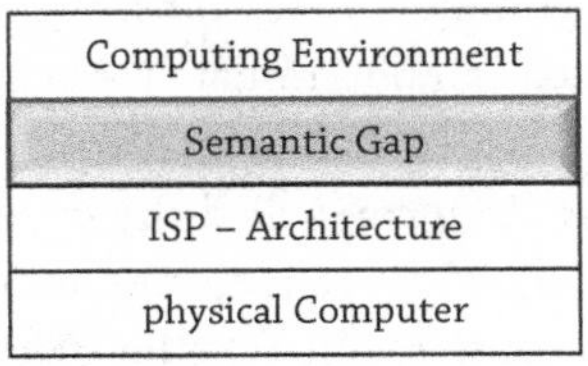

| Computing Environment |
| Semantic Gap |
| ISP – Architecture |
| physical Computer |

FIGURE 3.5 The "semantic gap."

and manipulating it. The System/370 provided nothing in support of these concepts. So also in the case of the data structure called "structure" in PL/1 and Algol 68 or "record" in Pascal and Ada.

In PL/1 and other languages the procedure call is quite complex, entailing saving the state of the calling procedure, allocating dynamically storage space in memory for local variables of the called procedure, transmitting arguments from the calling to the called procedure, and initiating the execution of the latter. There was virtually nothing in the 370 instruction set that provided support to the semantics of procedure calls.

This linguistic semantic gap was perhaps the most visible instance of the phenomenon, but it was not the only one. There was also a chasm between the demands imposed by operating systems and the prevalent architectural style. For instance, nothing in the 370 architecture recognized the *process* concept—which has a very definite, yet subtle, meaning in the context of operating systems. It signifies a program or a subprogram in a state of execution. The process is the dynamic *alter ego* to the static program held in memory. And one of the fundamental responsibilities of an operating system was then (and still is) process management: the creation of a process, its suspension and (reawakening), its termination; and the synchronization of concurrently active communicating processes. Computer scientists, most notably Edsger Dijkstra and C. A. R. Hoare, had invented concepts—semaphores, critical sections, monitors, and send–receive protocols—that were "household" concepts in the operating systems literature. Yet none had any reflection in the 370 architecture design.

These examples should suffice to convey the essence of Myers's argument and his discontent with the nature and extent of the semantic gap. And, to reiterate, there had been others, since the 1960s, who had this same sense of dissatisfaction with the state of the architectural paradigm. Myers's important contribution was to articulate these thoughts systematically and thoroughly—and to emphasize that these were *live* issues even in 1982.

3.15 "THE WORM TURNETH"

And yet, by the time the second edition of Myers's *Advances in Computer Architecture* appeared (1982), there were signs of a turning of the worm, in both the industrial and academic sectors. The topic of LD architecture had emerged as a vibrant and visible discipline within the larger canvas of computer architecture. Like an effective military campaign, the semantic gap was being breached on different fronts.

One approach followed the general philosophy embodied in the Burroughs machine architectures: to elevate the abstraction level of ISP architectures so that the semantic gap between the computing environment and the architecture was significantly narrowed, making the mapping between HLL programs and the executable object code simpler and more efficient. The idea was to make more visible the correspondence between programming language concepts and architectural concepts.

One of the most prominent exemplars of this approach, coinciding temporally with the emergence of object-oriented programming and the programming language Ada (Chapter 2, Sections 2.12, 2.14), was Intel Corporation's iAPX 432 "general data processor," circa 1981,[73] a 32-bit microprocessor. Myers—by then employed by Intel so, perhaps, not the most objective of observers—regarded this as possibly "the most significant architectural development in the last 5–10 years."[74] The central architectural concept was an entity called object, which, as the name suggests, mirrored loosely the abstract data type concept. Memory in the iAPX 432 architectural level was a *network of objects*, each an instance of a small number of object types. For instance, an "instruction object" was a procedure or subprogram; a "domain object" referred to the "access environment" of instruction objects; and a "context object" would be created during a subprogram call to specify the run-time environment of the subprogram along with its local variables. In the 432, objects and the operations defined on them were directly supported by the hardware. An iAPX 432 program would be composed of domain, instruction, and context objects.

There was, of course, much more of note in the 432 architecture, but its philosophy of raising the architectural level to reduce the semantic gap should be evident to the reader. Needless to say, registers were banished from this architecture: All iAPX 432 instructions took their operands from memory and left their results in memory. To use a taxonomic term, the Intel 432 manifested a memory-to-memory architecture. It was not a register machine.

But, amid all the discussion of approaches to language-directed architecture design—for instance, microprogramming UHMs continued to interest computer scientists and computer designers—the approach that most caught the broader architectural imagination was, in many ways, as audacious in concept as had been the UHM concept almost a decade earlier. This was the concept of the *reduced instruction set computer*. And its audacity lay in challenging a continuing and unexamined trend in the evolution of computer architectures: the trend of ever-growing complexity. Here the worm turned in a most unexpected way.

3.16   REVERSING AN EVOLUTIONARY TREND: A RISCY MOVEMENT

In 1980, David Patterson of the University of California, Berkeley, and David Ditzel of Bell Laboratories published a paper titled "The Case for a Reduced Instruction Set Computer" in *Computer Architecture News*.[75] Like others such as John Iliffe and Glenford Myers, Patterson and Ditzel were discontented with the state of computer architecture, but their discontent lay in a different direction. It lay in their observation of a historical trend: that architectures had been evolved in complexity over time, where complexity was characterized by such parameters as the size of a computer's instruction set, the functional power of individual instructions, the variety of data types, and the number of distinct addressing modes (that is, ways in which data objects could be referenced by instructions.)

They coined the term complex instruction set computer (CISC) to describe such machines.

On the other hand, they argued, given the trend in physical technology—specifically, the emergence of VLSI, which enabled an entire processor and memory to be placed on a single IC chip, and the evolving nature of computing environments, wherein almost all programming was done using high-level programming languages—there were neither theoretical nor practical justifications for this trend toward increasing complexity to continue. Rather, more cost-effective computers with simple architectures could be built taking advantage of both the technological and programming realities. They named such a machine the reduced instruction set computer (RISC). Thus were born two new concepts, two new terms, and a new way of classifying computer architectures—though the RISC class was hypothetical, as there were yet no such computers.

Here, then, was an unexpected and rather ironic twist to the Scylla–Charybdis story of architecture. For what Patterson and Ditzel were contemplating was that designing ISP-level architectures did not *have* to be a battle between the Scylla of software and the Charybdis of hardware. It could be a cooperative endeavor between the two.

But their proposal was also unexpected, indeed quite radical in another direction. Anyone even moderately familiar with evolutionary phenomena in nature or culture would realize that evolution carries with it the idea of growth of complexity. A fundamental reason for this is, as widely believed, that the environment in which an evolving entity resides itself changes and can become, in some sense, more complex. Thus, whether by natural selection in the case of living objects[76] or by design in the case of artifacts,[77] the entity must necessarily become more complex to not just survive but thrive in the changing environment.

Patterson and Ditzel were advancing a reversal of this complexity trend: They were suggesting that architectures do not have to evolve toward increasing complexity because both the material basis, Charybdis, and the environmental condition, Scylla, did not deem this necessary.

Let us call these three claims—the evolutionary trend toward increasing complexity, the technology–programming environment situation, and the claim of cost effectiveness of RISCs—the initial *RISC postulates*.

The first postulate was empirically based. One needed only to examine particular "lines" of computers—mainframes (e.g., the IBM System/360 and System/370 families or the Burroughs B5000/B6000/B7000 series), minicomputers (e.g., the DEC PDP-11/VAX-11 family), or microprocessors (e.g., the Motorola MC6800/MC68000 or Intel's 8080/8086/iAPX 432 series) to find evidence in support of this postulate.

But what were the specific *causes* of this trend? One was that because the computing environment was dominated by HLLs, new instructions had been introduced to support the efficient implementation of HLL programs—so as to reduce the semantic gap. Laudable as this goal was, an unintended consequence was to increase the size of the instruction set and the functionality (and thus the complexity) of individual instructions.

Moreover, as biological evolution tells us, even within a given environment artifacts compete for dominance. Such competition acts as a selection pressure so that only those artifacts "survive" that perform best in the given environment. In practical terms this puts pressure on the architect to design machines that are more cost effective than their competitors, for instance, by making them faster for the same or moderately increased cost.

One way architects and system designers effected this was to produce powerful instructions that did the job of a programmed sequence of simpler instructions—that is, by migrating functions from software to the physical computer. The idea was to reduce the number of instruction fetches to be performed to complete a functional task. Migrating functions across the architectural interface between software and hardware naturally increased the size of the instruction set as well as the functional complexity of the new instructions.

The initial RISC manifesto was, however, rather fuzzy in delineating the character of a RISC computer. But this was not unusual in the larger scheme of things. Initial proposals serve as visions, and so are often vague, incomplete, ambiguous. The purpose of a manifesto is to initiate and stimulate a conversation on a topic on which there had never been a prior conversation because the topic had not existed before. This was what happened with the initial RISC postulates proposed by Patterson and Ditzel. An animated debate was immediately spawned that came to be called the RISC–CISC controversy, a controversy that ranged over the next 5 years.[78] Lines were drawn. At Stanford University, led by John Hennessey,[79] at the IBM Watson Research Center led by George Radin,[80] and at Berkeley, led by Patterson,[81] projects were launched and in the process, the original, intuitive notion of the reduced instruction set computer was refined, over time, into a small set of reasonably clear characteristics:[82]

(1) All instructions are register–register type except LOAD and STORE (which transfers from memory to registers and from registers to memory, respectively).
(2) All instructions (except, possibly, LOAD and STORE) consume a single processor cycle.
(3) Fixed-length and fixed-format instructions that do not cross memory word boundaries.
(4) Relatively few operations and addressing modes.
(5) Use of hardwired rather than microprogrammed control.
(6) Use of certain specific processor implementation and appropriate compiler implementation techniques.

It is not so important for the reader to be familiar with these and other associated technical details characterizing RISCs. What is striking was the imperative nature of this refined RISC concept. This became a clarion call for architects to reverse the complexity trend. Scientifically speaking the refinement produced what we may call The

RISC Hypothesis: Machines that satisfy the RISC features will be more cost effective than machines that do not satisfy them.

But for this hypothesis to be properly scientific a definition of cost-effectiveness was needed. Patterson and his collaborators offered the following as defining criteria:

The speed of the resulting processor, that is, the execution time of individual instructions and the throughput time of instruction streams; ease of compilation; the time to design, verify, and implement the architecture; and the time to test and debug the resulting physical machine.[83]

3.17 TAKING "EXPERIMENTAL COMPUTER ARCHITECTURE" SERIOUSLY

But a hypothesis—to count as a genuine scientific proposition—must be testable. More precisely, if one is a follower of philosopher of science Karl Popper, for something to count as a scientific proposition it must be so couched as to be potentially falsifiable.[84]

Conducting falsifiable experiments on architectural hypotheses had never been particularly common. If the term "experimental computer architecture" was at all used, it referred to the empirical study of architectures: that is, implement a proposed architecture, as a prototype or a fully operational, one-of-a-kind machine and observe its behavior and performance "in the field." An example par excellence of this modus operandi was the development at Cambridge University of the CAP computer, initiated in 1970 and completed circa 1979.[85] The analogy was that of the anthropologist who observes a culture or community in situ rather than that of the laboratory scientist conducting experiments in vitro.

But the RISC hypothesis *qua* hypothesis put forth by its proponents demanded "proper" experiments in the traditional scientific sense. This was what Patterson and his Berkeley colleagues did. Their "experimental apparatus" comprised a simulated version of the machine they were developing, called RISC-I,[86] along with a collection of commercially available and, in the early 1980s, widely used physical CISC computers. They identified metrics to measure design time, execution time, and ease of compilation. The experimental "subjects" were a set of "benchmark" programs written in the programming language C and compiled using a portable C compiler that generated object codes for all the testbed machines—real and simulated. The results of the experiment suggested to Patterson et al. that for the given benchmark programs and testbed computers, RISC-I "outperformed" the CISC computers.[87]

One of the failures of what is often called experimental computer science is that the experiments that are performed, when they are, and as reported by one set of investigators, are almost never repeated by others. This, of course, is anathema to the culture of the natural sciences, for which replicability of experiments is part of its very creed, reaching back certainly to the 17th century, if not earlier. The methodology of natural science insists that an experiment performed in some laboratory at one time must be replicable somewhere else at some other time. This does not mean that experiments are inevitably replicated. But at the very least, the description of the experimental setup and the conduct of the experiment must be precise enough that a fellow scientist can conduct the same experiment. (In fact, physicist and science studies scholar John Ziman has pointed out that experiments in the natural sciences are actually far less replicated than one is led to believe.[88])

Still, this ideal of replicable in vitro experiments is fundamental to the scientific ethos, and the failure to observe this ideal has haunted the more reflective computer scientists over the years. It continues to do so even now, well into the 21st century,[89] and it remains a damning indictment against computer science as a science.

But, as we have just seen, in vitro experimentation was certainly performed to test the RISC hypothesis: a manifestation of an experimental consciousness. An even more striking, indeed remarkable, feature of this consciousness was that though the same experiments carried out by Patterson et al. were not replicated, their conclusions were experimentally challenged by a group of researchers at Carnegie-Mellon University in 1985.[90] More precisely, deploying a combination of qualitative and quantitative experiments, critical analysis, and simulation, and using the same benchmarks used by the Berkely group but deploying different performance measures, Robert Colwell, Douglas Jensen, and their associates produced evidence that *refuted* the RISC hypothesis.

This did not by any means banish the RISC concept from architectural discourse: The RISC style would prevail as part of this discourse in the practice of computer architecture and design. But what Colwell et al. did was cast doubt on the general validity of the RISC hypothesis. This was taking the notion of experimental computer science seriously.

3.18   THE TAXONOMY MOVEMENT IN ARCHITECTURE

Ever since the beginning of science itself observers of nature have attempted to impose order on the seemingly chaotic diversity of observed objects—not only by proposing unitary laws but also by *classifying* the objects of interest. A *taxonomic system* is a system of rules or principles whereby entities in the universe of interest are classified in a particular way. Taxonomy, strictly speaking, is the theory and practice of classification. Inventing a taxonomic system for some universe of objects is a scientific endeavor in its own right. The histories of biology and chemistry, among the natural sciences in particular, are in substantial part stories of the search for classification systems for biological and chemical objects, respectively.[91]

Within computer science we find ample evidence of this instinct to classify. The ACM, for example, has classified computing topics according to a scheme for its review journal *Computing Reviews*. But within its constituent disciplines, computer architecture is especially noteworthy for the scope and extent of attempts to create serious taxonomic systems—especially for PMS-level architectures. Architectural theorists and researchers have shown a marked fascination with taxonomy, especially in the

1980s. Indeed, it seems reasonable to refer to a *taxonomy movement* in computer architecture between 1981 and 1990. One reason for this is the quite spectacular diversity of computers, and because architectures are abstractions of physical computers it behooved architectural researchers to create architectural taxonomies for classifying physical computers. But also, this fascination surely reflected the architectural researchers' desire to elevate computer architecture's scientific persona.

The sophisticated state of biological taxonomy (or *systematics,* to use the formal term) offers some guidelines as to what a taxonomic system should be like.[92]

Given a universe of basic objects, a *taxonomic character* is a property of these objects. A *taxon* is a named subset of this universe of objects that are sufficiently distinct with respect to a specific set of taxonomic characters from objects belonging to some other taxon. The simplest (nonhierarchic) taxonomic system will have only two kinds of entities: the objects to be classified and a single set of *taxa* (plural of taxon) among which the objects are distributed so that every object is placed in exactly one taxon. The set of such taxa constitutes a *category.*[93] In more elaborate systems several categories may exist, each with its own set of taxa. The categories form a hierarchy so that each object will appear in exactly one taxon in each category. The categories are *ranked* from high to low, depending on their places in the hierarchy. So, as Figure 3.6 shows, the taxonomic system forms a *tree* structure with the objects as "leaves" and the overall system the "root."

In the realm of computer architecture the basic objects that are usually classified are PMS-level architectures.[94]

As mentioned, the 1980s witnessed something like a taxonomy movement for (mostly) von Neumann–style computer architectures. (Of course, the partitioning of architectures into "von Neumann" and "non–von Neumann" *styles* was itself an exercise in classification.) But the earliest, simplest, and thus still most frequently cited classification scheme was proposed by Michael Flynn (then of Northwestern University, later of Stanford University).[95] Flynn's concern was to construct a scheme that took into account the newly emerging vista of computer systems that used some versions of parallel processing to enhance performance. Using the vocabulary introduced earlier—Flynn himself did not deploy the formal vocabulary of taxonomy theory—we can represent Flynn's system in terms of six taxonomic characters:

Instruction memory (IM);
Data memory (DM);
Control unit (CU);
Processing unit (PU);
Instruction stream (IS);
Data stream (DS).

The distinction between IM and DM was a logical or functional one because in von Neumann machines both instructions and data are held in the same physical memory. IS referred to an ordered stream of instructions under the charge of a single

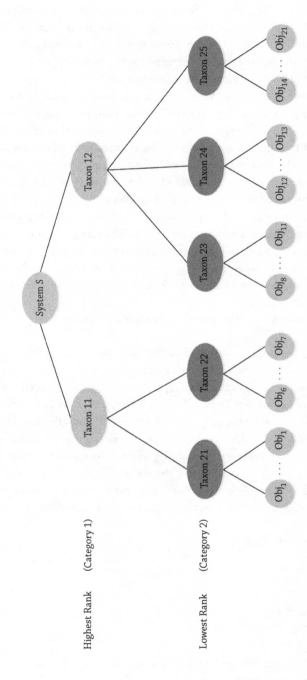

FIGURE 3.6 A hierarchic taxonomic system.

CU (Figure 3.7). DS was somewhat ambiguously defined but one interpretation is the data traffic exchanged between a single DM and a single PU (Figure 3.8).

Flynn's nonhierarchic system comprised a single category composed of four taxa:

SISD: Single instruction stream, single data stream.
SIMD: Single instruction stream, multiple data stream.
MISD: Multiple instruction stream, single data stream.
MIMD: Multiple instruction stream, multiple data stream.

So each taxon is composed of one or more copies of the elemental structures shown in Figures 3.7 and 3.8. If the root of the tree is denoted as a "computer system," Flynn's scheme, reconstructed into a two-level hierarchy, would look like Figure 3.9.

Flynn's system was not seriously questioned until the 1980s. By then some researchers were pointing to perceived weaknesses of the scheme. Most glaring, perhaps, was that no computer architecture could be "naturally" or obviously placed in the MISD taxon. Another criticism was that the other three taxa were too "coarse grained": for instance, the 8-bit Intel 8080 microprocessor, the 32-bit DEC VAX 8600 "supermini," and the

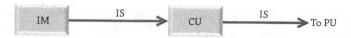

FIGURE 3.7 A single instruction stream.

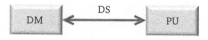

FIGURE 3.8 A single data stream.

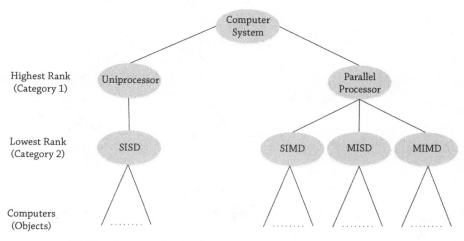

FIGURE 3.9 Flynn's taxonomy in hierarchic form.

64-bit Cray-1 "supercomputer" all belonged to the SISD taxon, although in very significant ways they were very different. Furthermore, Flynn's system did not cater—at least not in an unforced manner—to the type of parallel processing architecture called pipelining.

An indicator of the interest of researchers in architectural taxonomy was an international workshop on computer architecture taxonomy—arguably the first of its kind—held in Nuremberg, (then West) Germany, in 1981.[96] But perhaps the first serious critique of Flynn's system was mounted by Kai Hwang and Faye Briggs of the University of Southern California in 1984.[97] They dispensed with the MISD taxon as being redundant, and refined each of the other three taxa, resulting in a system of six taxa:

SISD-S: SISD with single functional unit.
SISD-M: SISD with multiple functional units.
SIMD-W: SIMD with word-slice memory access.
SIMD-B: SIMD with bit-slice memory access.
MIMD-L: MIMD with loosely coupled processing.
MIMD-T: MIMD with tightly coupled processing

It is not necessary for the reader to grasp the significance of these refinements except to note that the Hwang–Briggs system differentiated between nonparallelism and parallelism at the PMS level, an obviously important differentiation. The taxonomic system is shown in Figure 3.10.

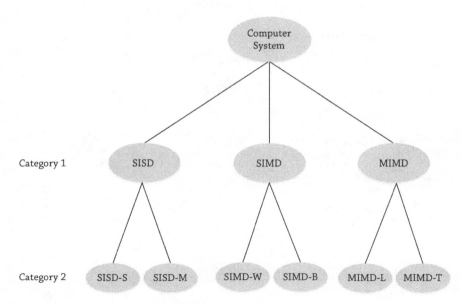

FIGURE 3.10 Hwang–Briggs taxonomy in hierarchic form.

The various other taxonomic systems proposed between 1981 and 1990 reveal, above all else, the many different lenses through which the same architectures could be perceived. These disparate lenses produced different sets of TCs and thus different taxonomies. An early proposal reaching back to 1977 was due to Wolfgang Händler of Friedrich Alexander Universität of Erlangen-Nürnberg, called the Erlangen classification system (ECS). This system was *morphological* and assigned a "metric" to PMS-level architectures based on key structural features such as CUs, ALUs, logic circuits, and so on.[98] And later systems, in the proper spirit of science, purported to improve earlier ones. If Hwang and Briggs strove to critique and improve on Flynn, David Skillicorn of Queen's University in Kingston, Ontario, attempted, in 1988, to improve on both.[99] His system was far more nuanced, drawing on six TCs that related to instruction processors (IPs) and data processors (DPs) and to *switches* (interconnections). Skillicorn took great care in defining his taxonomic characters; using his six TCs he identified a single (nonhierarchic) category of 28 taxa. His system enabled the differentiation among different kinds of von Neumann computers as well as the distinctiveness of other types of non–von Neumann machine architectures.

Sometimes, the taxonomic systems originated in concepts from entirely unrelated domains. For example, in 1987, Wolfgang Giloi of the Technical University of Berlin presented a system inspired by *formal grammars* of the kind familiar to programming language designers and language theorists.[100] Using a BNF-like metanotation, a very tiny fragment of Giloi's "architectural grammar" would look like this:

```
<computer_architecture> ::= (<operational_principles>, <hardware_structures>)
<operational_principles> ::= (<information_structure>, <control_structure>)
        . . . . . . .   . . . . .
<cooperation_rules> ::= <synchronous_cooperation> | <asynchronous_cooperation>
```

So a specific architecture would be described by a "syntactic tree" in the manner of the structure of a sentence or statement in a program.

An altogether different inspiration led the present author to a hierarchical system intended as an improvement of Skillicorn's scheme. The inspiration came from chemical notation and how it facilitated representation of atoms, radicals, and molecules of arbitrary complexity.[101]

This system was built from seven taxa referred to as "atoms," which represented concepts such as interleaved and noninterleaved memories, pipelined and nonpipelined processing units, instruction interpretation and instruction execution units, and cache memories. Atoms of the same kind could combine to form more complex "atomic radicals" (e.g., to represent multiple memory modules). Atomic radicals could combine to form still more complex "nonatomic radicals" (for example, combinations of instruction interpretation units and cache memories). The latter could combine to form "molecules" representing still more complex architectural entities. Finally molecules could combine into "macromolecules" representing complete PMS-level architectures. The rules of composition were defined by a collection of syntactic rules expressible in BNF notation, and by use of these rules Flynn's three meaningful taxa and all six of the Hwang–Briggs taxa could be represented. And because of the hierarchical nature of this system, even higher categories of taxa could be constructed.

Coincidentally, this taxonomic system was not the only one inspired by chemical notations and concepts. In 1981, Roger Hockney of the University of Reading was also similarly motivated, though his had a very different outcome than this author's scheme.

3.19   THE SIGNIFICANCE OF THE TAXONOMY MOVEMENT

From a 21st-century perspective Flynn's taxonomy must be judged the most successful; it is the only one of the many proposals put forth in the '80s that has prevailed. The others ultimately proved to be inconsequential. But the discipline of history is as much concerned with "*why* did it happen?" questions as "*what* happened?" and "*how* did we get here from there?" types of questions.

The significance of the taxonomy movement of the '80s in architecture lay in this *why* question. Let us understand that constructing taxonomic systems for architectures was a purely *nontechnological* project. And so *why* did this movement come into being at all because it had no consequence for practical computer design?

Its significance lay, I believe (both as a historian and one who participated in the movement), in a deep-seated desire all scientists harbor to demystify phenomena; and one aspect of this demystification is to classify and categorize a population of diverse objects or phenomena according to some *ordering principle*. So it is in biology; so it is in chemistry; so it is in geology and paleontology and astronomy.

For computer scientists involved in the taxonomy movement in architecture, the inspiration came from the remarkable success of taxonomic systems in biology and chemistry in particular in building sophisticated and *useful* (from expository and explanatory points of view) systems. The computer scientists' hope was that taxonomic research would lend a certain kind of *scientific aura* to computer architecture: Their hope was that the rules of their taxonomic systems would serve the nearest thing to *laws* of architecture. The taxonomy movement of the '80s was ultimately an aspiration for a "pure" science of architecture.

NOTES

1. C. C. Foster, 1970. *Computer Architecture.* New York: Van Nostrand-Rheinhold.

2. G. M. Amdahl, G. A. Blaauw, and F. P. Brooks, Jr., 1964. "Architecture of the IBM System/ 360," *IBM Journal of Research & Development*, 8, 2, pp. 87–101.

3. Anon., 1968. "Curriculum 68," *Communications of the ACM*, 11, 3, pp. 151–197.

4. Here, for example, are the titles of some well-known texts and monographs of the time: H. W. Gschwind, 1967. *Design of Digital Computers.* New York: Springer-Verlag; C. G. Bell and A. Newell, 1971. *Computer Structures: Readings and Examples.* New York: McGraw-Hill; Y. Chu, 1971. *Computer Organization and Microprogramming.* Englewood Cliffs, NJ: Prentice-Hall; E. I. Organick, 1973. *Computer Systems Organization: The B5700/B6700 Series.* New York: Academic Press.

5. M. J. Flynn, 1974. "Trends and Problems in Computer Organization," pp. 3–10 in *Information Processing 74 (Proceedings of the 1974 IFIP Congress).* Amsterdam: North-Holland.

6. M. V. Wilkes, 1975. *Time-Sharing Computer Systems* (3rd ed.). London: MacDonald & Jane's/ New York: American Elsevier.

7. See, e.g., K. J. Thurber, 1976. *Large Scale Computer Architecture.* Rochelle Park, NJ: Hayden; A. B. Salisbury, 1976. *Microprogrammable Computer Architectures.* New York: Elsevier; R. W. Doran, 1979. *Computer Architecture: A Structured Approach.* New York: Academic Press; H. S. Stone, 1980. *Introduction to Computer Architecture.* Chicago: SRA; J. B. Dennis, S. H. Fuller, W. B. Ackerman, R. J. Swan, and K.-S. Wang, 1979. "Research Directions in Computer Architecture," pp. 514–556 in P. Wegner (ed.), *Research Directions in Software Technology.* Cambridge, MA: MIT Press.

8. J. von Neumann, 1945. "First Draft of a Report on EDVAC," pp. 355–364 in B. Randell (ed.), *Origins of Digital Computers.* New York: Springer-Verlag; A. W. Burks, H. H. Goldstine, and J. von Neumann, 1946. "Preliminary Discussion of the Logical Design of an Electronic Computing Instrument." Unpublished report. This article appears in Bell and Newell, 1981, pp. 92–119.

9. U. Agüero and S. Dasgupta, 1987. "A Plausibility Driven Approach for Computer Architecture Design," *Communications of the ACM,* 30, 11, pp. 922–932.

10. For example, an excellent text by Andrew Tanenbaum of Vrie University, Amsterdam, was titled *Structured Computer Organization* (Englewood Cliffs, NJ: Prentice-Hall, 1984). Another fine textbook by V. C. Hamachar, Z. G. Vranesic, and S. G. Zaky of the University of Toronto was titled *Computer Organization* (2nd ed.). (New York: McGraw-Hill

11. For a discussion of subparadigms in the context of computer science, see S. Dasgupta, 2014. *It Began with Babbage: The Genesis of Computer Science.* New York: Oxford University Press, chapter 15.

12. Amdahl, Blaauw, and Brooks, 1964.

13. IBM, 1964. "IBM System/360 Principles of Operations," Form A22-6821-0. New York: IBM Corporation.

14. G. A. Blaauw and F. P. Brooks, Jr., 1964. "The Structure of SYSTEM/360. Part I—Outline of Logical Structure," *IBM Systems Journal,* 3, 2 & 3, pp. 119–136.

15. C. G. Bell and A. Newell, 1970. "The PMS and ISP Description System for Computer Structures," *Proceedings of the AFIPS Spring Joint Computer Conference,* 36, pp. 351–374.

16. C. G. Bell and A. Newell, 1971. *Computer Structures: Readings and Examples.* New York: McGraw-Hill.

17. See, e.g., Blaauw and Brooks, 1964, figure 1, p. 120.

18. M. R. Barbacci, G. E. Barnes, R. G. Cattell, and D. P. Siewiorek, 1979. "The ISPS Computer Description Language," Department of Computer Science, Carnegie-Mellon University, Pittsburg, PA; M. R. Barbacci, 1981. "Instruction Set Processor Specifications (ISPS): The Notation and Its Applications," *IEEE Transactions on Computers,* C-30, 1, pp. 24–40.

19. This architecture description is adapted from Barbacci et al., 1979.

20. C. G. Bell, R. Cady, H. McFarland, B. A. Delagi, J. F. O'Loughlin, R. Noonan, and W. A. Wulf, 1970. "A New Architecture for Minicomputers—the DEC PDP-11," *Proceedings of the AFIPS Spring Joint Computer Conference,* 36, pp. 657–675.

21. A. D. Falkoff, K. E. Iverson, and E. H. Sussenguth, 1964. "A Formal Description of SYSTEM/360," *IBM Systems Journal*, 3, 2 & 3, pp. 198–262.

22. K. E. Iverson, 1962. *A Programming Language*. New York: Wiley.

23. M. V. Wilkes and R. M. Needham, 1979. *The Cambridge CAP Computer and Its Operating System*. New York: North-Holland.

24. S. Rosen, 1969. "Electronic Computers: A Historical Survey," *Computing Surveys*, 1, 1, pp. 7–36.

25. G. E. Moore, 1965. "Cramming More Components Onto Integrated Circuits," *Electronics*, April 19, pp. 114–117. Reprinted, pp. 82–85 in *Proceedings of the IEEE*, 86, 1.

26. G. E. Moore, 1979. "VLSI: Some Fundamental Challenges," *IEEE Spectrum*, April, pp. 30–37.

27. Compare, e.g., the numbers given in the following contemporary publications: D. P. Siewiorek, C. G. Bell, and A. Newell, 1982. *Computer Structures: Principles and Examples*. New York: McGraw-Hill; and S. Muroga, 1982. *VLSI System Design*. New York: Wiley.

28. C. C. Foster, 1972. "A View of Computer Architecture," *Communications of the ACM*, 15, 7, pp. 557–565.

29. Ibid., p. 558.

30. Blaauw and Brooks, 1964, p. 120.

31. M. V. Wilkes, 1951. "The Best Way to Design an Automatic Calculating Machine." *Report of the Manchester University Computer Inaugural Conference*, Manchester: University of Manchester. Reprinted, pp. 266–270 in E. E. Swartzlander (ed.), 1976. *Computer Design Development: Principal Papers*. Rochelle Park, NJ: Hayden. I have described the history of this invention in S. Dasgupta, 2014. *It Began with Babbage: The Genesis of Computer Science*. New York: Oxford University Press, pp. 178–189.

32. M. V. Wilkes, 1969. "The Growth of Interest in Microprogramming: A Literature Survey," *Computing Surveys*, 1, 3, pp. 139–145; S. S. Husson, 1970. *Microprogramming: Principles and Practices*. Englewood Cliffs, NJ: Prentice-Hall.

33. H. Weber, 1967. "A Microprogrammed Implementation of EULER on IBM 360/30," *Communications of the ACM*, 10, pp. 579–588; H. W. Lawson, 1968. "Programming Language-Oriented Instruction Streams," *IEEE Transactions on Computers*, C-17, pp. 743–747.

34. R. F. Rosin, 1969. "Contemporary Concepts of Microprogramming and Emulation," *Computing Surveys*, 1, 4, pp. 197–212.

35. E. G. Wallach, 1972. "Emulation: A Survey," *Honeywell Computer Journal*, 6, 4, pp. 287–297.

36. Rosin, 1969, p. 210.

37. Ibid.

38. Ibid., p. 207.

39. M. J. Flynn and R. F. Rosin, 1971. "Microprogramming: An Introduction and Viewpoint," *IEEE Transactions on Computers*, C-20, 7, pp. 727–731; R. W. Cook and M. J. Flynn, 1970. "System Design of a Dynamic Microprocessor," *IEEE Transactions on Computers*, C-19, 3, pp. 213–222; A. B. Tucker and M. J. Flynn, 1971. "Dynamic Microprogramming: Processor Organization and Programming," *Communications of the ACM*, 14, 4, pp. 240–250; R. F. Rosin, G. Frieder, and R. H. Eckhouse, Jr., 1972. "An Environment for Research in Microprogramming and Emulation," *Communications of the ACM*, 15, 8, pp. 748–760.

40. Microdata, 1970. *Microprogramming Handbook* (2nd ed.). Santa Ana, CA: Microdata Corporation; H. W. Lawson and B. K. Smith, 1971. "Functional Characteristics of a Multilingual Processor," *IEEE Transactions on Computers*, C-20, 7, pp. 732–742; W. T. Wilner, 1972. "Design of the Burroughs B1700," *Proceedings of the AFIPS Fall Joint Computer Conference*. Montvale,

NJ: AFIPS Press; H. W. Lawson and B. Malm, 1973. "A Flexible Asynchronous Microprocessor," *BIT*, 13, pp. 165–176; Varian, 1975. *Varian Microprogramming Guide*. Irvine, CA: Varian Data Machines; Nanodata, 1979. *The QM-1 Hardware Level User's Manual*. Williamsburg, NY: Nanodata Corporation.

41. P. Kornerup and B. D. Shriver, Sr., 1975. "An Overview of the MATHILDA System," *ACM SIGMICRO Newsletter*, 5, 4, pp. 25–53; C. J. Neuhauser, 1977. "Emmy System Processor—Principles of Operation," Computer Systems Laboratory Technical Note TN-114. Stanford, CA: Stanford University; H. Hagiwara, S. Tomita, et al., 1980. "A Dynamically Microprogrammable Computer with Low-Level Parallelism," *IEEE Transactions on Computers*, C-29, 7, pp. 577–595; S. Tomita, K. Shibayama, et al., 1983. "A User-Microprogrammable Local Host Computer with Low-Level Parallelism," pp. 151–159 in *Proceedings of the 10th Annual International Symposium on Computer Architecture*. Los Angeles, CA: IEEE Computer Society Press.

42. A. B. Salisbury, 1976. *Microprogrammable Computer Architecture*. New York: American Elsevier.

43. E. I. Organick and J. A. Hinds, 1978. *Interpreting Machines: Architecture and Programming of the B1700/B1800 Series*. New York: North-Holland.

44. S. S. Husson, 1970. *Microprogramming: Principles and Practices*. Englewood Cliffs, NJ: Prentice-Hall.

45. T. S. Kuhn, 2012. *The Structure of Scientific Revolutions* (4th ed.). Chicago: University of Chicago Press.

46. H. A. Simon, 1996. *The Sciences of the Artificial* (3rd ed.). Cambridge, MA: MIT Press, pp. 27–30.

47. Wilkes, 1969.

48. Rosin, 1969.

49. S. Dasgupta, 1979. "The Organization of Microprogram Stores," *ACM Computing Surveys*, 12, 3, pp. 295–324.

50. Ibid., p. 295. Italics added.

51. Simon, 1996, chapter 5.

52. S. J. Schwartz, 1968. "An Algorithm for Minimizing Read-Only Memories for Machine Control," pp. 28–33 in *Proceedings of the IEEE 10th Annual Symposium on Switching and Automata Theory*. New York: IEEE.

53. A. Grasselli and U. Montanari, 1970. "On the Minimization of Read-Only Memories in Microprogrammed Digital Computers," *IEEE Transactions on Computers*, C-19, 11, pp. 1111–1114; S. R. Das, D. K. Banerji, and A. Chattopadhyay, 1973. "On Control Memory Minimization in Microprogrammed Digital Computers," *IEEE Transactions on Computers*, C-22, 9, pp. 845–848.

54. C. Montangero, 1974. "An Approach to the Optimal Specification of Read-Only Memories in Microprogrammed Digital Computers," *IEEE Transactions on Computers*, C-23, 4, pp. 375–389

55. See Dasgupta, 1979, pp. 52–53.

56. T. Jayasri and D. Basu, 1976. "An Approach to Organizing Microinstructions Which Minimizes the Width of Control Store Words," *IEEE Transactions on Computers*, C-25, 5, pp. 514–521.

57. J. L. Baer and B. Koyama, 1977. "On the Minimization of the Width of the Control Memory of Microprogrammed Processors," Tech. Report 77-08-01, Department of Computer Science, University of Washington.

58. C. B. Silo, Jr., J. H. Pugley, and B. A. Jemg, 1981. "Control Memory Width Optimization Using Multiple-Valued Circuits." *IEEE Transactions on Computers*, C-30, 2, pp. 148–153.

59. Burroughs Corporation, 1961. *The Descriptor—A Definition of the B5000 Information Processing System*. Detroit, MI: Burroughs Corporation; Burroughs Corporation, 1969. *B6500 System Reference Manual (001)*. Detroit, MI: Burroughs Corporation; E. I. Organick, 1973. *Computer System Organization: The B5700/B6700 Series*. New York: Academic Press; R. W. Doran, 1979. *Computer Architecture: A Structured Approach*. New York: Academic Press.

60. A. C. D. Haley, 1962. "The KDF.9 Computer System," pp. 108–129 in *Proceedings of the 1962 Fall Joint Computer Conference*. Washington, DC: Spartan.

61. B. Randell and L. J. Russell, 1964. *Algol 60 Implementation*. New York: Academic Press.

62. J. K. Buckle, 1978. *The ICL 2900 Series*. London: Macmillan.

63. S. H. Lavington, 1998. *A History of Manchester Computers* (2nd ed.). Swindon, UK: The British Computer Society, pp. 46–49.

64. Quoted in D. Morris and R. N. Ibbett, 1979. *The MU5 Computer System*. London: Macmillan, p. 3.

65. J. K. Iliffe, 1972. *Basic Machine Principles* (2nd ed.). London: Macdonald/ New York: American Elsevier.

66. Ibid., p. 3.

67. Ibid., p. 8.

68. Ibid., p. 7.

69. R. S. Barton, 1961. "A New Approach to the Functional Design of a Digital Computer," *Proceedings of the Western Jouint Computer Conference*, Los Angeles, pp. 393–396; R. S. Barton, 1970. "Ideas for Computer Systems Organization: A Personal Survey," pp. 7–16 in J. T. Tou (ed.). *Software Engineering, Volume I*. New York: Academic Press.

70. G. J. Myers, 1982. *Advances in Computer Architecture* (2nd ed.). New York: Wiley.

71. U. O. Gagliardi, 1973. "Report of Workshop 4—Software-Related Advances in Computer Hardware," pp. 99–120 in *Proceedings of a Symposium on the High Cost of Software*. Menlo Park, CA: Stanford Research Institute. Cited by Myers, 1982, p. 17.

72. Myers, 1982, p. 18.

73. Intel Corporation, 1981. *The iAPX-432 GDP Architecture Reference Manual*. Santa Clara, CA: Intel Corporation.

74. Myers, 1982, p. 325.

75. D. A. Patterson and D. Ditzel, 1980. "The Case for a Reduced Instruction Set Computer," *Computer Architecture News*, 8, 6, pp. 25–33.

76. See, e.g., J. T. Bonner, 1988. *The Evolution of Complexity by Natural Selection*. Princeton, NJ: Princeton University Press. For a dissenting view see D. W. McShea, 1997. "Complexity in Evolution: A Skeptical Assessment," *Philosophica*, 59, 1, pp. 79–112.

77. See, e.g., S. Dasgupta, 1997. "Technology and Complexity," *Philosophica*, 59, 1, pp. 113–140.

78. W. D. Strecker and D. W. Clark, 1980. "Comments on 'The Case for the Reduced Instruction Set Computer' by Patterson and Ditzel," *Computer Architecture News*, 8, 6, pp. 34–38; R. P. Colwell, C. Y. Hitchcock, E. D. Jensen et al., 1985. "Computers, Complexity and Controversy," *Computer*, 18, 9, pp. 8–20.

79. J. L. Hennessy, N. Jouppi, and S. Przybylski, 1982. "MIPS: A Microprocessor Architecture," pp. 17–22 in *Proceedings of the 15th Annual Workshop on Microprogramming*. Los Angeles: IEEE Computer Society Press; J. L. Hennessy, N. Jouppi, and S. Przybylski, 1983. "Design of a High-Performance VLSI Processor," pp. 33–54 in *Proceedings of the 3rd Caltech Conference on VLSI*. Pasadena, CA: California Institute of Technology.

80. G. Radin, 1982. "The IBM 801 Minicomputer," pp.39–47 in *Proceedings of the ACM Symposium on Architectural Support for Programming Languages and Operating Systems.* New York: ACM.

81. D. A. Patterson and C. Sequin, 1980. "Design Considerations for Single-Chip Computers of the Future," *IEEE Transactions on Computers,* C-29, 2, pp. 108–116; D. A. Patterson and C. Sequin, 1981. "RISC-I: Reduced Instruction Set Computer," pp. 443–458 in *Proceedings of the 8th Annual International Symposiumon Computer Architecture.* New York: IEEE Computer Society Press; D. A. Patterson and C. Sequin, 1982. "A VLSI RISC," *Computer,* 15, 9, pp. 8–21; M. G. H. Katevenis, 1985. *Reduced Instruction Set Computer Architecture for VLSI.* Cambridge, MA: MIT Press.

82. D. A. Patterson, 1985. "Reduced Instruction Set Computers," *Communications of the ACM,* 28, 1; Colwell et al., 1985.

83. Patterson, 1985.

84. K. R. Popper, 1968. *The Logic of Scientific Discovery.* New York: Harper & Row.

85. M. V. Wilkes and R. M. Needham, 1979. *The Cambridge CAP Computer and Its Operating System.* New York: North-Holland.

86. Patterson and Sequin, 1981.

87. D. A. Patterson and R. Piepho, 1982. "RISC Assessment: A High Level Language Experiment," pp. 3–8 in *Proceedings of the 9th Annual International Symposium on Computer Architecture.* Los Angeles: IEEE Computer Society Press.

88. J. Ziman, 2000. *Real Science: What It Is and What It Means.* Cambridge: Cambridge University Press.

89. W. F. Tichy, 1988. "Should Computer Scientists Experiment More?" *Computer,* 21, 5, pp. 32–40; R. T. Snodgrass, 2010. "'Ergalics': A Nature Science of Computation." Version Feb. 12: http://www.cs.arizona.edu/projects/focal/erga. Retrieved Sept. 8, 2014; D. Lamiere, 2013. "Should Computer Scientists Run Experiments?" http://lamire.me/blog/archives/2013/07/10/should-computer-scientists-run-experiments. Retrieved Jan.20, 2015.

90. Colwell et al., 1985.

91. For biology, see, e.g., E. Mayr, 1982. *The Growth of Biological Thought.* Cambridge, MA: Belknap Press of Harvard University Press. For chemistry, see, e.g., J. R. Partington [1957] 1989. *A Short History of Chemistry* (3rd ed.). New York: Dover Publications.

92. See, e.g., H. H. Ross, 1974. *Biological Systematics.* Reading, MA: Addison-Wesley; M. Ruse, 1973. *The Philosophy of Biology.* London: Hutchinson University Library.

93. Ruse, 1973, chapter 7.

94. To this writer's knowledge, the first attempt to map the taxonomy of computer architectures onto these standard biologically rooted taxonomic terms was the present author: S. Dasgupta, 1989. *Computer Architecture: A Modern Synthesis. Volume 2: Advanced Topics.* New York: Wiley, chapter 2.

95. M. J. Flynn, 1966. "Very High Speed Computing Systems." *Proceedings of the IEEE,* 54, 12, pp. 1901–1909.

96. G. A. Blaauw and W. Händler (eds.), 1981. *International Workshop on Taxonomy in Computer Architecture.* Nuremberg, Germany: Friedrich Alexander Universität, Erlangen, Nuremberg.

97. K. Hwang and F. Briggs, 1984. *Computer Architecture and Parallel Processing.* New York: McGraw-Hill.

98. W. Händler, 1977. "The Impact of Classification Schemes on Computer Architecture," *Proceedings of the 1977 International Conference on Parallel Processing,* pp. 7–15; W. Händler, 1981.

"Standards, Classification and Taxonomy: Experiences with ECS," pp. 39–75 in Blauuw and Händler, 1981.

99. D. B. Skillicorn, 1988. "A Taxonomy for Computer Architecture," *Computer,* 21, 11, pp. 46–57.

100. W. K. Giloi, 1983. "Towards a Taxonomy of Computer Architectures Based on the Machine Data Type View," pp. 6–15 in *Proceedings of the 10th Annual International Symposium on Computer Architecture.* Los Alamitos, CA: IEEE Computer Society Press.

101. S. Dasgupta, 1990. "A Hierarchical Taxonomic System for Computer Architectures," *Computer,* 22, 3, pp. 64–74.

4

GETTING TO KNOW PARALLELISM

Every morning the first thing that X does is make tea for herself. She *first* turns on the stove and *then while* the stove ring is heating up, she pours water from the faucet into the kettle. She *then* places the kettle on the stove ring, now nicely hot, and *while* the water is being heated she puts tea bags into the teapot; she *then* pours milk from the milk carton into a milk jug and *then* puts the milk jug into the microwave oven. *After* the water starts to boil she pours water into teapot. And *while* the tea "gathers strength" in the teapot, she presses the time button on the microwave to start warming the milk. *After* the milk is warmed she *first* pours tea from the teapot into a teacup and *then* adds milk from the warmed milk jug to the tea in the cup.

This tiny, humdrum, comforting, domestic scenario X enacts every morning has many (but not all) of the ingredients of a situation that involves the scope and limits of *parallel processing*. More precisely, the art of making tea as practiced by X entails a *blend* of both sequential and parallel events.

We note that certain events *can* take place in parallel (or concurrently) because they do not *interfere* with one another; for example, the heating of the stove and the pouring of water into the kettle. But other events *must* be sequentially ordered either because they interfere with one another or because one event must complete before the other can begin. The kettle can be placed on the stove ring only after it has been filled with water; water can be poured into the teapot only after the water has boiled.

But notice also that there is some flexibility in the ordering of X's actions. She can defer turning on the stove until after the kettle is placed on the stove ring; she can alter the ordering of pouring water into the teapot and placing teabags into the pot; she could defer warming the milk in the microwave until the tea has brewed. Indeed, the ultimate deferment of tasks would be to impose a *totally* sequential ordering on all the activities in the environment:

Pour water into kettle → Switch on stove → Put kettle on stove → Heat kettle → Put teabag in teacup → Pour water into teacup → Warm milk in microwave → Brew tea in cup → Pour milk into cup

Something else might be noticed: In the preceding totally sequential ordered situation, nine "steps" are required, each of some duration that differs from one step to another. But in Figure 4.1 there is a critical sequential path that shows the minimal number of steps needed—in this case six steps. Thus the minimal number of steps needed to make tea in this scenario is constrained by the sequential component of the process. There still remains, however, a certain amount of flexibility in accommodating the other steps in Figure 4.1.

X's experience in the art of tea making is, of course, not unique to her. We all navigate such worlds that offer the *promise* of parallelism. Sometimes we are conscious of this promise, sometimes we are not. Sometimes we exploit this promise because it seems natural or instinctive to do so: While driving to work, we manipulate the steering wheel, brake, and accelerator, watch the road ahead and through the rearview and side mirrors, selects roads, and make turnings in a seamless web of sequential and parallel actions. Other times we consciously impose parallelism in our actions—as in X's case—for the sake of efficiency: to "speed up" one's work, to "increase throughput."

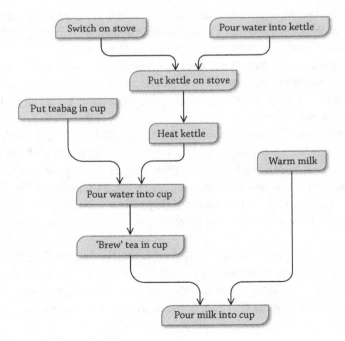

FIGURE 4.1 The parallel art of making tea.

4.2   THE PROMISE OF PARALLELISM IN COMPUTATIONAL ARTIFACTS

Exploiting the promise of parallelism has always been an element of the computer scientist's consciousness, long before the making of a discipline called computer science. After all, the task of a computational artifact was always to perform a computation as quickly as possible, and parallelism offered an obvious means of doing this. In the 1950s, engineers were devising binary arithmetic circuits that would perform addition and subtraction on the bit-wise pairs of inputs to the circuit in parallel.[1] And after the invention of microprogramming in 1951, engineers were designing microprogrammed control units in which the format of each microword in the control memory could allow two or more microoperations to be activated in parallel (see Chapter 3, Sections 3.11, 3.12).[2] So computers at the lowest levels of abstraction—at the logic circuit and microarchitecture levels—were routinely exploiting parallelism.

But it was not until the 1960s that the promise of parallelism began to be explored at higher levels of abstraction within a computer system. It really began when mainframe computers such as the IBM System/360 unshackled input and output (I/O) processing from the stranglehold of the central processing unit (CPU). I/O were assigned to independent *I/O processors*—IBM famously, and rather confusingly, called them "channels"[3]—whose actions, once initiated by the CPU, would go about their own businesses, while the CPU went about *its* own business, potentially in parallel.

A major reason for the uncoupling of I/O from CPU was that I/O were so much slower than the "real" computing performed by the CPU and so the CPU should not have to wait, impatiently twiddling its thumbs (so to speak) while input or output was going on. Thus this unshackling of I/O from CPU processing offered a promise of parallelism.

Of course, this made sense only if main memory could hold, simultaneously, two or more programs so that while one program was involved with I/O activity (reading or writing files), thus engaging the I/O processors, another program could engage the CPU in performing computation. In other words, the programs in main memory shared the CPU and I/O processors among themselves in a time-interleaved fashion but allowing for all the programs to be active at any time. This was the principle of *multiprogramming* that emerged as a topic of enormous interest in the 1960s for both academic computer scientists and industrial computer systems designers.[4]

The principle of multiprogramming and the uncoupling of I/O and CPU processing together offered the promise of parallel processing. But the realization of this promise lay with the *operating system* that, after all, is by design the computational agent responsible for controlling, managing, and allocating a computer's resources so as to enhance their efficient usage—and thus the overall efficiency of the computer system—and also, vitally, make the user's interaction with the computer more congenial.

Thus parallelism first became a phenomenon of interest to programmers (if not to computer designers) by way of operating systems design. It was the systems programmers, builders of operating systems, who first paid attention, within the software community, to the intriguing problem of parallel programming.

And it was Edsger Dijkstra (yet again) of the Technische Universiteit Eindhoven (THE) in the Netherlands who shed the first light on this problem. In a seminal paper written in 1965—a technical report in his celebrated "EWD" series (see Chapter 2, Section 2.2)—Dijkstra offered a certain vision of parallel programming that set the agenda that would occupy computer scientists over the next two decades.[5] Mastering parallelism certainly had many facets, and parallel programming was but one. But it was a hugely important and influential facet; and the striving for such mastery began with Dijkstra.

His vision was first presented theoretically in his 1965 report and then was cast more practically in his 1968 paper on the design of a multiprogramming system then being developed at THE.[6] As a (systems) programmer, Dijkstra imagined the whole multiprogramming system as a society of *sequential* processes, each running on its own abstract (or virtual) processor. For instance, every distinct user program in a state of execution would be a distinct sequential process; each I/O device in operation was a distinct sequential process.

The idea of a *process* was vital to this vision. Dijkstra himself was (surprisingly, given his meticulous style of exposition and his usual attention to detail) rather casual about the meaning of this concept. What he did imply was that a process was not a computational artifact "in repose"; it was not a static entity but rather an entity in a state of activity. A process is a dynamic entity: " . . . what happens during . . . a program execution is called a sequential process."[7]

The concept of a process has remained somewhat elusive. Later in the 1970s, other computer scientists tried to articulate the concept more precisely.[8] Thus, in 1975, Maurice Wilkes of Cambridge University, along Dijkstra's lines, suggested that if a process is "executed" by a dedicated virtual processor, the process is "the path of such a virtual processor."[9]

According to Dijkstra's vision, then, the building blocks of a parallel processing system are sequential processes. But what makes the ensemble interesting and complicated is that these processes are not entirely independent of one another but, rather, they are "loosely coupled": Most of the time they go about their own businesses but every now and then they *interact* with one another. Such interaction is necessary for the system as a whole to fulfill its goal. So the sequential processes must interact in a cooperative fashion.

Dijkstra's most consequential contributions in his 1965 paper were his notions of *critical region* and *semaphore*. (Dijkstra named the former "critical section," but here we use the later, more commonly used, term.)

The early literature on parallel programming drew on a number of "toy" illustrative problems that gradually became standard "textbook exemplars." One of these is the following. Two independent sequential processes, Q_1 and Q_2, are running simultaneously on their respective virtual processors. At some point in their executions each needs to access and modify a data object D in a shared memory. For the sake of clarity, let us assume that the code segments executed by Q_1 and Q_2 to access and change D are, respectively,

---

Q_1:   . . . . .

     $Reg_1 \leftarrow D$;

     $Reg_1 \leftarrow Reg_1 + 1$';

     $D \leftarrow Reg_1$;

     . . . . . .

Q_2 :   . . . . . .

     $Reg_2 \leftarrow D$;

     $Reg_2 \leftarrow Reg_2 + 2$;

     $D \leftarrow Reg_2$;

     . . . . . .

Here, Reg_1 and Reg_2 are, respectively, registers private to the two virtual processors.

Now, because Q_1 and Q_2 are running on virtual processors, we can make no assumptions about their relative speeds. They are, essentially, asynchronous parallel processes. So it is quite possible that the instructions in these code segments may interleave in time, leading possibly to unpredictable values read from and written into the shared data object D. For example, assuming that initially $D = 0$, depending on the nature of the interleaving of instructions from the two processes, the values of D can be 1, 2, or 3 after both code segments have executed.

This is clearly an unsatisfactory situation. To prevent such arbitrary interleaving Dijkstra stipulated that each of these code segments must execute as *indivisible* units. Such indivisible units acting on a data object—or more generally on some shared memory—is a *critical region*.

This, then, is an example of what Dijkstra called the mutual-exclusion problem. By implementing the two code segments as critical regions CR1 and CR2, D will be accessed and modified by the critical regions in a mutually exclusive manner. Regardless of which critical region executes first, the value of D after both have executed will always be 3.

So how should critical regions be implemented? Commonsensically, some sort of locking mechanism must be used. Dijkstra proposed a nonnegative integer data type that he called *semaphore* for this purpose. Given a semaphore σ, the operation $P(\sigma)$ will block the calling process unless σ has a positive value. The execution of the P operation reduces the value of σ by 1. The operation $V(\sigma)$ will cause a process previously blocked on that semaphore to be released and increases the value of σ by 1.[10]

FIGURE 4.2 The producer–consumer problem.

So, for the mutual-exclusion problem, assuming that initially $\sigma = 1$, the two critical regions can be programmed as follows:

Q1:   . . . . . .

P(σ);

{critical region CR1}

      $Reg1 \leftarrow D$;

      $Reg1 \leftarrow Reg1 + 1$;

      $D \leftarrow Reg1$

V(σ);

. . . .

Q2:   . . . .

P(σ);

{critical region CR2}

      $Reg2 \leftarrow D$;

      $Reg2 \leftarrow Reg2 + 2$;

      $D \leftarrow Reg2$

V(σ)

. . . . . .

The mutual-exclusion problem depicted here was less of cooperation than of competition between two sequential processes vying for a shared resource. But processes can also *really* cooperate. An example is the producer–consumer problem. Here, a cyclic process Q3 produces and deposits a value into a shared "buffer" B that is then consumed by another cyclic process Q4 (Figure 4.2). The two processes run freely and asynchronously at their own speeds, but there is a point at which they must be synchronized so that Q3 does not deposit a new value into B before the previous value has been consumed by Q4. Conversely, Q4 must not consume the same value twice.

In this situation two semaphores $\sigma1$ and $\sigma2$ are required. Initially, suppose $\sigma1 = 1$, $\sigma2 = 0$. Then Q3, Q4 will be correctly synchronized as follows.

Q3:   **repeat forever**

. . . . . .

P($\sigma3$);

{critical region CR3}

    Produce value and deposit it in B

$V(\sigma 4)$

. . . .

end repeat

$Q4$: **repeat forever**

. . . .

$P(\sigma 4)$

{critical region CR4}

Consume value from B

$V(\sigma 3)$

. . . .

end repeat

$Q3$ and $Q4$ will then interleave in their respective critical regions, the one cyclically depositing a new value into the buffer that is then cyclically consumed by the other. Data are produced and consumed in the correct sequence.

4.4 AN ARCHITECTURAL ASSUMPTION UNDERLYING THE NEW DISCOURSE

We will not exaggerate if we say that Dijkstra's 1965 paper on "Cooperating Sequential Processes" was as consequential for parallel programming as his "Notes on Structured Programming" was for sequential programming (see Chapter 2, Section 2.2). As in the latter case Dijkstra laid the foundation for a brand new discourse (if not an overarching paradigm). Perhaps the only difference was that Dijkstra's vision of parallel programming took slightly longer to seep into the computing community's consciousness. The new discourse properly took shape in the 1970s.

Note though that underlying this particular discourse was a certain model of the underlying computer architecture. Dijkstra's processors, virtual though they were, were grounded in a hard dose of architectural reality. The assumption was that these sequential processes running in parallel on their respective virtual processors would interact by way of shared memory. After all, the raison d'être for critical regions was to discipline the manipulation of data objects located in a shared memory. In other words, the assumption was a *shared-memory* model of computer architecture.

4.5 HOW TO COMPOSE PARALLEL PROGRAMS

Any practical discussion of computational processes must, of course, come to grips with the problem of language: how to describe, how to specify a computational process. So it was in the new discourse on parallel programming: What kinds of notation would express the nuances of parallel programs? How should one compose them?

Dijkstra (characteristically) worried over this question. Writing in the Algol 60 era he proposed an extension of the Algol notation for the sequential statement **"begin. . . end."** Dijkstra's original notation of 1965 was adapted in a slightly modified form in 1972 by the Dane Per Brinch Hansen, then at the California Institute of Technology,[11] and because he used this construct extensively in his influential textbook *Operating Systems Principles* (1973), we use his notation rather than Dijkstra's.[12]

Brinch Hansen's *concurrent statement* took the form

$$\textbf{cobegin } S_1; S_2; \ldots ; Sn \textbf{ coend,}$$

where $S_1, S_2, \ldots , Sn$ are themselves other statements (including, possibly, nested concurrent statements). This stipulated that $S_1, S_2, \ldots , Sn$ can be executed concurrently. The statement terminates when all of the Si's have terminated execution; then only will the statement following the concurrent statement begin execution.

As an elementary example, our tea-making scenario of Section 4.1 can be described as follows:

---

begin
    cobegin
        switch on stove;
        pour water into kettle
    coend;
    cobegin
        place kettle on stove;
        put tea bags in cup;
    coend;
    heat kettle;
    cobegin
        pour water from kettle into cup;
        warm milk in microwave oven
    coend;
    brew tea in cup;
    pour milk from into cup
end

Of course, underlying this composition is the Dijkstravian assumption that each of the actions within a concurrent statement is performed by its own dedicated "virtual processor": a "virtual person," so to speak. The composition describes a parallel processing system in its most pristine form, in which the only limiting constraints on parallelism are "data dependencies" between actions (see Figure 4.1). If and when there are physical constraints—for example, all the virtual persons are concretely embedded in a single "real person"—then this composition may not work.

Per Brinch Hansen was, of course, only one of many contributors to the new discourse on parallel programming. But his contribution was especially compelling for a number of reasons.

First, his research papers, essays, and published lectures on parallel programming through the '70s especially and into the '80s used a judicious blend of theoretical and concrete ideas. He served as a bridge between the theorists of parallel programming and practitioners of parallel systems programming, and this surely granted him a certain universality.[13]

Second, Brinch Hansen authored between 1973 and 1982 three highly accessible graduate-level books on parallel programming that together bore the stamp of his authority on the topic.[14]

Finally, between 1975 and 1987, he designed an evolutionary trio of parallel programming languages including Concurrent Pascal (on which more later); the latter was likely the first fully implemented language of its kind.[15] Each of these languages and their implementations served as an experiment to test certain parallel programming concepts. Like Nicklaus Wirth and Maurice Wilkes, Brinch Hansen was an exponent par excellence of experimental computer science—in his case, in the realm of parallel programming.

4.6 "MONITORY" POLICY

The concurrent statement was a beginning. But the crux of the matter—as the mutual-exclusion and producer–consumer problems illustrated—was how to compose asynchronous parallel processes that occasionally interact by means of a shared memory. Dijkstra had identified critical regions and semaphores as the means to achieve this goal, but these were "low-level" mechanisms. For Per Brinch Hansen and C. A. R. Hoare, working simultaneously and occasionally communicating with each other—a case of cooperative cognitive processes!—the question was this: how to represent critical regions in a high-level programming language?

By the early 1970s, Pascal had been invented (see Chapter 1, Section 1.10) and for many, including Brinch Hansen and Hoare, it was the language of choice. For the Pascalites, the natural instinct was to extend Pascal to encompass parallelism. The concurrent statement **cobegin . . . coend** was one such extension.

For Hoare, a shared memory was an instance of a shared resource. For example, several processes running independently may need to access a single peripheral device—the latter, a shared resource.[16] Following Dijkstra's line of thinking, Hoare envisioned the execution of a parallel program as a set of processes running freely, independently, and simultaneously. But each process needs, occasionally, to enter its critical region involving a shared resource. When it does so, it must have exclusive control of that resource. To effect this in a Pascal-like language, Hoare proposed the construct

with r do C

to mean a critical region *C* associated with a resource *r*.

He went further. A process may face a situation wherein it enters its critical region only if its shared resource satisfies a certain condition—resulting from the action of another concurrent process. To represent this situation Hoare extended the **with** construct to one that Brinch Hansen would term a *conditional critical region*:[17]

$$\textbf{with } r \textbf{ when } B \textbf{ do } C$$

Here, the critical region *C* associated with resource *r* is not entered until the Boolean condition *B* is satisfied.

As an example, Hoare considered a generalization of the producer–consumer problem called the bounded-buffer problem:[18]

A process *Q1* produces a sequence of values of type *T* that are deposited in a circular *buffer* of *N* slots. These values are consumed by another process *Q2* in the same sequence in which they are produced. *Q1* and *Q2* run asynchronously and in parallel. Now assume the variables:

counter: An integer variable that keeps count of the number of values in *buffer*.
writeptr: If *count* < *N* then this points to the first available empty slot in *buffer*;
   otherwise, it has the same value as *readptr*.
readptr: If *count* > 0 then this points to the slot in *buffer* from where the next
   value is to be consumed; otherwise, it has the same value as *writeptr*.
buffer: An array of *N* slots each to hold a value of type *T*.

The shared resource is then represented by a Pascal **record** type:

B: **record** *writeptr, readptr, counter:* integer;
          buffer: **array** 0 .. *N* − 1 **of** *T*
    end

All the variables in *B* are initialized to zero. The producer and consumer processes can be depicted by the concurrent statement

$$\textbf{cobegin } Q1; Q2 \textbf{ coend}$$

The conditional critical regions inside *Q1* and *Q2* will be, respectively, of the following forms:

Q1:
. . . .
with *B* **when** *counter* < *N* **do**
       begin

```
            buffer [writeptr] := 'next value';
            writeptr := ( writeptr + 1 ) mod N;
            counter := counter + 1
    end
..........
Q2:
...........
with B when counter > 0 do
    begin
            'this value' := buffer [readptr];
            readptr := (readptr + 1) mod N;
            counter := counter − 1
    end
.............
```

Publishing in the same year (1972), Brinch Hansen offered the constructs

$$\textbf{var } v : \textbf{shared } T$$
$$\textbf{region } v \textbf{ do } S$$

to signify, respectively, a shared variable of a certain type and a critical region associated with that variable.[19] He also proposed a "synchronizing primitive" **await** (analogous to Hoare's **when** primitive) that would *delay* a process until a shared variable satisfies some Boolean condition. So Hoare's conditional critical region would be represented as

$$\textbf{region } v \textbf{ do begin} \ldots \textbf{ await } B; \ldots \textbf{ end}$$

Yet even as Hoare's and Brinch Hansen's respective papers on conditional critical regions were being published (in 1972), there were signs of discontent. Indeed, in 1971, at the same seminar held in Queen's University, Belfast, in which Hoare presented his ideas about shared resources and conditional critical regions, Brinch Hansen read a paper on a topic quite unusual for a research conference. The paper was titled "An Outline of a Course on Operating System Principles."[20] The paper was really about a book-in-progress. And here, talking about conditional critical regions, Brinch Hansen pointed out that the descriptive elegance of the construct had a price: one abstracted away and ignored the sequence in which processes waiting to enter their critical regions were scheduled. In the case of processes that are largely independent and "loosely coupled" so that simultaneous requests for resources are relatively rare, this is not a serious problem. But, Brinch Hansen pointed out, ignoring scheduling is an unrealistic abstraction in the case of heavily used shared resources. Rather, the operating system responsible for managing concurrency among processes would need to identify processes competing for shared resources and control the scheduling of such resources among the competing processes.

And he went on to introduce, rather cryptically, a new concept he called a *monitor* as a resolution of this problem.

The book-in-progress was published as *Operating System Principles* (1973).[21] As noted before, this was the first of a trio of books Brinch Hansen wrote between 1973 and 1982 on the general topic of parallel programming. If we call this book a textbook, we do so in the older (19th- to early 20th-century) sense when the makers of a scientific subject wrote comprehensive textbooks that expressed their personal, idiosyncratic synthesis of the subject and included, often, new, previously unpublished material.[22] Like its illustrious predecessors, *Operating System Principles* bore the stamp of originality and the authority of its author. Few books on operating systems of this ilk had been written before. And certainly no book on parallel programming of this kind had been previously published.

One of its original contributions was an extensive discussion of the monitor concept. The adoption of this word for the purpose at hand was perhaps slightly unfortunate, for monitor was already known in a different sense in the operating systems discourse. Robert Rosin of the State University of New York, Buffalo, writing in 1969, identified "monitor system" as synonymous to "operating system," "executive system," and "supervisory system."[23] In other words, a monitor, in this established sense, *was* an operating system.

But in the hands of, first Brinch Hansen and then Hoare, a monitor came to have a very different meaning: in Brinch Hansen's view, it was a high level parallel programming construct of special use as, in the words of Hoare, "an operating system *structuring* concept."[24]

The monitor idea drew its sustenance—as did objects, as did abstract data types (Chapter 2, Sections 2.10, 2.12, 2.14)—from the SIMULA class concept invented by Ole-Johan Dahl and Kristian Nygaard. As Brinch Hansen announced in 1973, he would use the word monitor to denote "a shared variable and the set of meaningful operations on it." However, its purpose was to control resource *scheduling* among individual processes.[25] He justified the use of the term by noting its functional affinity to low-level software often called "basic monitor," used to control physical resource sharing in computer systems.[26]

In other words, a basic monitor in Brinch Hansen's language is the means by which an operating system, written in assembly language (thus utilizing a machine's hardware-implemented architectural features), schedules the machine's *physical* resources. Brinch Hansen appropriated the term for "higher levels of programming" as a new construct for a high-level parallel programming language that could be used to represent parallel programs executing on a shared-memory computer.

We recall (from Chapter 2, Section 2.10) that in Simula 67, a class is composed of a data type and the meaningful operations defined on it are expressed as a package. Given such a declaration and the declaration of a variable as an instance of that class, the compiler can check whether an operation specified by the programmer on that variable is valid. Brinch Hansen wanted this compile-time checking capability to extend to operations or procedures on *shared* variables and also to guarantee that two or more processes wanting to perform computations on such a shared variable could do so only in a mutually exclusive manner.

And so he proposed the *shared class* as a high-level construct for representing critical regions:

---

shared class $T = v_1: T_1; v_2 : T_2; \ldots ; v_m: T_m;$
    procedure $P_1 (\ldots)$ **begin** ... **end**;
    . . . .
    procedure $P_n (\ldots)$ **begin** ... **end**;
    begin S_0 **end**

This defines a data structure of type T consisting of components $v_1, \ldots, v_m$ of types T_1, $\ldots, T_m$ respectively, a collection of procedures $P_1, \ldots, P_n$ that can operate on this data structure, and an initializing statement S_0. Thus, given the declaration of a variable of this shared-class type,

$$\textbf{var}\ v : T$$

storage would be allocated for the components $v_1, \ldots, v_m$ and the initializing statement S_0 executed on v. A call of a procedure P_i on v, denoted as

$$v. P_i (\ldots),$$

will enable P_i to refer to the components of v as well as to P_i's local variables and the actual parameters specified in the call. The procedures $P_1, \ldots, P_n$ on v of type T are not only the only procedures defined on v, but simultaneous calls of these procedures will exclude one another in time.[27]

The shared-class construct is one way of realizing a monitor. Consider, as an example, representation of the bounded-buffer problem described at the beginning of this section. As a shared-class-based monitor it might be represented as follows:

---

shared class $B =$
buffer : **array** $0.. \ N - 1$ **of** T;
writeptr, readptr : $0 .. N - 1$;
counter : $0 .. N$;
procedure *send* $(m: T)$;
    begin
        await *counter* $<$ N;
        buffer (*writeptr*) := m;

```
            writeptr := (writeptr + 1)mod N;
            counter := counter + 1;
     end
procedure receive (var m : T);
     begin
            await counterer > 0;
            m := buffer (readptr);
            readptr := (readptr + 1) mod N;
            counter := counter − 1;
     end;
begin writeptr := 0; readptr := 0; counter := 0 end
```

Now, given the declaration

$$\text{var } b : B; \ t : T,$$

the buffer variable b and the message variable t can be accessed by a process by way of the procedure calls:

$$b. \ send \ (t)$$
$$b. \ receive \ (t)$$

So, in effect, a monitor "pulls out" the individual critical regions defined on some shared resource from individual concurrent processes and *centralizes* it into a single entity. Scheduling access to a shared resource is defined by the constructs specified in the monitor procedures.

The monitor—or, more generally, the parallel programming—story is really a two-part one. The first, recounted in this chapter thus far, tells of the fundamental concepts leading to the emergence of the monitor concept. Per Brinch Hansen was certainly a leading protagonist in this story so far but he was not in splendid isolation. As I have remarked, the monitor concept evolved from a confluence of ideas emanating from Brinch Hansen and C. A. R. Hoare in particular but also, given his pioneering role in this story, Edsger Dijkstra.[28]

In the same year *Operating System Principles* was published (1973), Hoare—already in possession of a manuscript copy of the book according to Brinch Hansen[29]—published a paper in the British *Computer Journal* wherein the monitor was used to structure a paging system.[30] A year later, Hoare published a comprehensive "tutorial" on monitors in *Communications of the ACM* with several illustrative examples of the use of monitors in operating systems situations.[31] Thus it would seem that other than Brinch Hansen's book Hoare's two papers were the first on monitors to appear in the standard, journals-based research literature. And although Hoare announced in the

"Abstract" heading the paper that it "develops Brinch Hansen's concept of a monitor,"[32] there is no further attribution of Brinch Hansen's priority in this research. Rather, the reader is told in the final acknowledgments section that "The development of the monitor concept is due to frequent discussions and communications with E. W. Dijkstra and P. Brinch Hansen."[33] This only further emphasized that the first part of the monitor story entailed a web of interacting cognitive processes involving Brinch Hansen, Hoare, and Dijkstra.

As the examples in this chapter show, once Pascal was invented it became the linguistic bedrock for the specification of all the concepts in parallel programming in the hands of Brinch Hansen and Hoare. Constructs to denote critical regions, conditional critical regions, synchronizing primitives, shared classes, and monitors were "add-ons" to Pascal (although the spirit of Simula 67 hovered benevolently above). The second part of the monitor story is, unsurprisingly, about the emergence of the first high-level programming languages designed expressly for parallel programming. As unsurprisingly, for someone like Brinch Hansen the foundation of such an endeavor was Pascal itself.

4.7 THE "PASCALIZATION" OF PARALLEL PROGRAMS

The paper heralding the advent of Brinch Hansen's programming language Concurrent Pascal, published in 1975,[34] was a stylistic masterwork of the same quality as Nicklaus Wirth's 1973 book *Systematic Programming* (see Chapter 1, Section 1.10). A reader of the latter was led into Pascal without seeming to be aware of it because of Wirth's focus on programming concepts ahead of notation. So also, Brinch Hansen, in part I of the paper, introduced his reader to the language by way of the concepts of parallel programming he was interested in before bringing forth, in part II, the notation in which these concepts were cloaked.

This was one facet of his expository style. The other was the progressive development of a single example of a concurrent system, first conceptually, utilizing a combination of text and elegant pictorial diagrams, then describing the system in the language itself.

But for those who were cognizant of the evolving thoughts voiced by Dijkstra, Wirth, Hoare, Dahl, Nygaard, and Brinch Hansen over the previous 7 or 8 years about structured programming, hierarchical design, cooperating asynchronous parallel processes, abstract data types, classes, monitors, and processes, reading this paper may well have been akin to a homecoming experience: they would find themselves in a familiar environment. What was new was how Brinch Hansen absorbed these elements and synthesized them into a coherent programming language wrapped in a Pascalian framework.

To give the reader a general feeling for the nature of a Concurrent Pascal program, and thus parallel programming in a shared-memory environment—consider Brinch Hansen's running example of a spooling system, perhaps the earliest and conceptually the simplest task an operating system may perform: Data are read from an input device to a disk; a job reads from this disk, performs some computation on it, and writes the result onto another disk; these data are then output in some appropriate form. Except for the peripheral devices the others are all taken to be abstract.

So here we find two of the three major system types recognized in Concurrent Pascal: processes and monitors. The peripherals are assumed to be monitors implemented in hardware. The third principal system type is the class as understood in the Simula 67 sense. The difference between a class and a monitor—both are abstract data types—is that the latter are shared between processes whereas the former is accessible to only one process. For example a disk buffer (conceived as a monitor) is given the illusion of possessing its own private disk. This illusion is created by way of a "virtual disk" to which the disk buffer is given exclusive access. The virtual disk is, thus, a class.

So at the most abstract level the spooling system is declared as

```
var   b1, b2: diskbuffer;
      read: input;
      do: job;
      write: output
```

So b1 and b2 are system components of monitor-type diskbuffer; read, do, and write are system components of process types input, job, and output, respectively. Each of these process types will, of course, have to be defined and declared; they happen to be nonterminating cyclic processes, for example,

```
type job =
      process (input, output: diskbuffer)
            var . . . .
            cycle .
              . . . . .
            end
```

Ignoring the details of these processes, monitors, and classes that emerge during the hierarchical or stepwise refinement of the system, a complete Concurrent Pascal program will be composed of a hierarchical set of system types. The "highest" system type is a predefined process type called initialization process, an instance of which is created when the concurrent program is loaded for execution. The initialization process defines all system types and their instances (i.e., system components) and initializes the system components. For example,

```
type =
      "declaration of system types as process, monitor, and class types"
      . . . . .
      diskbuffer = monitor ( . . . );
end;
var
"declaration of system components as instances of system types"
      . . . . . .
      b1, b2: diskbuffer;
   begin
   init  . . . .
      read (b1);
      do (b1, b2);
      write (b2);
   . . . . .
   end
```

Concurrent Pascal was the first fully designed and implemented parallel programming language. Brinch Hansen's student at the California Institute of Technology, Alfred Hartmann, implemented a Concurrent Pascal compiler (written in Pascal) in 1974–1975 for a DEC PDP 11/45 minicomputer.[35] In the spirit of the times in compiler thinking, the code generated by the compiler was for a virtual Concurrent Pascal machine designed by Brinch Hansen, which was then interpreted by a PDP 11/45.[36]

At the same time Hartmann was implementing the language, Brinch Hansen, as befitting an experimental computer scientist, designed and implemented Solo, a single-user operating system for the PDP 11/45. This was the first operating system written in Concurrent Pascal.[37] Two years later, Brinch Hansen—by then at the University of Southern California—published his book *The Architecture of Concurrent Programs* (1977).[38] This book offered a synthesis of his thoughts on parallel programming, an exposition on Concurrent Pascal, and descriptions of Solo and other system utilities all written in the language. As far as parallel programming in a shared-memory environment was concerned, this book was his *opus*.

4.8   BRINCH HANSEN'S CREATIVITY

Some creativity researchers maintain that the originality of an artifact is Janus-faced. On the one hand, looking back in time from its moment of creation, an artifact is deemed *historically original* (H-original) if no other functionally identical artifact had existed prior to this.[39] Correspondingly, the creator (or artificer) of the H-original artifact is judged to be *H-creative*. On the other hand, an artifact is deemed to the *consequentially original* (C-original) if it affects "future history"—that is, it shapes the development of artifacts that come after the artifact of concern, or it has epistemic, cultural, social, economic, or technological consequences.[40] The artificer is, accordingly, judged to be *C-creative*.

There is, of course, a certain asymmetry between H-originality–H-creativity and C-originality–C-creativity. Usually, judgment of H-originality can be effected almost immediately after the artifact has been created because only the antecedents of that artifact need be considered. But although judgment of C-originality might be anticipated at the time of artifact creation (for example, the consequence of the stored-program concept was grasped by many computer pioneers almost as soon as it emerged in the mid-1940s), more often than not, assessment of C-originality demands the elapse of time following the time of creation of that artifact. (A striking example was the invention of microprogramming in 1951: Its influence and effect were not felt until over a decade later.[41])

Brinch Hansen, reflecting back on the invention of Concurrent Pascal in 1993, almost two decades after its creation, would comment that what he was most proud of was to show that one can write "nontrivial concurrent programs" in a "secure" language, something that no one had demonstrated before.[42]

So he clearly believed in the H-originality of the language. And most observers would agree with him. Concurrent Pascal's H-originality—and Brinch Hansen's H-creativity—is incontestable. But his remark is interesting also because his claim to what we are calling H-originality for the language lay not so much in the specifics of the language but because it demonstrated the feasibility of writing concurrent programs in a high-level language. He would go on to say that "The particular paradigm we chose (monitors) was a detail only."[43] By his own reckoning the significance of Concurrent Pascal lay in its status as a proof of concept.

But his throwaway remark that the monitor paradigm "was a detail only" is worth noting. In the realm of parallel programming, as we have seen, the monitor concept was of major significance. Brinch Hansen was (at the very least) a coinventor of the concept (along with Hoare and, possibly, Dijkstra). In the specific domain of parallel programming he co-created something akin to a paradigm. And in this lay Brinch Hansen's C-creativity, for several "monitor languages" (to use his term) came into existence between 1976 and 1990, as he himself listed.[44] Not all were influenced by Concurrent Pascal, he noted, but several were. More generally, insofar as Concurrent Pascal was the first monitor-based language, no subsequent design of parallel or modular programming language could *not* have been influenced to some degree by Concurrent Pascal. The language was thus C-original, and Brinch Hansen, C-creative.

4.9 THE PROSPECT OF THE USER AS "PARALLEL PROGRAMMER"

Strictly speaking, who a "user" is is context-specific. For example, the systems programmer is a user of the "bare" machine as defined by its ISP-level and PMS-level

architectures (see Chapter 3, Sections 3.4, 3.5). But in ordinary computational discourse the user is one who is given a complete computer system (the physical machine, its operating system, language compilers, and other software tools) and uses this system to solve his or her problems. It is this sense of user that is relevant here.

In the 1970s and 1980s the primary class of users were application programmers concerned with solving scientific, engineering, mathematical, informatic, and business-related problems. Some wrote application programs on behalf of clients, others for themselves. Some were professionals, others amateurs. But by and large application programmers had a common expectation: that the systems they would use would hide from their sight all the nasty realities of *how* their system did what they did on the behalf of the users. Indeed, ever since the first assembly language was created (by David Wheeler of Cambridge University) in 1949,[45] the mission of computer architects, language designers, and systems programmers was to deliver to the user community an abstraction: a virtual computer. The intricacies underlying the innards of the actual computer system were to be safely tucked away from the user's sight.

But even as application programmers wrote their programs in high-level languages, and relied on the magic of virtual processors, virtual memory, and virtual I/O to provide them the illusion of an ideal user interface, there was another, concurrent movement afoot. For, by the beginning of the '70s, computer architects and hardware designers were producing a certain class of very expensive, very high-performance computers whose whole purpose was to solve computation-intensive scientific and engineering problems.

In the short history of computing, the fastest machines in the world at any given time were dubbed "supercomputers." These were the machines that, through a blend of the state-of-the-art semiconductor technology, physical packaging, and ingenious architectural design, performed at the limits of performance at any point of historical time. The first half of the 1980s witnessed a small confederacy of such machines including the Fujitsu VP-200 (1983), Hitachi's S-810/20 (1983), Cray Research's series of machines—Cray-1M (1983), Cray-X-MP (1983), and Cray-2 (1984)—and the CDC Cyber 205 (1982).[46]

These designs did not emerge de novo. They were no doubt influenced by their predecessors, the first generation of supercomputers developed in the 1960s and early 1970s. One subclass of these earlier high-performance machines were dedicated to synchronous parallel programming that exploited the "natural" parallelism of certain types of computational problems. They were generically called *vector-array computers*. And the question of enormous interest to both designers and users of these systems was this: Who should be responsible for mapping the natural parallelism of computation-intensive problems onto such machines?

One option was to give the user full control of this mapping: a scenario of the user as parallel programmer.

For us to appreciate the nature of "natural parallelism let us consider as a simple example the following mathematical problem:

We are given two n-element vectors:

$$A = [a_1, a_2, \ldots, a_n],$$
$$B = [b_1, b_2, \ldots, b_n].$$

To computer their "inner product," we have

$$C = a_1 * b_1 + a_2 * b_2 + \cdots + a_n * b_n.$$

Each of the products $a_i * b_i$ can be calculated independently and in parallel (if resources are available) before they are summed. Moreover, the summation of the products can also be organized to exploit natural parallelism. For example, perform the parallel additions:

$$a_1 * b_1 + a_2 * b_2,$$
$$a_3 * b_3 + a_4 * b_4,$$
$$\ldots \ldots$$
$$a_{n-1} * b_{n-1} + a_n * b_n;$$

then add pairs of these results in parallel and so on.

Calculation of the inner products of vector pairs is an ubiquitous operation in many scientific computations. And, as can be seen from this example, its characteristic is that some sort of an identical operation (multiplications or additions) is required to be performed on independent sets of variables. So if such multiple identical operations could be performed by independent multiple processing units then one can effect such computations by parallel processing means. Array–vector computers were intended precisely for this purpose. In Michael Flynn's taxonomy (see Chapter 3, Section 3.19) they constitute the SIMD taxon.

From about 1966 through 1980 a number of experimental and commercial vector-array computers were designed, built, and became operational. Arguably the most notable was the ILLIAC IV (circa 1966–1972), developed at the University of Illinois, originally under the leadership of Daniel Slotnick,[47] but there were also ICL's Distributed Array Processor (circa 1973),[48] Goodyear's STARAN (circa 1974)[49] and Massively Parallel Processor (circa 1980),[50] and the Burroughs Scientific Processor (circa 1980).[51]

For a sense of the structural complexity of such machines, consider the ILLIAC IV, certainly one of the most studied vector–array computers. A program would be distributed across 64 memory modules connected to a single, nonpipelined instruction interpretation unit. The program's data are also distributed across 64 memory modules, but here each is connected to its own dedicated nonpipelined execution unit, so that there are 64 pairs of memory module–execution unit combinations.

4.10   THE IDEA OF A VECTOR PROGRAMMING LANGUAGE

High-level languages for user programming of array–vector computers—generically called vector programming languages—never seemed to receive the kind of attention that parallel languages for systems programming did. Perhaps one reason was the conviction on the part of some that users should be shielded from the perils of parallel programming, that they should be allowed to program "as usual" in sequential fashion and leave the task of program parallelization to the compiler.

But there were others who wanted to give the users, concerned with exploiting as much parallel processing out of their programs in the interest of enhancing performance, the right tools to exercise this freedom; hence research into the design of vector programming languages. Some were (oxymoronically) high-level, machine-specific languages. The ILLIAC IV researchers, for instance, designed and implemented two such languages named CFD (circa 1975) and Glypnir (circa 1975), the former Fortran-like, the latter based on Algol.[52] But in keeping with the "proper" spirit of high-level programming, Ron Perrott of Queen's University Belfast designed a machine-*independent* vector programming language called Actus (circa 1979), as yet another descendant of Pascal.[53]

As one might guess, the dominant data type in a vector programming language would be the array. Of course Pascal and other high-level languages had this as a structured data type, but where Actus differed lay in that parallel actions that could be performed on array elements. For example, given the declaration

var *matrixa*: **array** [1. . *m*, 1 . . *n*] **of** real,

its elements could be referenced in Actus individually (as in Pascal), but by changing this declaration to

var *matrixb* : **array** [1 : *m*, 1 . . *n*] **of** real,

the *m* elements of *matrixb* along any one of the *n* columns could be accessed in parallel. Moreover, subarrays other than rows or columns could also be referenced together. And Pascal statement types (such as assignments and conditionals) were extended to cater for parallelism. For example, consider the following fragment:

```
const diagindices = 1 : 50;
var maindiag : array [1 : 50] of integer;
var matixc : array [1 : 50, 1 . . 50] of integer;
var vectork : array [1:50] of integer;
var vectorx : array [1 : 50] of integer;

. . . . . . . .
```

maindiag [1 : 50] := *diagindices*;
vectork [1 : 50] := *matrixc* [1 : 50, *maindiag* [1 : 50]];
vectorx [1 : 50] := *matrixc* [1 : 50, 50];
if any (*vectork* [1 : 50] > *vectorx* [1 : 50])
then *vectork* [1 : 50] := *vectork* [1 : 50]–1.

This first declares an array of constants 1, . . . ,50; these constants will be assigned to the elements of *maindiag*; the main diagonal elements of *matrixc* will then be assigned to the array *vectork*; then all 50 elements of column 50 of *matrixc* will be assigned to *vectorx*; and, finally, if any one of the elements of *vectork* is greater than the corresponding elements of *vectorx*, each of the *vectork* elements will be decremented by one.

However, as Ron Perrott, Actus's designer cautioned, the luxury afforded the programmer of being able to specify and control synchronous parallelism in this fashion came with a significant price. For even though an Actus-like language was machine independent in its syntax and semantics, users must be cognizant of the actual target computer for such programs. The target array–vector architecture would impose *constraints* on a problem's natural parallelism. Users would have to map the problem structure onto the target computer's architecture. The clarity of the resulting program could well be seriously obscured, leading to difficulties in its readability, understandability, and modifiability.[54]

4.11   THE COMPILER–WRITER AS PARALLELIZER

If the user of a vector-array computer were to be shielded from the minutiae of parallel processing then, as we have noted, the burden of parallel programming would fall on the compiler—more precisely on the compiler–writer. The user would write his or her program in the normal sequential fashion in a sequential language, and it would then be the compiler–writer's task to dissect the program and expose its parallelism in a form suitable for execution on the target array–vector computer.

So this was the other side of the parallel programming debate. And as befitting one of the pioneers—some would say *the* pioneers—of synchronous parallel computing, the ILLIAC IV researchers hedged their bets. On one front they studied the possibility of programmers doing their own parallel programming using the languages CFD and Glypnir. On the other front they investigated strategies for automatically detecting parallelism in sequential programs and vectorizing the sequential code.

This investigation was led by David Kuck of the University of Illinois, the home of the ILLIAC IV project. Over much of the 1970s Kuck and his colleagues explored and explicated the theory and practice of vectorizing sequential programs, culminating in a system called PARAFRASE (circa 1980).[55]

PARAFRASE was really a compiler "preprocessor" that performed transformations on source programs written in Fortran. It was also "retargetable" in the sense that its repertoire of transformation strategies encompassed machine-independent, partially machine-dependent, and machine-specific strategies. The vectorizer's output was thus Fortran code functionally equivalent to the input source program. This output would then be input to a compiler implemented for some specific array–vector machine.

4.12 THE FUNDAMENTAL RULE OF PARALLELISM, REVISITED

We have already encountered, first in the context of microprogramming (Chapter 3, Sections 3.11, 3.12), then in the context of asynchronous concurrency (Sections 4.3, 4.5), what we may call the *fundamental rule of parallelism*: Tasks can be activated in parallel as long as there are no resource dependencies between the tasks.

Kuck and his colleagues analyzed this principle in the context of array–vector processing. Thus, given a statement sequence

$$S_1 ; S_2$$

if an input variable of S_2 is also an output variable of S_1 they called it a "direct data dependency" (*dd*) relation between S_1 and S_2. If an output variable of S_2 is also an input variable of S_1, this is a "data antidependency" (*da*) relation between S_1 and S_2. Finally, if S_1 and S_2 share an output variable then the relationship between them is an output dependency (*od*). If any of these conditions prevailed S_1 and S_2 could not be parallelized.

These conditions, in a manner rediscovered by computer scientists investigating parallelism of different kinds, were first articulated formally in the context of parallel programs by Arthur Bernstein, then of General Electric Research and Development Center, in 1966.[56]

So the theoretical foundation for automatic parallelism detection was a structure called a "data dependency graph" (DDG)—drawn from mathematical graph theory—that would show, for some sequential source program segment, the presence or absence of the *dd, da*, and *od* relationships between pairs of statements. As an example, consider the following sequence of assignment statements:

$$S_1{:}\ a \leftarrow a + 1;$$
$$S_2{:}\ b \leftarrow a + c;$$
$$S_3{:}\ b \leftarrow e - c;$$
$$S_4{:}\ e \leftarrow a + f;$$

The corresponding DDG is shown in Figure 4.3. In the jargon of graph theory this is an "acyclic" graph (that is, free of cycles or loops). On the other hand, consider the following nested loop:

$$\textbf{for } i \leftarrow 1 \textbf{ until } 10 \textbf{ do}$$
$$\textbf{for } j \leftarrow 1 \textbf{ until } 10 \textbf{ do}$$
$$\text{S1: } a(i, j) \leftarrow a(i, j) + b(i, j);$$
$$\text{S2: } d(i, j) \leftarrow a(i, j) + e$$
$$\textbf{end for}$$
$$\textbf{end for}$$

The DDG of the inner-loop body will be a "cyclic" DDG, as shown in Figure 4.4.

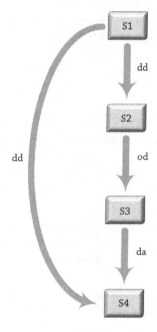

FIGURE 4.3 An acyclic DDG.

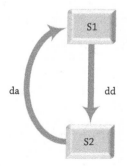

FIGURE 4.4 A cyclic DDG.

PARAFRASE was, in fact, more than a detector of such data dependencies. It contained a catalog of strategies for transforming sequential code containing dependencies into sequences that were data independent so that they could not only be parallelized but also improved or (in the jargon of compiler theory) optimized in some sense. To give an example, consider the strategy called "loop distribution." We have the following loop:

$$\textbf{for } i \leftarrow 1 \textbf{ until } 10 \textbf{ do}$$
$$\text{S1: } a(i) \leftarrow b(i) + c(i);$$
$$\text{S2: } d(i) \leftarrow r(i) \: / \: a(i);$$
$$\text{S3: } x(i) \leftarrow y(i) \: ^* \: w$$
$$\textbf{end for}$$

There is a *dd* relationship between S1 and S2 but S3 is quite independent of S1 and S2. So if the loop is "distributed" then we obtain

---

$$\textbf{for } i \leftarrow 1 \textbf{ until } 10 \textbf{ do}$$
$$\text{S1: } a(i) \leftarrow b(i) + c(i);$$
$$\text{S2: } d(i) \leftarrow r(i) \: / \: a(i)$$
$$\textbf{end for};$$
$$\textbf{for } i \leftarrow 1 \textbf{ until } 10 \textbf{ do}$$
$$\text{S3: } x(i) \leftarrow y(i) \: ^* \: w$$
$$\textbf{end for}$$

The second loop can be performed by a single vector operation. Moreover, this would enhance what was called the program segment's *locality* characteristic. That is, references to all the arrays in the first loop will be completed before control transfers to the second loop. At program execution time this reduces the amount of data traffic between memory modules and processors.

4.13   EFFECTING PARALLELISM THROUGH PIPELINING

As we have seen, the phenomenon of parallelism is abstraction sensitive: A computational artifact that appears sequential at one level of abstraction may be found in reality to be teeming with parallel activity at a lower abstraction level. Abstraction, by design, creates illusions; refinement dispels illusions.

And so, when computer architects call a computer a *uniprocessor*, the implication is that it can process only one task at a time. That is, there is a single processing unit that accepts a single instruction stream, one instruction at a time, and processes that instruction before accepting the next instruction in the input stream. In other words, the ICycle (see Chapter 3, Section 3.5) is assumed to be an *indivisible* process.

A uniprocessor, then, corresponds to the SISD taxon in Michael Flynn's taxonomy (see Chapter 3, Section 3.19). But not all uniprocessors (or SISD computers) *really* process one instruction at a time. Not all ICycles are really indivisible processes. After all, an ICycle typically consists of six sequential steps:

Instruction fetch;

Instruction decode;

Operands address calculation;

Operand fetch;

Instruction execute;

Update program counter for accessing next instruction.

Assume that each of these steps is performed by a distinct processing resource within the processor—let us call them stages—denoted IFETCH, IDECODE, OADDR, OFETCH, IEXEC, and UPCTR, respectively. Consider an instruction stream $I_1, I_2, \ldots,$ to be processed sequentially by the ICycle. Each instruction I_j will "flow" through the six stages of the processor, and each stage will execute its designated step; at the end of the final stage I_j will have been processed. The stages then form a *pipeline* through which I_j flows. But at one time, I_j will occupy only one of the six stages. Suppose it is occupying UPCTR; in that case each preceding stage can be kept busy executing its designated step on instructions that follow I_j in the instruction stream. In the ideal situation, there can be six instructions, $I_j, I_j + 1, I_j + 2, \ldots, I_j + 5$ in the pipeline simultaneously. If I_j is "occupying" UPCTR, $I_j + 1$ will occupy IEXEC, $I_j + 2$ will occupy OFETCH, and so on. And when the pipeline is full, and assuming there are no dependencies between the instructions in the pipeline, a six-stage pipeline can increase the processing unit's throughput sixfold.

This, then, is an instance of an *instruction pipeline*.

But consider an individual floating-point arithmetic unit that adds two floating-point numbers $N_1 = <E_1, M_1>$, $N_2 = <E_2, M_2>$, where E_i, M_i ($i = 1, 2$) are the exponents and mantissa, respectively, of number N_i. Then the addition of N_1, N_2 would typically involve the following sequence of operations:

Comparison of exponents E_1, E_2;

Alignment of mantissas M_1, M_2;

Addition of the mantissas;

Checking for the number of leading zeros in the result;

Normalizing the result to eliminate leading zeros.

So if each of these operations is performed by a distinct stage in the arithmetic unit—let us call them COMP, ALIGN, ADD, ZERO, and NORM, respectively—then add instructions from the instruction stream could be fed to this *arithmetic pipeline*. A single floating-point unit could process up to (theoretically) five floating-point arithmetic operations concurrently.

The idea of effecting parallelism through pipelining certainly reaches back to the mid-1960s, although other roughly equivalent terms such as "instruction overlap" were used and related schemes such as "instruction lookahead" were implemented. In 1964, the same year that the IBM System/360 family was introduced, Control Data Corporation (CDC) released the CDC 6600, the supercomputer of its time, which manifested a form of processor pipelining in that its CPU comprised 10 independent functional units to perform different kinds of arithmetic operations, logical operations, and shift operations. Thus instructions in the instruction stream could be parceled out to these parallel functional units that would perform their tasks concurrently.[57] So, in effect, multiple instructions from the same instruction stream could execute in parallel.

But we would not be remiss if we select January 1967 as seminal in the context of pipelining. That month, an issue of the *IBM Journal of Research and Development* (at the time called the *IBM Journal*) was published that was devoted entirely to the design of the IBM System/360, Model 91, at the time the high-end member of the System/360 family. This issue included a paper by D. W. Anderson, F. J. Sparacio, and R. M. Tomasulo on the architecture of a multistage instruction pipeline in the 360/91.[58] Interestingly, the term pipeline did not appear in this paper; the preferred term was "assembly line."

So the seeds of pipelining were planted in the mid-to-late 1960s. But their fruits were most abundantly evident in the following two decades, both in the realms of experimental computers and commercial systems.

Perhaps the most prominent of the experimental machines that explored pipelining extensively was the MU5, a powerful system designed and built between circa 1968 and 1974 at the University of Manchester as the fifth of their series of innovative experimental systems.[59] In the commercial sector, pipelining became ubiquitous across all class lines (so to speak), ranging from the microprocessor, such as the Intel 8086 (circa 1978), through superminis, for example, DEC's VAX 8600 (1985), through very high-performance machines, like Texas Instrument's Advanced Scientific Computer, TI-ASC (circa 1972), to the later Cray series of supercomputers, Cray-1 (1978), Cray X-MP (circa 1983), and Cray-2 (circa 1986).

Effecting parallelism through pipelining of course meant that the fundamental principle of parallelism must be met (Sections 4.3, 4.5): There must be no conflicts in data resources or functional unit resources between instructions in the instruction stream. But pipelining is attended with its own genre of problems because it entails single instruction streams within which parallelism must be detected. The smooth flow of such an instruction stream through a pipeline will be often rudely disrupted by "eddies" and "whirlpools": conditional and unconditional branches and program loops. Furthermore, scheduling of an instruction stream through the pipeline has to be achieved through hardware, and *timing* becomes an issue. The design of pipelines is further exacerbated by the variations in the nature of the individual instructions in the pipeline (e.g., differences in instruction formats results in variations in the ICycle process itself). Further complexity arises if there are multiple functional units to which different instructions in the pipeline are parceled out in the execution phase.

Thus, stimulated no doubt by the ubiquity of pipelining in different computer classes, much theoretical research ensued in the 1970s, not only on the logic of pipelined parallel processing but also in analytical model building to elicit performance characteristics of pipelined systems. It is interesting to note that in the book *Computer Structures: Readings and Examples* (1972), authored and compiled by Gordon Bell and Allen Newell of Carnegie-Mellon University, "pipelining" does not even appear in the otherwise comprehensive index.[60] How things changed over the next decade! A comprehensive survey article, the first of its kind on the topic, by C. V. Ramamoorthy of the University of California, Berkeley, and H. F. Li of the University of Illinois appeared in *ACM Computing Surveys* in 1977.[61] Derrick Morris and Roland Ibbett, two members of the MU5 design team at Manchester University, published a book-length monograph on the computer in 1979 that included two detailed chapters on the MU5 pipelines.[62] The first comprehensive textbooks on computer architecture, Harold Stone's *Introduction to Computer Architecture* (1980) and Jean-Loup Baer's *Computer Systems Architecture* (1980) contained valuable chapters on pipelining.[63] When Daniel Siewiorek of Carnegie-Mellon University joined Gordon Bell and Allen Newell in publishing *Computer Structures: Principles and Examples* (1982), a much-revised and updated edition of the Bell–Newell volume of a decade earlier, pipelining was not only in the index, an entire section of the book was devoted to pipelined computers.[64] What was probably the first book devoted entirely to parallel processing systems and architecture, *Computer Architecture and Parallel Processing* (1984) by Kai Hwang and Faye Briggs of the University of Southern California, contained valuable chapters on pipelining.[65] And, as the ultimate signal of a specialized topic of "having arrived," Peter Kogge of IBM published his authoritative treatise, *Architecture of Pipelined Computers* in 1981.[66]

4.14   PARALLEL MICROPROGRAMMING: DÉJÀ VU WITH A TWIST

As we already noted, in the 1970s, with the advent of writable control memories and the emergence of universal host machines, the prospect of user microprogramming (also called "dynamic" microprogramming) became an intriguing, indeed, exciting, possibility for some researchers (Chapter 3, Sections 3.9, 3.10). One outcome of this was the idea of high-level microprogramming[67]—that is, writing microprograms (or *firmware*, a termed coined by Ascher Opler of IBM in 1967[68]) in a high-level language, then compiling them into an executable microcode, and loading into writable control memory ready for interpreting user-defined architectures.

A new discourse was thus initiated in the mid-1970s under the banner of what came to be called *firmware engineering,* a discourse that gained much traction over the next decade,[69] and loosely defined as being concerned with the scientific principles underlying microprogramming and applying these principles to the firmware development process.[70] We have already noted the emergence of a science of the artificial pertaining to the *architectural* face of microprogramming (see Chapter 3, Sections 3.11, 3.12); the discourse of firmware engineering addressed its *programming* face.

Some people of that period versed in programming methodology and parallel programming, reading the emerging literature on firmware engineering, may well have experienced a sense of déjà vu. It might have seemed to them that all the issues, debates, and solutions from the realm of software were reappearing in the firmware domain. They might even have concluded, at first sight, that this was a simple and relatively uninteresting case of *knowledge transfer* from one level of computational abstraction to another: from a higher to a lower level. But if they had persisted in their reading they would also have realized that this was déjà vu with a twist—a few twists, in fact. They would have realized that the knowledge transfer was not straightforward at all. For the lower level of abstraction—the microprogram level—had its own idiosyncrasies, its own quirkiness, especially in coping with parallelism. If parallel programming was a discourse of its own then parallel microprogramming lay at the intersection of this discourse with the discourse of firmware engineering.

Therefore researchers in parallel microprogramming had to integrate the nuances of parallelism in microarchitectures (see Chapter 3, Sections 3.9–3.12) with the accumulated understanding of both asynchronous and synchronous parallel programming and of parallel programming languages of the kinds described earlier in this chapter.

The crux of the matter was the inerasable fact that microprograms were *machine specific.* There were no virtual processes, processors, or memory resources at this level of abstraction. The resources manipulated by a microprogram were *physical* (memories, registers, data paths, functional units, etc.); moreover, they were physical resources in a *particular* machine—the host machine and its microarchitecture on which a target, higher-level system would be implemented (Chapter 3, Sections 3.9, 3.10). This machine specificity meant that any dream of high-level microprograms completely abstracted from the host-machine microarchitecture was just that—a dream.

And this machine-specificity imperative was what demarcated microprograms from programs and parallel microprogramming from parallel programming.

Let us recall the fundamental principle of parallelism (Section 4;12): that tasks can be done in parallel as long as there are no resource conflicts between them. Bernstein's conditions must be met.

In microprograms the basic task is the microoperation (MO) that, in the formal literature on microprogramming, was represented as the 4-tuple:

$$Mi = <\ OPi,\ INi,\ OUTi,\ Ui\ >,$$

where OPi is the operation performed, INi is the set of data input resources to OPi, $OUTi$ is the set of data output resources from OPi, and Ui is the operational unit that performs Opi. (For simplicity's sake this representation ignores the matter of *timing* of MOs.)

Imagine a sequential microprogram S where each statement in the program is an MO. Two MOs Mi and Mj such that Mi *precedes* Mj in S, ($Mi < Mj$) can be executed in parallel ($Mi \parallel Mj$) from within the same microinstruction (MI) if Bernstein's conditions are met:

$$INi \cap OUTj = \Phi,$$
$$INj \cap OUTi = \Phi,$$
$$OUTi \cap OUTj = \Phi,$$

and also

$$Ui \cap Uj = \Phi.$$

Here Φ is the empty set. Let us collectively call these the *resource rule for microparallelism*.

But this ignores the all-important factor of timing. For the execution of microinstructions are controlled by a clock cycle called a microcycle. And usually (for reasons we will not enter into here) the microcycle is longer than the time to execute most MOs. Thus, different MOs may be timed to execute in different *phases* of the microcycle. Some MOs may need a single phase, others two or more phases, still others, the whole microcycle. At any rate, let us term the phase or phases of the microcycle in which a given MO Mi executes its *time attribute Ti*. So, in addition to the resource rule, we also have the *timing rule*:

$$Ti \cap Tj = \Phi.$$

There is still something else: MOs are encoded in particular *fields* of the microword (Chapter 3, Section 3.13). Clearly, if two MOs are encoded in the same field, they *cannot* be activated in parallel. Denoting by Fi, Fj the fields in which MOs Mi, Mj (where $Mi < Mj$ in S) are encoded, in addition to the resource and timing rules the *field rule* must also be satisfied:

$$Fi \cap Fj = \Phi.$$

These then collectively define the fundamental rule of microparallelism. Collectively they add the "twist" to any sense of déjà vu the seasoned researcher on parallel programming might experience when considering parallel microprogramming.

And, as we have observed in the case with synchronous parallel programming, a debate ensued: Should the burden of parallel microprogramming fall on the high-level language microprogrammer or should it be assigned to a *microcode compiler*?

The former path demanded the availability of high-level microprogramming languages with constructs for describing the intricacies of microparallelism à la Concurrent Pascal.[71] But it was the latter path that attracted the most attention. The general sentiment was that microprograms would be—should be—written in a high-level language but in sequential form. A compiler would first generate an "intermediate" sequential microcode for the target microarchitecture, possibly with some microcode "optimization" (a euphemism for code improvement) that would eliminate redundancies in the microcode; and then the "back end" of the compiler would detect parallelism within the sequential, optimized microcode and *compact* the code into horizontal microprograms.

Thus was born the twin problems of "vertical" (that is, sequential) microcode optimization and "horizontal" microcode compaction.

4.15   MICROCODE COMPACTION: AN IDENTITY OF ITS OWN

1971–1981 was an interesting decade for these two classes of problems. The pioneering theoretical work on vertical optimization was done in 1971 by Richard Kleir and C. V. Ramamoorthy, then of the University of Texas, Austin,[72] who explored how program optimization techniques performed by compilers could be adapted to microprograms.[73] Kleir and Ramamorthy's paper was undoubtedly significant in that what had hitherto been a severely empirical field of investigation [as evidenced by Samir Husson's *Microprogramming: Principles and Practices* (1970), the first book on microprogramming[74]] was subject, for the first time, to formal, mathematical treatment.

Microprogram optimization as theory was transformed into the domain of a practical artifact in 1976 by David Patterson of the University of California, Berkeley [who would shortly thereafter be one of the originators of the RISC movement in computer architecture (see Chapter 3, Sections 3.16, 3.17)]. In his PhD dissertation (done at the University of California, Los Angeles) Patterson developed a microprogramming system called STRUM that, among other things, generated microcode for the Burroughs D microprogrammable computer. Patterson listed a number of optimization techniques incorporated in STRUM.[75]

Vertical optimization is relevant to this story only because, first, Kleir and Ramamoorthy formalized the representation of microprograms, and then Patterson constructed a tool for optimizing vertical (sequential) microcode. But compared with that of vertical optimization, the problem of (horizontal) compaction undoubtedly earned much greater attention. This was probably because the compaction problem entailed issues that were not simply "spin-offs" of ideas from the domain of software. Microcode compaction, as a class of parallelism problems, rather like pipelining, acquired an identity of its own.

Thus, between 1974 and 1982, a spate of papers offered algorithmic solutions to the compaction problem, the differences lying in the specific assumptions about microarchitecture governing the underlying formal models and the strategies used in the algorithms. This was indeed *transfer of knowledge*, in part from graph theory, in part from compiler code optimization theory. But the idiosyncracies of microprogram-ming injected their own problematics so that research into microcode compaction and the automatic production of parallel (horizontal) microprograms also *generated new knowledge* that contributed a distinctive understanding of the nature of low-level parallelism.

The earliest studies adopted a divide-and-conquer policy. A sequential program was assumed to have been partitioned into *straight-line microprogram* (SLM) segments. (Such partitioning was well known in compiler research wherein straight-line program segments were termed "basic blocks.")

Precisely: A SLM was defined as a sequence of MOs with no entry point except the first MO in the sequence and no branch exit except (possibly) the last MO in the sequence.

So these early researchers—Stephen Yau and his colleagues at Northwestern University,[76] the present writer and his collaborators at the University of Alberta,[77] Masahiro Tsuchiya and his associates at the University of Texas, Austin,[78] and Graham Wood at Edinburgh University[79]—concentrated on the so-called local compaction problem, confining analysis to within SLMs.

By 1976, the more challenging problem of extending parallelism detection and com-paction beyond individual SLMs—the global compaction problem—was well under way, undertaken first by the present writer at the University of Alberta,[80] followed soon after by Mario Tokoro and his coworkers at Keio University[81] and Joseph Fisher at New York University.[82]

The ultimate goal of microcode compaction was to minimize the time to execute a microprogram. Making the simplifying assumption that each horizontal microin-struction in the microprogram was executed in one microcycle—this was not always true, for instance, accessing main memory may have required multiple microcycles—the compaction goal would become the minimization of the control memory *space* re-quired to hold the microprogram.

However, satisfying such an optimization goal would, in general, be exorbitantly expensive in computational time. The algorithms invented by the various researchers were thus all heuristic algorithms: They might, for some input sequential microcode, produce the minimal output but they were never guaranteed to do so.

For example, one local compaction algorithm, which was due to C. V. Ramamoorthy, Masahiro Tsuchiya, and Mario Gonzales, first constructed a dependency graph and then used this to identify a *lower bound* on the number of microinstructions that would be required. These microinstructions formed a "critical path" through the dependency graph. (This is of the same nature as the critical path in the tea-making scenario of Section 4.1.) Each remaining MO was then placed in the earliest possible microinstruc-tion without violating the MO dependency relations.[83]

Another local compaction algorithm, which was due to the present writer, did not actually build a dependency graph. Rather, MOs from the input SLM were added in the order they appeared in the SLM to a list of (initially empty) microinstructions. Each MO was placed in the earliest possible nonempty microinstruction without violating the MO dependency relationships. If this was not possible—if the MO could not be "packed" into an existing microinstruction—then it was placed in a newly created microinstruction at the "head" of the "current" microinstruction sequence, if possible; otherwise, a new microinstruction with this MO was created at the "tail" of the current microinstruction sequence.[84]

As an example of global compaction, a technique devised by Joseph Fisher, then of New York University, identified paths (called by Fisher traces) through a dependency graph representing the microprogram as a sequence of MOs that *may* be executed when the microprogram is activated. Using probability estimates of which paths would be followed at branch points, the algorithm first selected the most likely trace through the dependency graph. In effect, such a trace denotes the straight-line sequence of MOs most likely to be executed. This "extended" SLM is then compacted by some compaction algorithm. The next most likely trace is then selected and compacted. This procedure is followed until all possible traces through a dependency graph have been exhausted.[85] The reader will realize that the resulting size of the parallelized microprogram, in general, will be larger than if the individual SLMs had been locally compacted. However, the average execution time of the resulting compacted microprogram is likely to be less.

4.16 DATA FLOW: APOSTATE OF PARALLELISM

Thus far this chapter has told a story of parallelism as it unfolded through the 1970s and the early years of the 1980s. As we have seen, the phenomenon of parallelism got to be known and understood at several levels of abstraction: at the user level, at the operating systems level, at the level of PMS architecture, at the ICycle level, and at the microarchitectural level. The phenomenon entailed both asynchronous and synchronous concurrency; both abstract time and real, physical time; and virtual resources as well as real resources. Threading through all these faces of parallelism were variations of the fundamental rule of parallelism.

But this rule itself was grounded in an assumption about the underlying machine architecture: that, regardless of the level of abstraction, the constraint imposed on parallel processes was the constraint of *shared resources*—primarily shared memory and data objects, but also shared functional units and data paths. The constraints were imposed by the von Neumann style of computing.

But now imagine a world that flagrantly violates the rules of the von Neumann universe; a world without the tyranny of the ICycle; a world without memory, without flow of control between instructions; indeed, without instructions that lord over data; without anything resembling an instruction stream. Imagine instead a world in which *data rule*. In which instead of a program showing control flowing from one instruction to another, the program depicts *data flowing* from one operation to another.

This was precisely what Jack Dennis of the Massachusetts Institute of Technology began to imagine in the early–mid-1970s.[86] Dennis believed that the kind of "massively parallel processing" needed for very high-performance computing demanded a departure from the general tradition of von Neumann computations: It demanded a new *style* of computing altogether, one that could exploit parallelism in its most pristine, most natural state, unfettered by the constraints of "conventional" parallel computers. He called this new style *data flow* to contrast it with the "control-flow" nature of the von Neumann style.

Figures 4.5 and 4.6 illustrate the difference. The former shows a control-flow-constrained DDG for a program fragment in which each vertex of the DDG denotes an instruction and the edges signify data dependencies between instructions. Such a DDG renders explicit all the natural parallelism between *instructions* within the framework of von Neumann architectures.

But as we have seen, supporting such "fine-grain" parallelism in a multiprocessor system would necessitate all kinds of overheads: allocating to, or scheduling individual

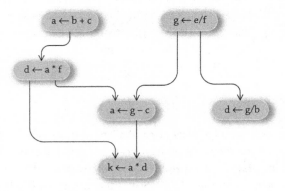

FIGURE 4.5 An acyclic control-flow DDG.

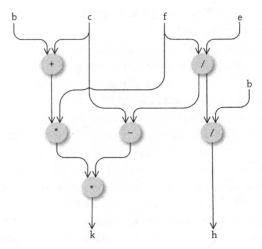

FIGURE 4.6 An acyclic DFG.

instructions on, individual processors, and synchronizing operations to preserve the data dependencies between instructions in the DDG.

Figure 4.6 shows a DDG that is functionally identical to the one in Figure 4.5, except that now the vertices of the graph are operations that do the actual computation. An operation "fires" whenever all its input operands are available. Synchronization between operations is achieved by the flow of *data* between operations. And now there is no need to assign the result of an expression evaluation to a variable in memory.

This was the idea behind the *data-flow computer* as conceived by Jack Dennis. But like most acts of creation, Dennis drew from others before him. In particular, in 1962, Carl Adam Petri submitted in a doctoral dissertation to the University of Bonn a powerful graph-theoretic formalism, called *Petri nets*, to model concurrent and distributed systems.[87] This would have profound theoretical influence on subsequent research on distributed computing, and we can certainly find its trace in the idea of data-flow representations. But the most direct influences on Dennis, as he acknowledged in a review paper of 1979,[88] were formalisms proposed in the late '60s by a number of researchers in the United States.[89]

Data flow really was an apostate in the world of practical parallel computing. Naturally, most of the early development of its ideas, in the mid–late 1970s, were due to Dennis and his students at MIT[90] but, as we subsequently see, before the end of the decade, a certain excitement about this style, engendered perhaps by its "cool" elegance, would spread beyond the reaches of Cambridge, Massachusetts.

Dennis invented a new language (both metaphorically and literally) to describe the nature of this new style of computing. Literally speaking, he and his coworkers created a new low-level (machine) language called *data-flow graph* (DFG), a specialized version of the DDG. Here, nodes (vertices) denoted entities called *actors* and *links*, and the edges (arcs) signified *paths* that carried *tokens* (representing data values) between nodes. In Dennis's original model, called *static* data flow, an actor could fire when there was a token in each of its input arcs and its output arcs were all empty. The result of the firing was that the actor *consumed* the input tokens—they disappeared into oblivion—and *produced* a new token on each output arc. This "firing semantics" meant that there was no need to write values into variables—a massive deviation from the von Neumann style.

A *link* was a vertex in a DFG with one input and two or more outputs; its function was simply to copy a token on its input arc to all its output arcs. A data-flow program described as a DFG would then be a machine language program in the data flow style.

DFGs can get quite complex very quickly so here the general nature of such a graph is depicted with a very simple example. Figure 4.7 shows the DFG corresponding to the following conditional expression:

$$Result \leftarrow \textbf{if } x{<}y \textbf{ then } u * v \textbf{ else } u * 6.$$

Figure 4.7 shows four actor types: the squares denote operators that perform basic arithmetic or logical operations and fires by consuming its input tokens and producing an output token. The boxes labeled 'SW' act as switches that allow the input from above

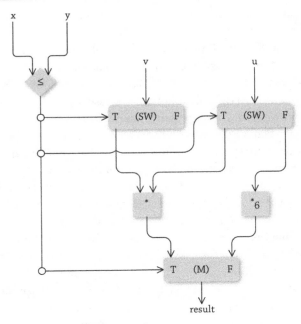

FIGURE 4.7 A DFG for the conditional expression.

to pass through to the output side as a function of a Boolean input at the side. The dia-mond is a "decider": It represents some predicate, and it fires by consuming its input tokens and producing a Boolean value as an output token as a function of its inputs. The box labeled 'M' is a "merge" actor: It accepts two input tokens on its "T" and "F" sides respectively, along with a control (Boolean) token. If the latter value is *true* the T input is transmitted as output; otherwise the F input is transmitted.

Using these semantic rules the reader will realize how the conditional expression is realized by this DFG. The potential parallelism and the inherent sequentialism are also evident.

4.17   FROM APOSTATE TO MOVEMENT

Apostasy either gets quickly punished or it gathers momentum and, if sufficiently sustaining, becomes a movement and even gains respectability. As the '70s gave way to the '80s data flow manifested the clear signs of a movement-in-progress. Small but unmistakable clusters of data-flow researchers and designers were soon in evidence: at the University of California, Irvine; at the University of Manchester; at the University of Utah; in Texas Instruments, Dallas; and, most spectacularly and visibly, as the Fifth Generation Computer Project in Japan.

But whether one likes it or not, some form of synchronization in a parallel pro-cessing environment cannot be avoided. Dennis's original idea was the static model in which only one token could occupy an arc at a time. But this seemingly innocuous principle had a price tag. Consider the DFG of Figure 4.6. The node (or actor) performing the "upper" multiplication can fire only if the output arc from it to the "lower" multiplication is token-free. Likewise, the subtraction actor can fire only if the output arc from it to the lower multiplication is token-free. So the lower-multiplication actor must *signal* back to the upper-multiplication and subtraction actors *acknowledging* that the tokens on its input arcs have been consumed.

Each instruction for such actors must then have places for accepting its operands and a place for holding the pointers to successor instructions; it must also have a place to accept acknowledgement signals. So in the static scheme an instruction's places must all be filled for the instruction to fire. The acknowledgment signal serves as a synchronizing operation.

There were other limitations of the static data-flow model, most notably the fact that it limited the amount of parallelism during execution of a DFG. Consider a subgraph S of a DFG D that represents a procedure to be invoked from two or more other parts of D. Suppose further that at two successive points of time S is called from two different places in D. If the input tokens have not all arrived at a particular node in the earlier invocation, the latter cannot proceed. However, the later invocation could proceed if its corresponding input tokens are all available; but in the static model the second invocation would not be permitted.

In his original (1974) paper Dennis acknowledged this problem, and he raised the idea of a *dynamic* data-flow model wherein several tokens could be present simultaneously on an arc because of multiple invocations of the relevant actor. Clearly, then, there would have to be some way of distinguishing tokens from different invocations: The solution was to attach different *tags* to the tokens, which is why the dynamic model also came to be called *tagged-token* data flow.

But Dennis and his group did not pursue this model any further; they focused on static data flow. The dynamic model was taken up by other groups, most notably, from circa 1978, by Arvind and his collaborators, first at the University of California, Irvine, then at MIT,[91] A. L. Davis of the University of Utah,[92] and John Gurd, Ian Watson, and their collaborators of the University of Manchester.[93]

Theory was one part of the data-flow movement. There was also the practical matter of testing theory by building machines and observing their performances. Moreover, such machines had to be programmed, and so this necessitated high-level programming languages. The various groups embarked on their respective designs of *data-flow languages*: VAL, developed by Dennis's group, Arvind et al.'s Id, and SISAL, first built at the Lawrence Livermore Laboratory circa 1983, but implemented also by the Gurd et al. group in Manchester.[94]

Oddly enough, neither Dennis nor Arvind seemed eager to actually build physical data-flow computers. The earliest static data-flow machines were a system called LAU developed in Toulouse circa 1977[95] and the Distributed Data Processor (DDP) built by Texas Instruments in Dallas circa 1980.[96]

The University of Manchester, of course, had a long history of machine building reaching back to the later 1940s,[97] so it was entirely in keeping with this tradition that

Gurd, Watson, and their collaborators would embark on building a data-flow computer. A prototype version of the Manchester Dataflow Computer became operational in 1981 and was probably the first dynamic data-flow machine to be fully implemented. Over the next several years this computer was extensively studied experimentally, as documented in a series of publications.[98]

But arguably the most publicized project related to data flow was the Japanese Fifth Generation Computer Project, announced in 1982, which centered around the development of data-flow machines for supporting knowledge-based computing. This project entailed the participation of several leading Japanese companies under the management of an organization called the Institute for New Generation Computer Technology (ICOT).[99]

A part of the appeal of the data-flow style undoubtedly lay on aesthetic grounds: the formal elegance of variables-absent, data-memory-absent systems. In principle this meant that data structures (such as arrays) could also be passed around as tokens. When input to an actor the data structure would be consumed and a new data structure would be produced as an output token.

So pronounced the theory. In reality it was entirely impractical to pass entire, possibly very large, data structures around as tokens. Rather, the solution that emerged was that data structures would be held in a memory ("structure store") and only *pointers* to them would serve as tokens.

The formal elegance of the data-flow style thus crumbled. The dream of memoryless data tokens and the nonnecessity of storing values in variables held in memory dissolved. Attention thus shifted to data-flow architectures containing structure stores. Both Arvind at MIT and Gurd and Watson in Manchester proposed careful solutions to the structure store problem,[100] but the brute reality could not be ignored: Data structures had to be stored in some memory; there had to be instructions to read from and write into memory. The von Neumann style peeped through like a pentimento in a painting.

Ultimately, as the 1980s came to an end—despite the various data-flow projects on four continents (in addition to the projects in the United States, the United Kingdom, and Japan, a project on the design and evaluation of a Manchester-style data-flow computer had been reported by Lalit Patnaik and his collaborators at the Indian Institute of Science, Bangalore[101])—the data-flow movement remained just that: a movement that never quite entered into the mainstream of parallel processing and very high-performance computing, which had been the aspiration of its pioneer Jack Dennis.[102] The von Neumann style remained in command.

NOTES

1. See, e.g., the classic work by R. K. Richards, 1955. *Arithmetic Operations in Digital Computers*. Princeton, NJ: Van Nostrand.

2. See S. S. Husson, 1970. *Microprogramming: Principles & Practices*. Englewood Cliffs, NJ: Prentice-Hall.

3. A. Padegs, 1964. "Channel Design Considerations," *IBM Systems Journal*, 3, 2 & 3, pp. 165–180.

4. For an early but brief exposition of multiprogramming see R. F. Rosin, 1969. "Supervisory and Monitor Systems," *Computing Surveys*, 1,1, pp. 37–54.

5. E. W. Dijkstra, 1965. "Cooperating Sequential Processes," Tech. Report EWD 123, Mathematics Department, Technische Universiteit Eindhoven, Eindhoven.

6. E. W. Dijkstra, 1968. "The Structure of the 'THE' Multiprogramming System," *Communications of the ACM*, 11, pp. 341–346. Reprinted, pp. 89–98 in E. Yourdon (ed.), 1982. *Writings of the Revolution: Selected Readings in Software Engineering*. New York: Yourdon Press.

7. Dijkstra, 1965.

8. Probably the most formal proposal was due to J. J. Horning and B. Randell, 1973. "Process Structuring," *ACM Computing Surveys*, 5, 1, pp. 5–30.

9. M. V. Wilkes, 1975. *Time-Sharing Computer Systems* (3rd ed.). London: Macdonald & Jane's/New York: American Elsevier, p. 37

10. P stands for the Dutch work *passern* [to pass] and V stands for *vrygeren* [to release]. G. R. Andrews and F. B. Schneider, 1983. "Concepts and Notations for Concurrent Programming," *ACM Computing Surveys*, 15, 1, pp. 3-44.

11. P. Brinch Hansen, 1972a. "Structured Multiprogramming," *Communications of the ACM*, 15, 7, pp. 574–578.

12. P. Brinch Hansen, 1973. *Operating Systems Principles*. Englewood Cliffs, NJ: Prentice-Hall.

13. See, e.g., a comprehensive (if not complete) collection of his most important papers on the topic in P. Brinch Hansen, 1996. *The Search for Simplicity: Essays in Parallel Programming*. Los Alamitos, CA: IEEE Computer Society Press.

14. Brinch Hansen, 1973; P. Brinch Hansen, 1977. *The Architecture of Concurrent Programming*. Englewood Cliffs, NJ: Prentice-Hall; P. Brinch Hansen, 1982. *Programming a Personal Computer*. Englewood Cliffs, NJ: Prentice-Hall.

15. P. Brinch Hansen, 1975. "The Programming Language Concurrent Pascal," *IEEE Transactions on Software Engineering*, 1, 2, pp. 199–207; P. Brinch Hansen, 1981. "The Design of Edison," *Software—Practice and Experience*, 11, 4, pp. 363–396; P. Brinch Hansen, 1987. "Joyce— A Programming Language for Distributed Systems," *Software—Practice and Experience*, 17, 1, pp. 29–50.

16. C. A. R. Hoare, 1972. "Towards a Theory of Parallel Programming," pp. 61–71 in C. A. R. Hoare and R. H. Perrot (eds.), *Operating Systems Techniques*. New York: Academic Press. Reprinted, pp. 202–214 in D. G. Gries (ed.), 1978. *Programming Methodology: A Collection of Articles by Members of IFIP WG2.3*. New York: Springer-Verlag. All page references are to the reprint.

17. Brinch Hansen, 1973.

18. Hoare, 1972, p. 209.

19. Brinch Hansen, 1972a.

20. P. Brinch Hansen, 1972b. "An Outline of a Course on Operating System Principles," pp. 29–36 in Hoare and Perrott, 1972.

21. Brinch Hansen, 1973.

22. I have in mind classic textbooks written by William James on psychology in the late 19th century and by Alfred Marshall on economics, Charles Sherrington on neurophysiology, and G. H. Hardy on pure mathematics in the early 20th century.

23. R. F. Rosin, 1969. "Supervisory and Monitor Systems," *Computing Surveys*, 1, 1, pp. 37–54.

24. C. A. R. Hoare, 1974. "Monitors: An Operating System Structuring Concept," *Communications of the ACM*, 17, 10, pp. 550–557.

25. Brinch Hansen, 1973, p. 121.

26. Ibid.

27. Ibid., p. 229.

28. Brinch Hansen, 1996, p. 349, acknowledges this.

29. Ibid., p. 361.

30. C. A. R. Hoare, 1973. "A Structured Paging System," *The Computer Journal*, 16, 8, pp. 209–214. A paging system is the most fundamental means for implementing virtual memory (the illusion of unlimited memory) within a computer system. The classic discussion on virtual memory is P. J. Denning, 1970. "Virtual Memory," *Computing Surveys*, 2, 3, pp. 153–190. See also Wilkes, 1975, chapter 4.

31. Hoare, 1974.

32. Ibid., p. 549.

33. Ibid., p. 557.

34. Brinch Hansen, 1975.

35. A. C. Hartmann, 1977. *A Concurrent Pascal Compiler for Minicomputers* (Lecture Notes in Computer Science 50). Berlin: Springer-Verlag.

36. Ibid., pp. 55–57.

37. P. Brinch Hansen, 1976. "The SOLO Operating System," *Software—Practice and Experience*, 6, 2, pp. 141–200.

38. Brinch Hansen, 1977.

39. The notion of H-originality is due to cognitive scientist Margaret Boden of the University of Sussex. See M. A. Boden, 1991. *The Creative Mind*. New York: Basic Books.

40. The notion of C-originality was formulated by the present author. See S. Dasgupta, 2011. "Contesting (Simonton's) Blind-Variation, Selective-Retention Theory of Creativity," *Creativity Research Journal*, 23, 2, pp. 166–182.

41. See S. Dasgupta, 2014. *It Began with Babbage: The Genesis of Computer Science*. New York: Oxford University Press, chapters 8 and 12.

42. P. Brinch Hansen, 1993. "Monitors and Concurrent Pascal: A Personal History." *SIGPLAN Notices*, 28, 3, pp. 1–35. Reprinted, pp. 343–421, in Brinch Hansen, 1996. The quote is on p. 392 of the reprint.

43. Ibid.

44. Brinch Hansen, 1996, p. 388.

45. See Dasgupta, 2014, chapter 9.

46. J. P. Riganati and P. B. Schneck, 1984. "Supercomputing," *Computer*, 17, 10, pp. 97–113.

47. G. H. Barnes, R. M. Brown, M. Kato et al., 1968. "The ILLIAC IV Computer," *IEEE Transactions on Computers*, C-18, 8, pp. 746–757; W. Bouknight, S. A. Denenberg, D. E. McIntyre et al., 1972. "The ILLIAC IV System," *Proceedings of the IEEE*, 60, 4, pp. 369–388; R. M. Hord, 1982. *The ILLIAC IV: The First Supercomputer*. Rockville, MD: Computer Science Press.

48. S. F. Reddaway, 1973. "DAP—A Distributed Array Processor." *Proceedings of the IEEE*, 60, 4, pp. 369–388.

49. K. E. Batcher, 1974. "STARAN Parallel Processor System," pp. 405–410 in *Proceedings of the National Computer Conference, Volume 43*. Montvale, NJ: AFIPS Press.

50. K. E. Batcher, 1980. "Design of a Massively Parallel Processor," *IEEE Transactions on Computers*, C-29, 9, pp. 836–840.

51. D. J. Kuck and R. A. Stokes, 1980. "The Burroughs Scientific Processor (BSP)," *IEEE Transactions on Computers*, C-31, 5, pp. 363–376.

52. K. Stevens, 1975. "CFD—A Fortran-like Language for the ILLIAC IV," *SIGPLAN Notices*, 3, pp. 72–80; D. H. Lawrie, T. Layman, D. Baer et al,. 1975. "Glypnir—A Programming Language for ILLIAC IV," *Communications of the ACM*, 18, 3, pp. 157–163. See also Hord, 1982.

53. R. H. Perrott, 1979. "A Language for Array and Vector Processors," *ACM Transactions on Programming Languages and Systems*, 1, 2, pp. 177–195.

54. R. H. Perrott, 1980. "Languages for Parallel Computers," pp. 255–282 in R. M. McKeag and R. Macnaghtan (eds.), *The Construction of Programs*. Cambridge: Cambridge University Press, p. 279.

55. D. J. Kuck, 1976. "Parallel Processing of Ordinary Programs," pp. 119–179 in M. Rubinoff and M. C. Yovits (eds.), *Advances in Computers, Volume 15*. New York: Academic Press; D. J. Kuck, 1977. "A Survey of Parallel Machine Organization and Programming," *ACM Computing Surveys*, 9, 1, pp. 29–60; D. J. Kuck, R. H. Kuhn, B. Leasure et al., 1980. "The Structure of an Advanced Retargetable Vectorizer," in *Proceedings of COMPSAC 80*. New York: IEEE Computer Society; D. J. Kuck, R. H. Kuhn, D. Padua et al., 1981. "Dependence Graphs and Compiler Organization," *Proceedings of the 8th Annual ACM Symposium on Principles of Proramming Languages* (POPL). Williamsburg, VA, pp. 207–218. See also D. J. Kuck, 1978. *The Structure of Computers and Computation, Volume 1*. New York: Wiley, chapter 2, section 2.4.

56. A. J. Bernstein, 1966. "Analysis of Programs for Parallel Processing," *IEEE Transactions on Computers*, EC-5, 10, pp. 757–763.

57. J. E. Thornton, 1964. "Parallel Operation in the Control Data 6600," pp. 33–40 in *Proceedings AFIPS Fall Joint Computer Conference, Volume 26, Part II*. Washington, DC: Spartan Books.

58. D. W. Anderson, F. J. Sparacio, and R. M. Tomasulo, 1967. "The IBM System/360 Model 91: Machine Philosophy and Instruction Handling," *IBM Journal*, January, pp. 8–24.

59. D. Morris and R. N. Ibbett, 1979. *The MU5 Computer System*. London: Macmillan; S. H. Lavington, 1998. *A History of Manchester Computers* (2nd ed.). Swindon, UK: British Computer Society.

60. C. G. Bell and A. Newell, 1972. *Computer Structures: Readings and Examples*. New York: McGraw-Hill.

61. C. V. Ramamoorty and H. F. Li, 1977. "Pipeline Architecture," *ACM Computing Surveys*, 9, 1, pp. 61–102.

62. Morris and Ibbett, 1979.

63. H. S. Stone, 1980. *Introduction to Computer Architecture*. Chicago: SRA; J-L. Baer, 1980. *Computer Systems Architecture*. Rockville, MD: Computer Science Press.

64. D. P. Siewiorek, C. G. Bell, and A. Newell, 1982. *Computer Structures: Principles and Examples*. New York: McGraw-Hill.

65. K. Hwang and F. A. Briggs, 1984. *Computer Architecture and Parallel Processing*. New York: McGraw-Hill.

66. P. M. Kogge, 1981. *Architecture of Pipelined Computers*. New York: McGraw-Hill.

67. S. Dasgupta, 1980. "Some Aspects of High Level Microprogramming," *ACM Computing Surveys*, 12, 3, pp. 295–324; S. Davidson, 1986. "Progress in High Level Microprogramming," *IEEE Software*, 3, 4, pp. 18–26.

68. A. Opler, 1967. "Fourth Generation Software," *Datamation*, 13, 1, pp. 22–24.

69. S. Dasgupta and B. D. Shriver, 1985. "Developments in Firmware Engineering," pp. 101–176 in M. C. Yovits (ed.), *Advances in Computers, Volume 24*. New York: Academic Press.

70. Ibid., p. 102.

71. Dasgupta, 1980.

72. R. L. Kleir and C. V. Ramamoorthy, 1971. "Optimization Strategies for Microprograms," *IEEE Transactions on Computers*, C-20, 7, pp. 783–795.

73. The literature on code optimization strategies performed by compilers for programming languages was already quite rich by the early 1970s. The authoritative text on the topic was A. V. Aho and J. D. Ullman, 1973. *The Theory of Parsing, Translation and Compiling, Volume 2*. Englewood Cliffs, NJ: Prentice-Hall.

74. Husson, 1970.

75. D. A. Patterson, 1976. "STRUM: A Structured Microprogram Development System for Correct Firmware," *IEEE Transactions on Computers*, C-25, 10, pp. 974–985.

76. S. S. Yau, A. Schowe, and M. Tsuchiya, 1974. "On Storage Optimization of Horizontal Microprograms," *Proceedings of the 7th Annual Microprogramming Workshop* (MICRO-7), Palo Alto, CA, pp. 98–106.

77. L. W. Jackson and S. Dasgupta, 1974. "The Identification of Parallel Microoperations," *Information Processing Letters*, 2, 6, pp. 180–184; S. Dasgupta and J. Tartar, 1976. "The Identification of Maximal Parallelism in Straight Line Microprograms," *IEEE Transactions on Computers*, C-25, 10, pp. 986–992.

78. C. V. Ramamoorthy and M. Tsuchiya, 1974. "A High Level Language for Horizontal Microprogramming," *IEEE Transactions on Computers*, C-23, 8, pp. 791–801; M. Tsuchiya and M. J. Gonzales, 1976. "Towards Optimization of Horizontal Microprograms," *IEEE Transactions on Computers*, C-25, 10, pp. 992–995.

79. W. G. Wood, 1978. "On the Packing of Microoperations Into Microinstructions," *Proceedings of the 11th Microprogramming Workshop* (MICRO-11), Pacific Grove, CA, pp. 51–55.

80. S. Dasgupta, 1977. "Parallelism in Loop-Free Microprograms," pp. 745–750 in *Information Processing 77* (Proceedings of the IFIP Congress, 1977). Amsterdam: North-Holland.

81. M. Tokoro, 1978. "A Technique for Global Optimization of Microprograms," *Proceedings of the 11th Microprogramming Workshop* (MICRO-11), Pacific Grove, CA, pp. 41–50; M. Tokoro, E. Tamura, and T. Takizuke, 1981. "Optimization of Microprograms," *IEEE Transactions on Computers*, C-30, pp. 491–504.

82. J. A. Fisher, 1981. "Trace Scheduling: A Technique for Global Microcode Compaction," *IEEE Transactions on Computers*, C-30, 7, pp. 478–490.

83. Ramamoorthy and Tsuchiya, 1974; Tsuchiya and Gonzales, 1976.

84. Dasgupta and Tartar, 1976.

85. Fisher, 1981; see also J. A. Fisher, D. Landskov, and B. D. Shriver, 1981. "Microcode Compaction: Looking Backward and Looking Forward," pp. 95–102 in *Proceedings, National Computer Conference*. Montvale, NJ: AFIPS Press.

86. J. B. Dennis, 1974. "First Version of a Data Flow Procedural Language," pp. 362–376 in *Proceedings Colloque sue le Programmation* (Lecture Notes on Computer Science, Volume 19). Berlin: Springer-Verlag.

87. C. A. Petri, 1962. "Kommunikation mit Automaten," PhD dissertation, University of Bonn.

88. J. B. Dennis, 1979. "The Varieties of Data Flow Computers." Computation Structures Group Memo 183–181. Cambridge, MA: MIT.

89. In particular, R. M. Karp and R. E. Miller, 1966. "Properties of a Model for Parallel Computation: Determinacy, Termination, Queuing," *SIAM Journal of Applied Mathematics*, 14, 6, pp. 1390–1411; D. A. Adams, 1968. "A Computational Model With Data Flow Sequencing,"

Tech. Report CS 117. Stanford University; J. E. Rodriguez, 1969. "A Graph Model for Parallel Computation," Report MAC-TR-64. Project MAC. Cambridge, MA: MIT.

90. J. B. Dennis and D. P. Misunas, 1974. "A Preliminary Architecture for a Basic Data Flow Computer," pp. 126–132 in *Proceedings of the 2nd Annual Symposium on Computer Architecture.* New York: ACM/IEEE, pp. 126–132; J. Rumbaugh, 1977. "A Data Flow Multiprocessor," *IEEE Transactions on Computers,* C-26, 2, pp. 138–146; J. B. Dennis, C. K. C. Leung, and D. P. Misunas, 1979. "A Highly Parallel Processor Using a Data Flow Machine Language," Computational Structures Group Memo 134–131. Cambridge, MA: MIT.

91. Arvind, K. P. Gostelow, and W. Plouffe, 1978. "An Asynchronous Programming Language and Computing Machine." Tech. Report 114a. Department of Information and Computer Science. Irvine: University of California.

92. A. L. Davis, 1978. "The Architecture and System Method of DDM1: A Recursively Structured Data Driven Machine," pp. 210–215 in *Proceedings, 5th Annual Symposium on Computer Architecture.* New York: ACM/IEEE.

93. J. R. Gurd, I. Watson, and J. R. Glauert, 1978. "A Multilayered Data Flow Computer Architecture," Internal report, Department of Computer Science, University of Manchester.

94. W. B. Ackerman, 1982. "Data Flow Languages," *Computer,* 15, 2, pp. 15–25; Arvind, Gostelow, and Plouffe, 1978; J. R. Gurd and W. Bohm, 1988. "Implicit Parallel Processing: SISAL on the Manchester Dataflow Computer," pp. 175–205 in G. Paul and G. S. Almasi (eds.), *Parallel Systems and Computation.* Amsterdam: North-Holland.

95. J. C. Syre, D. Comte, G. Durrieu et al., 1977. "LAU System—A Data-Driven Software/ Hardware System Based on Single Assignment," pp. 347–351 in M. Feilmeier (ed.), *Parallel Computers-Parallel Mathematics.* Amsterdam: North-Holland.

96. D. Johnson et al., 1980. "Automatic Partitioning of Programs in Multiprocessor Systems," pp. 175–178 in *Proceedings IEEE COMPCON.* New York: IEEE Press.

97. S. H. Lavington, 1998. *A History of Manchester Computers* (2nd ed.). London: The British Computer Society. See also, briefly, Dasgupta, 2014, pp. 125–126.

98. See, in particular, J. R. Gurd, C. C. Kirkham, and I. Watson, 1985. "The Manchester Prototype Dataflow Computer," *Communications of the ACM,* 28, 1, pp. 34–52; J. R. Gurd and C. C. Kirkham, 1986. "Data Flow: Achievements and Prospects," pp. 61–68 in H.-J. Kugler (ed.), *Information Processing 86* (Proceedings of the 1986 IFIP Congress). Amsterdam: North-Holland.

99. T. Moto-oka (ed.), 1982. *Fifth Generation Computer Systems.* Amsterdam: North-Holland. See also M. Amamiya, M. Tokesue, R. Hasegawa et al., 1986. "Implementation and Evaluation of a List Processing-Oriented Data Flow Machine," pp. 10–19 in *Proceedings of the 13th International Symposium on Computer Architecture.* Los Alamitos, CA: IEEE Computer Society Press.

100. Arvind and R. E. Thomas, 1981. "I-Structures: An Efficient Data Structure for Functional Languages," Tech. Report LCS/TM-178, Laboratory for Computer Science Cambridge, MA: MIT; Arvind and R. A. Ianuuci, 1986. "Two Fundamental Issues in Multiprocessing," CSG Memo 226–5. Laboratory for Computer Science. Cambridge, MA: MIT.

101. L. M. Patnaik, R. Govindarajan, and N. S. Ramadoss, 1986. "Design and Performance Evaluation of EXMAN: An EXtended MANchester Data Flow Computer," *IEEE Transactions on Computers,* C-35, 3, pp. 229–244.

102. J. B. Dennis, 1986. "Data Flow Ideas and Future Supercomputers," pp. 78–96 in N. Metropolis, D. H. Sharp, W. J. Worlton, and K. R. Ames (eds.), *Frontiers of Supercomputing.* Berkeley, CA: University of California Press.

5

VERY FORMAL AFFAIRS

5.1   MATHEMATICS ENVY

If social and behavioral scientists have harbored "physics envy" as some have wryly claimed—envy of its explanatory and predictive success—then computer scientists may be said to have suffered from "mathematics envy." Interestingly, this envy was less a characteristic of the pioneers of digital computing of the 1940s and 1950s, the people who shed first light on the design of digital electronic computers, the first programming languages, the first operating systems, the first language translators, and so on—though most of them were trained as mathematicians. They were too busy learning the heuristic principles of computational artifacts. Rather, it was in the 1960s when we first find signs of a kind of mathematics envy, at least in some segments of the embryonic computer science community. It was as if, having discovered (or invented) the heuristic principles of practical computational artifacts, some felt the need to understand the underlying "science" of these artifacts—by which they meant its underlying mathematics and logic.

Mathematics envy could be assuaged only by *thinking mathematically* about computational artifacts. Computer science would then be raised to the intellectual stature of, say, physics or indeed of mathematics itself if computer scientists could transform their discipline into a *mathematical science.*

5.2   THE PROMISES OF MATHEMATICAL THINKING

One cannot blame computer scientists who thought this way. The fact is, there is something about mathematics that situates it in a world of its own. "Mathematics is a unique aspect of human thought," wrote hyperprolific science (fact and fiction) writer Isaac Asimov.[1] And Asimov was by no means the first or only person to think so.

But wherein lies the uniqueness of mathematical thinking? Perhaps the answer is that for many people, mathematics offers the following *promises*:

(1) The *unearthliness* of mathematical objects.
(2) The *perfectness* and *exactness* of mathematical concepts.
(3) An inexorable *rigor* of mathematical reasoning.
(4) The *certainty* of mathematical knowledge.
(5) The *self-sufficiency* of the mathematical universe.

These promises are clearly enviable if they can be kept; usually, they *are* kept. And we detect a desire for such promissory notes in C. A. R. Hoare's assertion in 1969 about the exactness of computer programming in the sense that all the properties of a program can be excavated from the text itself by logical reasoning.[2]

As we have seen before (Chapter 1, Section 1.12), what was in 1969 something of a dream, crystallized for Hoare, into a manifesto as late as 1986. We also find evidence of this yearning for a mathematics of programming elsewhere. In 1985, at a Royal Society symposium on "Mathematical Logic and Programming Languages," Per Martin-Löf of the University of Stockholm presented a table showing correspondences between programming and mathematical concepts, suggesting the close kinship between mathematics and programming. For example:

PROGRAMMING	MATHEMATICS
Program	Function
Input	Argument
Output	Value
Sequential composition	Composition of functions
Conditional statement	Case-based definition
Iterative statement	Recursive definition
Data type	Set, type

But programming was not the only branch of computer science to be drawn to mathematical thinking in the 1960s. We will encounter other domains. And we will see that this kind of thinking became richer in its consequences in the 1970s and 1980s.

The discourse we are talking about here can best be described as *formal computer science*. At the core of this discourse is a modus operandi that is a legacy of Newtonian natural philosophy in the 17th century but adapted to the needs of computer science.

Computational artifacts (and associated phenomena) of interest are idealized, and abstract models are then constructed within *formal* (i.e., *axiomatic*) frameworks of the kind mathematicians and logicians are familiar with. The rules, laws, and theorems of such frameworks are used to manipulate the models with the intent of producing interesting insights pertaining to the models themselves.

5.3   "HARDYANS" AND "KNUTHIANS"

On the one hand, if the Newtonian style is properly respected then the results of formal manipulation would be interpreted in the context of, or mapped onto, the original real-world computational phenomena that prompted formalization in the first place, and insight and understanding of the latter would then be elicited. This is formalization as a means to an empirical end. On the other hand, if the Newtonian legacy is abandoned midway, if the formal, idealized model and its axiomatic domain become an almost self-contained world of their own, quite independent of the original real-world computational phenomena of interest, then this would be a case of formalization as an end in itself.

In fact, both these two modes of formal computer science came into being, the former in the 1950s, the latter in the 1960s.

On one hand was a milieu of abstract computational artifacts that obey formal laws; a world of definitions, axioms, rules of inference, theorems, and proofs. This milieu belonged squarely in the tradition of mathematics and logic. Computer scientists belonging to this milieu, like pure mathematicians and logicians, tended to dwell almost entirely within this milieu.

Here, I call them *Hardyans* after the distinguished English mathematician G. H. Hardy of Cambridge University, who took pride and comfort in the "uselessness" of the mathematics he had spent his whole life in doing; his sole justification *qua* mathematician was that he "added something to knowledge."[3]

The practitioners of the other tradition, in which the formal is a means to an empirical end, I call *Knuthians* after the computer scientist Donald Knuth of Stanford University, whose deep mathematical explorations of algorithms, programming, and language grammars were always infused with the desire to elicit practical insights.[4]

What Hardyans and Knuthians have shared was the value both have placed on abstraction and formalization and a conviction that formalism affords a rigor of reasoning and a depth of understanding of computational phenomena that elude other modes of doing computer science. Where Hardyans and Knuthians differed was in their respective *mentalities,* perhaps even *identities.* As Amnon Eden of the University of Essex has observed, those I am calling Hardyans seem to identify with mathematicians. More precisely, they have tended to regard computer science as a branch of mathematics.[5] In contrast, Knuthians have one foot in the mathematical domain and the other in the empirical domain; the value of a formal result is judged by the Knuthian in terms of its empirical consequence.

5.4   BUT WHAT DOES IT *MEAN*?

Semantics is one of those words (like "history") that refers both to a kind of entity and the systematic study of that entity. Semantics-as-entity refers to the relationship between language constructs and the objects they refer to: the *meanings* of words, terms, sentences, commands, statements, and the like. Semantics-as-discipline refers to the study of semantics-as-entity.

Where there is language there is semantics. Thus linguists, logicians, and philosophers have long been interested in semantics from their respective perspectives. As have anthropologists and literary scholars.

In the 1960s, computer scientists jumped on the semantics bandwagon. As designers and implementers of computer languages this was hardly surprising. The interest was not just intellectual curiosity; there was also a sound *practical* reason: Both language users and language implementers needed a clear, unambiguous comprehension of the meaning of the constructs of a programming language (or of a specification, a microprogramming, or a hardware description language) if implementers had to *correctly* translate a description in the language into executable code and if language users wished to write *provably correct* programs (or microprograms or hardware descriptions). This practical concern nicely complemented the purely intellectual, even philosophical, issue: What do the constituents of an invented language actually *mean*? And how do we *describe* this meaning?

Because the first computer languages were programming languages and because of their ubiquity in the design of computational artifacts, concern over the semantics of programming languages has dominated the semantics scene in computer science. Concerns with the meanings of other kinds of computational languages came later. And researchers of the "semantic question" in these latter domains—microrogramming and hardware description languages—followed the lead of research into programming language semantics.

5.5 THE VIENNESE MEANING OF MEANING

The philosophical school known as "logical positivism" was born in Vienna in the 1920s amid a group of philosophers and mathematicians called the "Vienna Circle." The core of logical positivism was the *verification principle of meaning*, a doctrine that proclaimed that a proposition has a meaning if and only if it can be logically or empirically verified.[6]

The vast philosophical discussion of (and disputation on) the verification principle is irrelevant here. What *is* significant for us is that the principle embodied the idea that meaning is intimately related to *knowing how* to verify propositions. This idea of "knowing how" found its way into a doctrine in the philosophy of science articulated most explicitly by the Nobel Laureate physicist Percy Bridgeman, circa 1927, called *operationalism*.[7] In the context of the physical sciences, operationalism asserts that the meaning of a physical concept is specified in terms of some set of experimental operations. For example, the meaning of the "length" of an object is determined by laying out a measuring rod along the side of the object and determining how many units of the measuring rod will cover the side.

Perhaps it was entirely coincidental that Vienna was also the locus of a theory of the meaning of programming languages called *operational semantics*. In 1968–1969, Peter

Lucas and his coworkers at the IBM Laboratory in Vienna developed a metalanguage for defining the semantics of the programming language PL/1. This metalanguage came to be known as the Vienna Definition Language (VDL).

Interestingly, the definitive publication on the VDL, coauthored in 1969 by Peter Lucas and Kurt Walk, never used this name.[8] Nor, indeed, did Lucas use this name in a later (1972) paper.[9] Perhaps this was modesty on his part. At any rate, in 1972, Peter Wegner of Brown University made the name widely known by way of his paper titled "The Vienna Definition Language," published in the *ACM Computing Surveys*.[10] Wegner's paper was crucial in clarifying the general nature of operational semantics as well as the specific nature of VDL. His paper became, de facto, the prime reference on both VDL and operational semantics.

As in the case of operationalism in physics à la Bridgeman, the meaning of a programming language according to the Vienna approach was defined operationally, but in a sense that computer scientists would have no problem in identifying with.

The method used for the definition of a programming language is based on the definition of an abstract machine described by the set of its states and its state transition function. A program together with its input data specifies an initial state of the machine, and the subsequent behavior of the machine is said to determine the *interpretation* of the program with respect to the input data.[11]

This would have made a lot of sense to both compiler writers and computer architects. After all, one of the techniques used in implementing a language was to postulate an abstract, "intermediate code machine" oriented toward the programming language and independent of any specific target physical computer. The compiler would translate the source program into object code for this abstract computer that would then be tailored for any particular physical machine (see, e.g., Chapter 3, Section 3.9). And as IBM had themselves pioneered, a microprogrammed emulator would create an abstract or virtual machine on a real-host machine that enabled programs intended for the emulated architecture to run on the real host (see Chapter 3, Sections 3.8–3.10). Both these scenarios entailed *abstract* computers that mediated between a programming language and its target physical machine.

The distinctive feature of the Vienna method was that the meaning of a language such as PL/1 was defined by formal rules governing the *interpretation* of a program written in the language by the abstract machine, rather than by rules governing the *compilation* of the program into the abstract machine object code. Thus Peter Wegner made the important distinction between two modes of operational semantics, one *interpretation oriented*, as the Vienna method was, the other, *compiler oriented*. In the latter, the meaning of a language is specified in terms of a set of translation rules. Compiler-oriented operational semantics was proposed by Donald Knuth in 1968.[12]

So how would interpreter-oriented operational semantics work? In Wegner's explication, suppose that X is a construct—a syntactic category such as a declaration or a statement—in the language of interest L. Then the Vienna approach associates with X a *state transformation* from the "current" state of an abstract, interpretive machine to a new (goal) state consistent with the syntax of *X*. Moreover, this interpretation of *X* would show the *path*—the state sequence—followed in the abstract machine in going from the current state to the goal state. In other words, the occurrence of a syntactic category *X* in a program would give rise to the execution of a sequence of operations in an abstract machine, manifested as a sequence of state transformations in the machine.

VDL, as designed by Lucas and his colleagues, was a metalanguage that rigorously and formally defined the interpreter-oriented operational semantics of PL/1. Of course, it could be used to specify the semantics of any language. Lucas and Walk paid tribute to the work of John McCarthy, then of MIT, and Peter Landin, at the time an independent researcher and later of Queen Mary College, London, a half-decade earlier on which the VDL approach was founded.[13]

Unfortunately, to deploy VDL, the metalanguage must itself be mastered. That is, the syntax and semantics of the metalanguage as well as the architecture of some abstract interpretive machine must be learned. Because of its elaborate mathematical structure, this was a severely nontrivial enterprise.[14]

To convey to the reader a whiff of the nature of the Vienna method, consider as an example the meaning of an infix arithmetic expression as might appear as the right part of an assignment statement:[15]

$$OPD1 \; opn \; OPD2.$$

Here *OPD1* and *OPD2* are operands and *opn* is an arithmetic operator. Simplifying considerably the actual VDL formalism, the meaning of this expression would be specified by the execution (by the appropriate abstract machine) of an instruction:

$$\underline{\text{eval-infix-expr}} \; (OPD1, OPD2, opn).$$

The interpretation of this instruction would, in turn, invoke the two instructions (in no specific order):

$$\underline{\text{eval-expr}} \; (OPD1, env),$$
$$\underline{\text{eval-expr}} \; (OPD2, env),$$

where *env* is the *environment* in which the expressions are interpreted. The interpretation of these two <u>eval-expr</u> instructions would "return" values of the respective expressions to the "parent" instruction <u>eval-infix-expr,</u> which would then execute *opn* on its arguments. The execution of the partially ordered set of three instructions (operations) thus constituted the operational semantics of the infix arithmetic expression.

A rather surprising use of VDL was explored at another IBM research center. Between 1974 and 1978, William Carter and his coworkers at the IBM T. J. Watson Research Center in Yorktown Heights, New York, published a series of papers on the construction and evolution of a *microcode certification system* (MCS).[16] Their approach was a textbook study in the application of operational semantics. It was also a study in the melding of language semantics and what has been called the "correctness problem in computer science."

The scenario postulated by Carter et al. was one that was well known in the computer design milieu: A computer can be viewed at several distinct abstraction levels (Chapter 3, Sections 3.2–3.6, 3.8). In particular, consider two representations of a computer A and μA. A describes the computer at the ISP level of architecture, and μA specifies the machine at the microarchitectural level—that is, a specification of the microprogrammable host machine together with the microcode interpreted by the host machine to implement A (Chapter 3, Sections 3.9, 3.10). The machine μA is of course at a lower abstraction level than is A.

The problem Carter et al. posed to themselves was to prove that A and μA were in some formal sense equivalent. That is, μA correctly implemented A. Toward this aim they drew on a theory of simulation put forward in 1971 by Robin Milner of Edinburgh University.[17] Informally, a simulation of a program P by another program P^* meant that whatever could be computed by P could also be computed by P^*. This meant that *any* sequence of state transformations performed by P could also be performed by P^*. So what Carter et al. desired to show was that μA could simulate A. Their approach entailed the following steps:

(1) Define operationally machine A.
(2) Define operationally machine μA.
(3) Establish a desired simulation relation R between A and μA.
(4) Prove that μA simulates A with respect to R.

In an earlier version of MCS Carter et al. used an extended version of VDL to specify both the higher-level architecture and the microarchitecture. MCS was an automated system that could then simulate the two machines and thereby show that the two were equivalent.

5.6   MEANING-AS-FUNCTIONAL DENOTATION

The prominent feature of operational semantics is that the operations performed by the abstract machine in interpreting a language construct elicits a computational *path*: a state sequence. In one sense this is attractive, for operationalism à la VDL is a rigorous formalization of our intuitive understanding of the meaning of a computer language.

But is it *necessary* to know the details of the path followed by the interpretative machine? Isn't this a case of overspecification? We may feel that the only states that matter are the initial and final states of a computation. How we arrive at a final state is irrelevant to the meaning of a language construct. Moreover, an operational definition of a language is abstract machine specific: This machine must be defined for an operational definition of a language to work. For example, to understand the semantics of PL/1, we must be thoroughly familiar with the abstract PL/1 machine.

Suppose, instead, that we abstract away the interpretative machine altogether; thus we abstract away the state sequence delineating the path from initial to final state. Suppose we focus on the language itself and we view a program in that language as a *mathematical function* from states to states. Suppose, then, we assign purely mathematical meaning (in terms of such mathematical objects as sets and functions) to language constructs in our language of interest:

$$F : S \to T,$$

where F is a function and S, T are sets of states ("state spaces").

This was the line of reasoning proposed in 1971 by logician Dana Scott, then of Oxford University (later of Carnegie-Mellon University), and computer scientist Christopher Strachey, also of Oxford, in their monograph "Toward a Mathematical Semantics for Computer Languages."[18] Again, the root of this line of thinking can be traced to John McCarthy's earlier work.[19] However, we would not be too remiss in signifying the Scott–Strachey paper as marking the foundations of what came to be called denotational semantics.

As sometimes observed in the history of science, the dissemination of a complex idea is often due to scientists other than the originators of the idea. (For example, Wolfgang Pauli's exposition of the special theory of relativity in 1921 was a major agent of dissemination of Albert Einstein's 1905 work.) We have noted this situation in the case of VDL by way of Wegner's monograph. Denotational semantics offers a similar story: The definitive discussion of the topic was the treatise by Joseph Stoy, also of Oxford University, titled *Denotational Semantics: The Scott–Strachey Approach to Programming Language Theory* (1977).[20]

To explicate the flavor of this approach consider a simple example.[21] But first, some notation is in order.

Let I = { $a, b, \ldots$ } be the set of integers.
Let *Ivar* denote the set of integer variables {$i_1, i_2, \ldots$ }.
Let *IExp* be the set of integer expressions {$e_1, e_2, \ldots$ }.

We suppose that each $e_i \in$ *IExp* consists only of integer variables or arithmetic expressions of the form $e_1 \ominus e_2$, where $\ominus$ is an arithmetic operator. Moreover, the semantics of $\ominus$ is also functionally defined as an ordered pair of integers:

$\ominus: I \times I.$

Let $\Sigma = \{\sigma_1, \sigma_2, \ldots\}$ denote the set of states, where a state is a function from *Ivar* to I. That is,

$\Sigma : Ivar \to I.$

Then the semantics of an integer expression is defined by a function

$SExp : IExp \to (\Sigma \to I).$

That is, given an integer expression $e \in IExp$, the value of SExp (e) is the function Σ, the set of states. So notationally, we can say that

for some state $\sigma \in \Sigma$, SExp (e) (σ) = a, where $a \in I$.

So, to summarize, the semantics of an integer expression is defined as

SExp (i) (σ) = σ (i) for a variable I;
SExp ($e_1 \ominus e_2$) (σ) = SExp (e_1) (σ) $\ominus$ SExp (e_2) (σ) otherwise.

The meanings of statements (such as the assignment, the conditional, or the iterative) are also defined as functions that map from statements to *their* values, which in turn are functions from states to states. That is, statements cause state transformations.

Suppose SStm is this function. Then if s is a statement then its meaning is given by

σ' = SStm (s) (σ)

That is, the effect of a statement is to transform the initial state σ to the final state σ'. And, of course, σ, σ' will determine the values of the variables before and after the execution of statement s, respectively.

5.7   THE AXIOMATIC APPROACH TO MEANING

We have already encountered the axiomatic approach founded as *Hoare logic*—or "Floyd–Hoare logic" (Chapter 1, Section 1.13). Here, the meanings of the constructs of a programming language are defined by a logical system of axioms and rules of inference (or "proof rules"). As Hoare told us in his foundational 1969 paper, his aim was to construct a logic that would ensure that the consequences of executing a program could be determined by "purely deductive reasoning."[22] It seems appropriate to label the axiomatic approach as "logicism" to contrast it to operationalism and functionalism.

Thus the axiomatic approach was not only intended to illuminate the semantics of a language (and of programs written in it) but also to provide a rigorous system to prove the correctness of programs. This then was logicism's advantage over operationalism and even functionalism.

In the 1970s and the early 1980s, the promise and scope of Hoare logic and the axiomatic approach were extensively and richly explored. We have already seen that the semantics of Pascal was defined by Hoare and Nicklaus Wirth in 1973 (Chapter 1, Section 1.13). In 1979, Saud Alagic and Michael Arbib of the University of Southern California proposed a proof rule for the contentious **goto** statement that Hoare and Wirth had not considered.[23] Indeed, the year before, their book *The Design of Well Structured and Correct Programs* (1978) appeared. This was an important contribution to the melding of structured programming (see Chapter 2, Sections 2.2–2.5), semantics, and program correctness issues.[24]

Of course, as we have seen (in Chapter 4), parallelism was much in the computer scientist's consciousness in the 1970s. And for the formalists the semantics of parallel programs evoked much interest. From the mid-1970s through end of the 1980s a profusion of publications—papers and books—addressing parallelism appeared.[25]

Susan Owicki of Stanford University and David Gries of Cornell University were among the earliest to propose an axiomatic semantics for parallel programs along the lines of Hoare logic.[26] In the spirit of the latter their aim was to offer a formalism that not only captured the semantics of parallel operations but also one that facilitated proving the correctness of parallel programs. Their proposal was confined to the shared-memory model of parallel processing (see Chapter 4, Section 4.4).

Toward this end they adopted two familiar language constructs for the parallel programming scenario:

$$\textbf{cobegin } S_1 \parallel S_2 \parallel \ldots \parallel S_n \textbf{ coend}$$
$$\textbf{await } B \textbf{ then } S$$

The first is a slight variant of Per Brinch Hansen's proposal of a parallel construct (Chapter 4, Section 4.5). To recapitulate, the execution of a **cobegin . . . coend** statement initiates the execution of all its component statements $S_1, \ldots, S_n$ in parallel; the statement terminates when all the S_i's have completed. In general, no assumption is made about the relative speed of the S_i's.

In the case of the **await**—where B is a Boolean expression—the assumption is that the statement S does not contain another **cobegin** or another **await** statement. The proof rule for this statement was proposed as

$$\{\,P \,\&\, B\,\}\,S\,\{Q\}$$

$$\{P\}\ \textbf{await } B \textbf{ then } S\ \{Q\}$$

The proof rule for the parallel statement was particularly interesting;

$$\{P_1\}\,S_1\,\{Q_1\},\ \{P_2\}\,S_2\,\{Q_2\},\ \ldots,\ \{P_n\}\,S_n\,\{Q_n\} \text{ are interference free}$$

$$\{\,P_1 \,\&\, P_2 \,\&\, \ldots \,\&\, P_n\}\ \textbf{cobegin } S_1 \parallel S_2 \parallel \ldots \parallel S_n \textbf{ coend}\ \{Q_1 \,\&\, Q_2 \,\&\, \ldots \,\&\, Q_n\}$$

That is, not only must the Hoare formulas $\{P_i\}\,S_i\,\{Q_i\}$ be proved true, their *proofs* must not interfere with one another.

The condition "interference free" was rigorously defined by Owicki and Gries. Stated informally, a statement Si does not interfere with a *proof* of $\{Pj\}$ Sj $\{Qj\}$ if the execution of Si has no effect on the truth of the precondition Pj of Sj and the postcondition Qj used in the proof of $\{Pj\}$ Sj $\{Qj\}$. So if each constituent statement in the **cobegin** statement does not interfere with the Hoare formulas of every other constituent statements then the proofs of all the Hoare formulas $\{P1\}$ $S1$ $\{Q1\}, \ldots, \{Pn\}$ Sn $\{Qn\}$ are interference free.

5.8   PLURALITY → COMPLEMENTARITY → UNITY

The plurality of approaches to semantics would have pleased postmodernists—had they been aware of it. They would have found it entirely natural that there was no "master narrative" governing the question, "What is semantics?"

Yet there were computer scientists who pondered the relationship among operationalism, functionalism, and logicism, for computer scientists, like all scientists, desire unity among diversity; they wish to unify. Scientists inherently are not inclined toward postmodernism.

Thus, according to James Donahue of the University of Chicago, writing in 1976, these three modes of semantics were not simply three points of view. Rather, they were *complementary* in the sense that they defined semantics at different levels of abstraction.[27]

For example, the operational approach defined semantics that was strongly suggestive of how a language may be implemented by prescribing sequences of state transformation resulting from the interpretation of a syntactic object. The denotational approach abstracted from state transformation sequences—and thus from any specific implementations—but retained the concept of states and defined meanings in terms of functions. The axiomatic approach abstracted still further in that states are only implicitly present, relying rather on relating syntactic objects to logical formulas. Furthermore, by design, logicism is more conducive to program design and verification than operationalism or functionalism.

But unity—in some sense—is what scientists most seek. Jaco de Bakker of the Mathematics Centrum, Amsterdam, discovered such unity between the denotational and the axiomatic. In his treatise *Mathematical Theory of Program Correctness* (1980), de Bakker showed how denotational semantics could be used to *define* not only programming languages but also as the metalanguage to specify logical assertions used in Hoare-style axomatic proof techniques. Denotational semantics was, in this sense, more fundamental than axiomatic proof techniques for the former provided a rigorous basis for establishing the *reliability* of axiomatic proof systems.[28] By reliabilty, de Bakker meant that the axiomatic proof system is *sound*—that is, *only* true Hoare formulas can be proved by the system—and *complete*—that is, *all* true formulas can be proved by the system.

5.9 CONSOLIDATION AND SPREAD OF THE FORMAL DESIGN PARADIGM

The leading spirits of the formalist school of programming were not content with inventing what de Bakker called a "mathematical theory of program correctness." In this they were Knuthians, not Hardyans (see Section 5.3). For they desired to integrate proof techniques into the program design process itself. Edsger Dijkstra and C. A. R. Hoare, for example, desired nothing less than a *calculus of programming* that would enable the creation of correct programs *by design*,[29] much like Boolean algebra was used in the construction of correctly functioning logic circuits. Others followed their lead. And in the decade 1976–1986 we witnessed the consolidation of what we may be justified in naming the *formal design paradigm*, not only in the realm of software artifacts but also in the hardware and firmware domains.

Here, I use the term *design paradigm* to mean the following:

(a) An abstract prescriptive model of the design process that (b) serves as a framework or schema for constructing practical methods, procedures, and tools for conducting design.[30]

A formal design paradigm is a design paradigm in which the prescriptive model is grounded in an axiomatic system.

This consolidation took the shape of an outstanding series of articles, texts, and monographs on formal software development.[31] And such was the enticing appeal of the formal paradigm—its promise of transforming design into a mathematically driven activity—that even as the paradigm was being consolidated in the realm of software, its influence was spreading to other domains: the design of computer architectures,[32] firmware (microprogramming),[33] and even digital circuits (hardware).[34] One might say that the formal design paradigm came to resemble a Thomas Kuhn–style scientific paradigm:[35] a core of fundamental elements surrounded by a collection of subparadigms pertaining to a variety of computational artifacts.

Perhaps the most exemplary project on the use of the formal design paradigm outside the software domain was executed by Werner Damm, then at the Technical University of Aachen, and his coworkers. Their system, called AADL/S* consisted of a computer architecture description language (AADL), a high-level microprogramming language (S*),[36] and a Hoare-style *microprogramming logic*[37] for proving that an architecture specified in AADL is correctly implemented by a microprogram written in S* on a lower level architecture. However, to manage the potential semantic gap between a target ISP-level architecture and a host microarchitecture, Damm et al. envisoned a rigorous stepwise-design-by-refinement process.

Figure 5.1 schematizes this approach. The design process would begin with a formal specification of both the target ISP-level architecture *T* and the host microarchitecture *H* in AADL.[38] The designer's aim is to develop a microprogram *M* that implements *T* on *H*. This would be effected by constructing, in top-down, hierarchical fashion, one or more

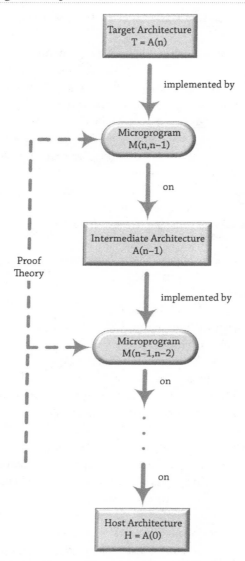

FIGURE 5.1 Hierarchical design of a provably correct target architecture.

intermediate architectures $A(n-1), A(n-2), \ldots, A(1)$, where $A(j)$ is more abstract than $A(j-1)$ and each architecture $A(j)$ is expressed in AADL. So each design step k would create a microprogram $M(k, k-1)$ in S^* that implements $A(k)$ on $A(k-1)$ (Figure 5.1) and at the same time uses the proof theory to prove that the microprogram $M(j, j-1)$ is correct.

5.10 TOWARD A FORMAL AGENDA FOR HARDWARE DESIGN

The 1960s were an extraordinarily fertile decade for the development of switching and finite automata theory—the theory for the design of digital logical circuits such as arithmetic units, registers, counters, shift registers, decoders, multiplexors, and memories—the functional building blocks of physical computers. Armed with such mathematical tools as Boolean algebra, group theory, coding theory, graph theory, and optimization theory, a mathematically rich foundation was established by the end of the decade for the design of logical circuits. Students and teachers of logic design were blessed with the availability of some outstanding introductory and advanced textbooks on the subject, some bearing the authorial authority of the very creators of the field.[39]

Thus researchers in this aspect of hardware design were well ensconced in a formal design environment, but it was an environment somewhat different from the formal design paradigm as defined here. In particular, the idea of proving the correctness in an axiomatic, Hoare-style manner was quite alien to their culture. Even as late as 1974, in an admirable historical survey of the theory and practice of logic design by Glen Langdon of IBM, there was much discussion of testing, simulation, and debugging, and even the use of flowcharts in the design process for logic circuits. The term "design verification" was used but the reference was to "check out and debug the design" but never to axiomatically prove design correctness.[40]

Yet even before Langdon's book appeared matters were changing. There arose a desire to automate the design of computers, to construct, at the very least, computer-aided design (CAD) tools and systems for supporting computer design. The basic schema envisioned was not unlike the well-established schema for compiler-based program translation. The computer (or some part thereof) would be specified formally at some level of abstraction (e.g., at the ISP level) and then have CAD tools aid the designer in transforming this specification into a low-level digital design.

Such formal specification demanded a computer hardware description language (CHDL) to capture the computer specification that would then be translated (compiled) into a low-level representation. This in turn would be input to either a simulator or a CAD system or (most ambitiously) to an automatic *synthesis* program that would produce an implementation of the target machine.

Crucial to this scenario was the construction of *formal models* of the desired target machine at the abstraction level of interest. Such models provide a theoretical basis for implementing simulators, emulators, or CAD tools. A new kind of computer scientist came into being: Their domain was computer design but they were not computer designers. They were, rather, theorists, modelers, and builders of computer design tools.

Such was the imperative for exploring this scenario—investigating the physical computer design process—that the International Federation for Information Processing (IFIP) under the umbrella of their Technical Committee on Digital Design (TC 10) established Working Group WG 10.2 devoted to "Digital System Description and Design Tools." Its first chairman and founder was Mario Barbacci of Carnegie-Mellon University, a prominent researcher in the domain of CHDLs [and who had designed and implemented the ISPS architecture description language (Chapter 3, Section 3.5)]. In 1973, the first IFIP International Conference on Computer Hardware Description Languages and Their Applications (CHDL-73) was held in New Brunswick, New Jersey.

This biennial conference became the principal international forum for exploring the formal design of digital systems.

The development of CHDLs reaches back to the 1960s.[41] But partly in reaction to the Babelic proliferation of CHDLs by the mid-1980s and partly because of its own needs the US Department of Defense (DoD) initiated, circa 1984, as part of its "Very High-Speed Integrated Circuit" (VHSIC) Program, the development of a DoD-wide standard CHDL—reminiscent of the Ada initiative (see Chapter 2, Section 2.12). This language came to called VHDL, an acronym for "VHSIC Hardware Description Language."[42]

5.11   VLSI DESIGN: A NEW EXCITEMENT

There was, of course, another technological imperative that drove the formal agenda in digital design. By the end of the 1970s integrated-circuit (IC) technology had evolved, following Gordon Moore's prediction (Chapter 3, Section 3.7), to the point that IC chips containing in excess of 100,000 "gates" (the most primitive building blocks in a digital system) were being fabricated. This was the era of very large scale integration: "VLSI" became a kind of household acronym, not just for hardware designers but for the entire computing community, such was its perceived impact. VLSI enabled entire computers to be implemented on a few chips or even, by clever design, on a single chip (Chapter 3, Section 3.6).

A seminal event in the development of VLSI "culture" was the publication, in 1980, of the book *Introduction to VLSI Systems* by Carver Mead of the California Institute of Technology and Lynn Conway of Xerox Palo Alto Research Center.[43] "Mead and Conway" became in the realm of VLSI design what "Knuth Volume 1" had been in the realm of algorithms a decade earlier: the de facto bible for VLSI designers.

What was especially original about Mead and Conway was its emphasis on what Dijkstra, some 15 years before, had called the "intellectual management of complexity." In the case of VLSI the complexity stemmed from the sheer circuit density of IC chips. And just as the concern with complexity in the software realm, though of a different nature, had evoked the twinned strategy of hierarchy and abstraction, so also concern with complexity in the realm of VLSI systems brought forth a quest for hierarchy and abstraction.

It was not that hardware designers were unfamiliar with the principles of hierarchy and abstraction. Far from the case. But much of the *excitement* generated by the Mead-Conway treatment of VLSI design seemed to have stemmed from the realization that hardware design was suddenly akin to software design. What programming theorists and methodologists had called "structured programming" as a way of coping with complexity in their realm became, in the minds of Carver Mead, Lynn Conway, and their associates, "structured design."

The differences lay in that one face of software was abstract whereas VLSI chips were uncompromisingly materialistic: A chip in its most abstract sense was a piece of two-dimensional *physical space*—"silicon real estate" as it was picturesquely called. And although the behavior of programs was governed by invented, artificial principles—as Hoare logic was—the behavior of VLSI chips was governed or constrained by the physical laws of semiconductors.

So such concepts as abstraction and hierarchy could not be "imported" lock, stock, and barrel from the programming to the VLSI realm. The concepts had to be reinterpreted and instantiated to make sense in the VLSI context.

At the most abstract level, the objective of VLSI design was a kind of physical floor planning: to lay out the circuit components on a given chip area, a given amount of available "real estate," in the most efficient manner possible. "Efficiency" in this context related to the physics of semiconductors but ultimately it entailed the minimization of *communication paths* between functional components placed on the chip, because the number and lengths of the interconnection wires were the constraining factors rather than the number of transistors on the chip. The Mead–Conway structured design philosophy was thus driven by these considerations. And its main elements were the following:

(a) The use of a systematic, hierarchic design approach involving both top-down and bottom-up strategies.

(b) The use of highly regular structures for subsystems and their composition into larger systems in a highly regular way. Regularity of structure both within a chip and of an ensemble of chips was a profoundly significant objective in VLSI implementation, whether the target artifact was a microprocessor, or complete microcomputer,[44] or a supercomputer.[45] Typical candidates for such regular subsystems were decoder matrices, shift registers, programmable logic arrays (PLAs) and read-only memories (ROMs).[46] The literature on VLSI design was soon illuminated by photomicrographs of microprocessor chips showing their highly regular susbsytems that were remarkably Mondrianesque in their aesthetics.

There does not seem to be anything particularly formal about the Mead–Conway design philosophy. But the very idea of composing components or subsystems into larger systems lended itself to formal investigation.

For example, James Rowson, Carver Mead's graduate student at the California Institute of Technology, completed a PhD dissertation in 1980—the same year as "Mead and Conway" was published—titled "Understanding Hierarchical Design."[47] A substantial portion of the "understanding" entailed exploring what Rowson called "the mathematics of hierarchies." The mathematics Rowson drew on was a branch of mathematical logic called "combinatory logic," developed in the 1930s and most closely associated with the name of Haskell Curry, a mathematical logician at Pennsylvania State University.[48] This logic was closely associated with another branch of mathematical logic called "lambda calculus" (or λ-calculus) also invented in the late 1930s by logician Alonzo Church of Princeton University (and Alan Turing's doctoral advisor).

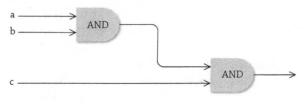

FIGURE 5.2 Composing a three-input AND gate from two two-input AND gates.

Stated *very* simply, combinatory logic is concerned with operators that take functions as arguments and return functions as values. Consider, for example, a two-input AND gate, a primitive building block in digital logic design. In Boolean algebra, given inputs *a* and *b* to this gate, the output would be defined by the Boolean expression *a and b*, where *and* is a Boolean operation. Now if the output of this AND gate is an input to another two-input AND gate and *c* is the other input to the latter (Figure 5.2), the Boolean expression for this structure would be (*a and b*) *and c*.

Using a notation that roughly corresponds to the notation of combinatory logic, in contrast, the first two-input AND gate would be expressed as

$$and\ a\ b.$$

Here, the *and* operator is defined to take an argument *a* and return a *function* that "ands" *a* to *its* argument *b*. The combination of the two AND gates of Figure 5.2 would be expressed as

$$and\ and\ a\ b\ c.$$

This composition of the two instances of the *and* operator represents the hierarchic composition of the two two-input AND gates into, in effect, a three-input AND gate. So the *structural* composition of circuit elements is expressed by a *functional* composition.

5.12   PROVING HARDWARE CORRECTNESS: AN ALIEN NOTION

But another intriguing appeal of the VLSI "movement" was to formalists interested in *proving* hardware correctness using Hoare-style and related logics—a very alien notion to the hardware design tradition.

Yet the idea of hardware correctness was problematic In a way, program correctness could not be. As we've noted, the goal of hardware design is the creation of wired structures laid out in physical space. Now, we can certainly specify the behavior of this circuit in a hardware description language. The following, for example, shows a specification of the behavior of the circuit in a VHDL-like notation:

---

entity full_adder **is**
    port
        (in1, in2, carry_in : **in bit**; sum, carry_out : **out bit**)
end full-adder;
architecture behavioral_description **of** full_adder **is**
    begin
        sum <= in1 **xor** carry_in **after** 30ns;
        carry_out <= (((in1 **and** in2) **or** (in1 **and** carry_in))
                or (in2 **and** carry_in) **after** 30ns
    end

As in the realm of microprograms we can then adapt Hoare logic and prove that this specification satisfies a given pair of preconditions and postconditions using an axiomatic proof theory for, say, VHDL. This approach was, in fact, used by Vijay Pitchumani and Edward Stabler of Syracuse University in 1984.[49]

But the fact is that the ultimate objective of hardware design is the creation of wired structures. A description in VHDL-like notation of the full-adder structure that would realize the previously mentioned behavior would look like this:

---

architecture structure of full_adder **is**
    component and_gate **port** (x, y: **in bit**; z : **out bit**);
    component or_gate **port** (x, y: **in bit**; z: **out bit**);
    component xor_gate **port** (x, y : **in bit**; z: **out bit**);
    --- internal signals
    signal sa, sb, sc, sd, se : **bit**
    begin
        A1: and_gate (in1, in2, sa);
        A2 : and_gate (in1, carry_in, sb);
        A3: and_gate (in2, carry_in, sd)
        O1: or_gate (sa, sb, sc);
        O2: or_gate (sc, sd, carry_out)
        X1 : xor_gate (in2, carry_in, se);
        X2: xor_gate (se, in1, sum)
    end

Notice that this includes declarations of the component types (rather analogous to the class concept in SIMULA), with their respective behaviors presumably well defined and known, and the interconnection wires ("signals" in VHDL terminology); and a specification of the instances of the component types (X_1, X_2, etc.) including the actual input and output ports.

The task of formal design then would be to prove that this structure exhibits the desired behavior described, say, in terms of preconditions and postconditions.

In 1983, Robert Shostak of the Stanford Research Institute adapted Robert Floyd's original technique of 1967 (Chapter 1, Section 1.13) to prove the correctness of such structures.[50] Input assertions were attached to each input wire of the circuit and output assertions to each output wire. Additional assertions were assigned to arbitrary internal wires, these denoting the initial states of components (such as the initial values of registers in the case of sequential circuits). The correctness criterion was defined thus:

If the inputs satisfy the input assertions and the initial states are assumed to hold and if each component behaves according to its defined characteristics, then the circuit outputs will satisfy the output assertions.

As in Floyd's original method (which analyzed programs in flowchart form) the proof procedure established paths through the circuit and identified *invariants* (that is, unchanging assertions) at key points in the paths and proving that these invariant assertions were correct.

Shostak was engaged in proving the correctness of existing circuit designs rather than in designing provable correct circuits. In 1986–1988 Michael Gordon (who had earlier written a monograph on denotational semantics[51]) and his coworkers at Edinburgh University developed an elegant method for the specification *and* verification of hardware using a formalism called *higher-order logic* (HOL).[52]

HOL extends "ordinary" (or first-order) logic by allowing variables to range over functions and predicates. That is, functions and predicates can be arguments and results of other functions and predicates. The payoff lay in that both behavior and structure of circuits could be represented as formulas in HOL. Furthermore, the laws of this logic could be used to draw inferences and prove assertions about circuits. A HOL, then, could serve both as a hardware description language and as a proof system for reasoning about hardware structures. This was in contrast to Hoare-style formalism in which the assertions about a program are in a different language from the programming language in which the program is specified.

The nature of this formalism can be glimpsed with a very simple example.[53] Consider an inverter, the output of which negates the input but is delayed by one time unit. Its behavior in HOL, the particular higher-order logic used by Gordon et al., may be specified by first representing the input (i) and output (o) signals by *functions* that map discrete time values to Boolean values. So the inverter's behavior can be expressed by the predicate

$$\text{INV}\ (i, o)\ \equiv \forall t.\ o(t + 1) = \neg\ i(t).$$

This can be read thus:

The truth-valued predicate INV with arguments i, o is defined as, "For all time units the output value at time $t + 1$ is equal to the negation of the input value at time t."

Suppose two inverters of this sort are connected in series by an internal wire p (Figure 5.3). Then the behavior of the device as a whole will be *constrained* by the *conjunction* of the behaviors of its constituents. This device can then be characterized by the predicate

$$\text{INV2 } (i, o) \equiv \exists p. \text{ INV } (i, p) \wedge \text{INV } (p, o).$$

This is read thus:

The predicate INV2 with arguments i, o is defined by this proposition: "There exists a value of p such that both predicates INV (i, p) *and* INV (p, o) are true."

We can see how the formal design paradigm works here. The specification of a hardware device would be composed hierarchically from the bottom up: given the behavioral specification of "more" primitive components $C_1, C_2, \ldots, C_n$ in the form of predicates, a structure S composed out of these components would be defined in terms of another predicate composed out of the more "primitive" predicates. S in turn may be a component of a still-larger structure S' and so on. Moreover—and this is where lay much of the appeal of this method—using the theory of the logic, the behavior of a structure S could be *inferred* from the behavior of the components and their interconnection structure. For instance, for the inverter system of Figure 5.3 this would entail proving that

$$\forall\, i, o : \text{INV2 } (i, o) \supset \forall\, t.\, o(t + 2) = i\,(t\,).$$

That is,

For all values of i, o if the predicate INV2 (i, o) is true then this implies that for all values of time t the output value of o at time $t + 2$ will equal the input value of i at time t.

In other words, the inverter series acts as a delay element.

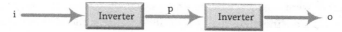

FIGURE 5.3 Two inverters connected in series.

The intellectual appeal of the formal paradigm was irresistible to many. We have witnessed C.A.R. Hoare's manifesto on "The Mathematics of Programming."[54] The appeal was rooted in a sort of mathematics envy (Section 5.1). And the response to this envy was to construct a mapping from the domain of computational artifact design onto the mathematical domain:

DESIGN	MATHEMATICS
Specification of requirements	Problem definition
Representation of the artifact	Theorem
Verification of representation against requirements	Proof

So if an artifactual design problem could be mapped in this fashion one could bring to bear the rich apparatus of mathematical reasoning and knowledge to derive provably correct solutions.

Incidentally, this was not only the computer scientist's dream. Theorists in (building) architecture and dwelling planning were contemporaneously pursuing a similar project. Probably the seminal work here was that of the University of California theorist Christopher Alexander who in 1964 gave voice to a similar mathematics envy in his influential monograph *Notes on the Synthesis of Form*[55]—which was even quoted by Per Brinch Hansen[56] in 1977. (Ironically, Alexander would later repudiate thoroughly his mathematical approach of *Notes*.) Over a decade after *Notes* was published Cambridge University architectural theorists Lionel March and Philip Steadman would write articles on the uses of boolean algebra and graph theory, respectively, on the design of "built forms."[57] And, influenced by formal computational theory, William Mitchell of MIT explored representations and a *calculus* of architectural form using predicate logic and formal language theory in his book *The Logic of Architecture* (1990).[58]

Returning to computer science, when David Gries of Cornell University wrote his elegant *The Science of Programming* (1986), the "science" he had in mind was not empirical science at all but mathematical logic: Part I of his three-part book was devoted to an exposition of axiomatic systems and formal logic.[59]

If the formalists believed that they had appeased their mathematical yearnings, that the mapping they had articulated from computational artifact design to mathematics would be acceptable to all, at least within their own community, they were mistaken. There were skeptics who chose to seriously contest the very premises of the formal paradigm. The unkindest cut of all was that some of these sceptics were nurtured in the same mathematical–logical milieu as the committed formalists themselves.

5.14   FORMALISM CONTESTED: A SOCIO-HISTORICAL CHALLENGE

In May 1979, Richard De Millo of Georgia Institute of Technology and Richard Lipton and Alan Perlis of Yale University—the latter one of the makers of the first age of computer science, a codesigner of Algol 60, and the first winner of the ACM Turing Award (de facto computer science's "Nobel")—published a paper in the *Communications of the ACM* bearing the title "Social Processes and Proofs of Theorems and Programs."[60]

The title was enormously significant. Here were three "mainstream" computer scientists, all thoroughly versed in mathematics and programming theory, willing—daring—to speak to the *social process* of doing mathematics and computer science. Practicing scientists are not particularly interested in the sociology or social history—or even the philosophy—of their particular sciences. Those were matters of little relevance to them.

And yet the paper was published in a time when actual scientific *practice* was being intensely scrutinized by historians, philosophers, and sociologists of science to determine what scientists actually do when they do science and how scientific knowledge actually grows. A major influence on this line of thinking was Thomas Kuhn's *The Structure of Scientific Revolutions*[61] but there were other more recent, post-Kuhnian influences such as *Laboratory Life* (1979) authored by sociologists Bruno Latour of the École Nationale Supérieure des Mines, Paris, and Steve Woolgar of Brunel University.[62] We would not be inaccurate in claiming that the paper by De Millo et al. must count as one of the earliest serious, scholarly contributions to the sociology and social history of computer science.

The issue De Millo et al. addressed was the relationship between programming and mathematics; more specifically, the perceived parallel between mathematical proofs and formal program verification. They began by recapitulating the mathematics envy-driven yearning to make programming "more like mathematics."[63]

A laudable goal—but "not in the way they seem to think."[64] For the formalist's vision of making programming "more mathematics like" was to deploy "a long chain of formal deductive logic" to certify that a program is functionally correct. Programs will then have been *proved* correct in exactly the same sense the formalists believed mathematical theorems are proved.

In theory, De Millo et al. admitted, mathematicians could indeed use long chains of logical deductions to prove theorems. But in practice this is not what mathematicians do. If a mathematical proof was indeed a long sequence of formal, inexorably logical deductions it would be quite unlikely that such proofs would ever be *read*, let alone accepted. In practice a mathematical proof is "a very different animal."[65] Mathematical proofs are more often proof *sketches*.

Moreover—and this was the essence of their skepticism—the mere existence of a proof of some theorem may not suffice; the proof must be *accepted* by the relevant community to be assimilated into mathematical knowledge. This acceptance of a proof depends on a *social process*—in the sense that the proof is studied, discussed, argued over, and criticized by members of the relevant community. As a result of such a process the proof may either be believed and the theorem accepted by the community—becoming part of the mathematics paradigm à la Kuhn—or it may be rejected. A theorem may even be only *tentatively* accepted and perhaps later discarded because of an error detected in the proof. De Millo et al. cited philosopher of science Imre Lakatos's now-classic discourse *Proofs and Refutations* (1976)[66] but then only recently published, in which Lakatos showed that a celebrated relationship in graph theory discovered by Leonhard Euler "was reformulated again and again after its first statement until it reached its final form."[67] They offered many other examples from the history of mathematics to illustrate the social dynamics of mathematical practice; and to point out that one cannot explicate mathematical concepts without taking into account the role of social processes in proof constructions. For example, they noted, mathematical errors are rectified by other mathematicians, not by formal logic.[68]

The process of proving programs, according to De Millo et al., is of quite a different nature. Program proofs—especially those generated by machines—are long, tedious sequences of logical deductions that look and are very different from mathematical proofs. They are almost never subject to the kind of social process mathematical proofs undergo. Thus their acceptance (or otherwise) through the process of criticism, argument, and analysis never actually happens. As De Millo et al. put it, one either believes program proofs as an act of faith or one does not.

Thus, from their socio-historical argument, in contrast to the formalists' analogy between program proofs and mathematical proofs, they offered an alternative analogy:

PROGRAMMING	MATHEMATICS
Specification	Theorem
Program	Proof
Verification	Imaginary Formal Demonstration

In other words the formalists map their verification strategy onto an imaginary version of the mathematician's proof strategy.

De Millo, Lipton, and Perlis struck another blow against the very possibility of program proofs. They noted the fundamental *discontinuity* between the world of small ("academic") algorithms and programs for which proofs of correctness have been provided—they cite C. A. R. Hoare's 1971 proof of a search program[69]—and the world of industrial software. The differences between these two worlds "are not differences in degree. They are differences in kind."[70] (In the more general context of software development, this schism had been observed by others, most notably, by Frank DeRemer and Hans Kron of the University of California, Santa Cruz, in a 1975 paper titled, catchingly, "Programming-in-the-Large Versus Programming-in-the-Small."[71]) Among the reasons for this, De Millo et al. noted, was that a large part of industrial code may consist of user interfaces and error messages—"ad hoc, informal structures that are by definition unverifiable."[72] Even more seriously, industrial software is constantly evolving: A large part of what would be called by the mid-1980s the "software life cycle" was given to software maintenance—meaning modification. This is in contrast to the general stability of small programs. There was no reason to suppose, De Millo et al noted, that verifying modified software would be any easier than verifying the original software.

This discontinuity between the world of relatively stable, small programs and the constantly evolving, unstable world of large software precluded even the possibility of the kind of social processes that attend mathematical practice.

As can be imagined, the paper by De Millo, Lipton, and Perlis provoked much discussion. The "ACM Forum" section of the November 1979 issue of *Communications of the ACM* carried some 10 letters supporting or refuting the De Millo et al. thesis. Most (but not all) supporters came from the industrial sector. And their general response was akin to a sigh of relief that at last someone (of impeccable academic credentials at that) had finally voiced what they had suspected all along, that the formalist emperor had no clothes.[73]

A rather well-tempered response from within the formalist camp was offered by William Scherlis and Dana Scott of Carnegie-Mellon University in 1983. They did not repudiate the thesis about the social process of proof acceptance in mathematics. But they objected to the proposition that computer-generated proofs were not—or could not—be subjected to a similar kind of social scrutiny. Their example was the celebrated and venerable *four-color problem* ("Can a finite planar map be colored with four colors so that no two adjacent regions have the same color?") and the equally celebrated but recent computer-aided proof discovered by mathematicians Wolfgang Haken and Kenneth Appel of the University of Illinois in 1977.[74] Scherlis and Scott documented how this proof underwent exactly the same kind of public debate and scrutiny traditional mathematical proofs go through.[75]

But it is worth noting that this computer-generated proof was for a celebrated, open mathematical problem; moreover, it was the first computer-generated proof produced in the mathematical realm. Thus the interest generated in the proof was twofold: because of the celebrity of the problem itself and because of the sheer novelty of the proof method. The social process De Millo et al. wrote about quite naturally carried into this domain. This did not imply that computer-generated correctness proofs for programs of less mathematical (or scientific) significance would engender similar scrutiny.

Despite the attention the paper by De Millo, Lipton, and Perlis received, it did not in any way dilute the subsequent efforts of the formalists. Indeed, some of the more influential books published thereafter on formal program design and verification paid no attention to the De Millo et al. paper.[76] A multiauthored volume titled *The Correctness Problem in Computer Science* (1981) edited by Robert Boyer and Strother Moore of SRI International contained five lengthy papers on various aspects of the correctness problem. Not a single one referred to the De Millo, Lipton, and Perlis paper. The formalists had a mission, and they were not willing to countenance any unpleasant obstacles.[77]

One of the tasks of philosophy is to clarify concepts, including the conceptual foundations of some field such as science. It was thus that a very different contestation, as forceful as that of De Millo et al., was put forth almost a decade after by a philosopher of science, James Fetzer of the University of Minnesota, Duluth. In September 1988, the *Communications of the ACM* published an article by Fetzer titled "Program Verification: The Very Idea."[78] This was probably the first publication in that journal on the philosophy of computer science that dealt with a topic that was *not* artificial intelligence.

Fetzer argued that the mere evidence of a social process for program proofs (assuming, *contra* De Millo et al., that such a social process exists) does not ensure the validity of the proof (or otherwise). On the other hand, he claimed, there *were* reasons for skepticism concerning formal program verification as a "completely reliable method" for ensuring the correct functioning of a program. Thus, he concluded, the arguments put forth by De Millo et al. were "bad" as a defense of positions that needed better supportive arguments.[79]

For Fetzer, being a philosopher in search for conceptual clarification, the problem was whether there are *logical* grounds for presuming programs are mathematical entities or whether mathematical theorems (and their proofs) and programs (and their proofs) are not altogether different kinds of entities. Fetzer's challenge was thus grounded in what philosophers would call ontological considerations—questions concerning the fundamental nature of objects, in this case, programs and theorems.

He pointed out that the term verification is used differently in mathematics on the one hand and in scientific discovery and ordinary reasoning on the other. In the realm of pure mathematics and logic a verification—a proof—is a demonstration that a given theorem or proposition can be deduced from some given set of axioms. Fetzer termed such propositions to be *absolutely verifiable*.

In contrast, in empirical science, or the real world, verification refers to the situations in which certain conclusions follow from some given set of premises in which the validity of the premises themselves may either be questioned or be unverifiable. Fetzer termed conclusions of this sort *relatively verifiable*.

Now a theorem is simply a symbol structure that has been deduced from some set of axioms by applying a chain of inference rules. So a theorem so deduced has no semantic significance. It is a purely syntactic entity. A theorem is what philosophers would call analytic statements—whose "truth" can be ascertained by the chain of deductions leading to the theorem. A program, on the other hand, has a meaning: It specifies a sequence of operations that can be, or are required to be, performed by a physical computer. (So Fetzer was suggesting that programs are not purely abstract but are rather what I called in the Prologue liminal artifacts.)

Further along these lines, Fetzer claimed that programs must be distinguished from algorithms: The latter are abstract logical structures that stipulate a function to produce outputs from given inputs. An algorithm, Fetzer claimed, is an executable entity only in the sense that its target computer is an abstract machine having no physical significance. (In my terminology of the Prologue, an algorithm is an abstract artifact.) A program, in contrast, requires the existence of a real processor or device occupying physical space, whose operations consume physical time.

Being abstract entities, algorithms, like mathematical theorems, may be absolutely verifiable. Being causal entities (that is, effecting physical events), programs can at best be subject to relative verification because the premises underlying such verification are, in the final analysis, dependent on empirical justification involving the physical system that executes the program.

More precisely, given a program Π, if Π is really an *algorithm,* the Hoare formula

$$\{Pre\}\ \Pi\ \{Post\}$$

can be absolutely verified from the axioms and proof rules governing the language in which Π is expressed. If, however, Π is really a *program* that will be compiled and executed on a physical machine, this Hoare formula will be relatively verifiable only because the underlying axioms and proof rules will depend on the causal properties of the physical target computer—its hardware and firmware. And *that* can be determined only through empirical means, not through formal arguments.

One must conclude, Fetzer stated, that programs can at best be treated just like empirical scientific statements, at best be treated as *conjectures* that can be *refuted* conclusively but never verified absolutely.

This idea of conjecture and refutation is, of course, at the heart of philosopher of science Karl Popper's celebrated theory of the scientific method.[80] Fetzer acknowledge this Popperian spirit of his conclusion, suggesting that programs bear a correspondence with the propositions of the empirical sciences rather than with mathematical theorems. This suggestion, of course, did not go down well with the formalists.[81]

Because Popper had a presence in Fetzer's critique, let us pursue this nexus a bit further.

Popper's concern was with the *logic* of how scientific knowledge grows, not with the sociology of science. Empirical science, he maintained, was a strictly logical process but of a particular kind. In *Objective Knowledge* (1972) he schematized this logic by the schema:[82]

$$P_1 \rightarrow TT \rightarrow EE \rightarrow P_2,$$

where **P1** is a scientific problem or a phenomenon, **TT** is a tentative theory (or hypothesis) that solves or explains **P1**, and **EE**, error elimination, is the critical attempt to falsify **TT**. The outcome is that a fresh problem or phenomenon **P2** emerges.

TT is in the nature of a bold conjecture; but *how* **TT** is obtained is not of logical concern in Popper's scheme of things. The logic lies in the critical attempt to refute **TT** rather than to prove it: on the one hand, one can never prove an empirical theory because there is no guarantee that such a statement about the empirical world that makes claims for universality (e.g., "all swans are white") will not be falsified by some later observation (such as a black swan). So if **EE** fails to falsify **TT**, the latter is a conjecture that is merely provisionally accepted. On the other hand, a single falsifying observation is enough to logically negate—refute—**TT** (e.g., the observation of a black swan). And so the process *evolves*; it is by way of this cycle of conjecture and refutation that scientific knowledge grows.

This was the direction of Fetzer's thinking. The new analogy, then, was one between program design (or, more generally, computational artifact design) and empirical science:

NATURAL SCIENCE	DESIGN
Problem/Phenomenon	Specification of Requirements
Hypothesis/Theory/Conjecture	Design of the Artifact
Critical Experiment/Observation	Critical Test
Error Elimination	Error Identification/Correction

5.16 DISCOVERING THE CONNECTEDNESS OF REALLY DIFFICULT PROBLEMS

Automata theory, concerned with what can or cannot be computed and constructed around the Turing machine, had developed into a rich Hardyan formal discipline (Section 5.3) by the end of the 1960s. Two authoritative works representative of the state of knowledge on automata theory were *Computation: Finite and Infinite Machines* (1967) by Marvin Minsky of MIT and *Formal Languages and Their Relation to Automata* (1969) by John Hopcroft of Cornell University and Jeffrey Ullman, then of Princeton University.[83]

But in the mid-1960s another remarkable topic evolved out of automata theory. This was termed *computational complexity* by its founders, Juri Hartmanis and Richard Stearns of the General Electric Research Laboratory in Schenectady, New York.[84] The journal in which their paper was published—*Transactions of the American Mathematical Society*—is worth noting, indicating, both to the authors and the journal editor, that computational complexity (and automata theory) were then perceived to belong to the realm of pure mathematics. Although since then the theory of computational complexity has been relocated in the computer scientist's milieu, the sense of a mathematical identity has prevailed ever since. The topic belongs unmistakably to the Hardyan wing of formal computer science.

Computational complexity belongs to our story here for a very specific reason. But to appreciate this a certain preamble is necessary.

The universe of computational complexity is the space of computational problems and comprises of two regions: one containing problems deemed *tractable* ("relatively easy"), the other consisting of problems deemed *intractable* ("very difficult"). In the language of complexity theory a tractable problem belongs to the *P*-class and intractable problems to the *NP*-class.

There are reasons why these letters were assigned, and they have to do with the relationship of problem difficulty with Turing machines. (All formal roads in computer science seem to lead to automata theory.) A "conventional" Turing machine of the sort Alan Turing conceived is also called a *deterministic* Turing machine (DTM) in that in each step of the machine's operation only one move can be made and only one state change is possible.[85] Following the lead of Michael Rabin and Dana Scott in 1959,[86] automata theorists invented, in the 1960s, the concept of a *nondeterministic* Turing machine (NDTM): Here, a finite number of alternative moves can be chosen from at each step of the machine's operation, leading to a finite number of alternative next states. One can think of a NDTM as starting with a given input string on its tape and executing all possible sequences of moves *in parallel* until the machine comes to a halt. It is as if a number of copies of the machine come into existence.[87]

Armed with these concepts of DTM and NDTM, *P* and *NP* can be more formally characterized: A *P*-problem is one for which a solution can be obtained on a DTM in *polynomial time*; that is, for a problem of size n, algorithms exist that solve the problem deterministically in time proportional to no more than that of a polynomial function of n. An example from the *P*-class is the sorting problem: To transform a randomly ordered array of n elements (say, numbers or character strings) into an ascending (or nondescending) sequence. Algorithms exist that solve the sorting problem in time of the order of $n \log_2 n$ [more concisely stated as $O(n \log_2 n)$].

An *NP*-problem is one for which no deterministic polynomial time algorithm is known to exist; such a problem *can* be solved in polynomial time on a NDTM, but its correctness can be *checked* in polynomial time on a DTM. An example is the "traveling-salesman problem": Given a set of cities connected by paths of given lengths, can a salesman, starting at a "base" city traverse all the cities and return to the base covering less than some specified distance? No polynomial time algorithm is known for this problem, but given an algorithm to solve it, it can be verified for correctness by way of a polynomial time algorithm.

But there is more to the *P–NP* story. First, given their definitions, it should be clear that the *NP*-space of problems contains the *P*-space of problems. In 1971, Stephen Cook of the University of Toronto introduced a concept called *NP*-completeness.[88] A problem Π is said to be *NP-complete* if Π is in *NP* and all other problems in *NP* can be transformed or *reduced* to Π in polynomial time. So, on the one hand, if Π is tractable, so are all other problems in *NP*; if, on the other hand, Π is intractable then so are all other problems in *NP*. In other words, there is a sense in which all the problems in *NP* are "equivalent" and are *connected* to one another. A remarkable discovery.

Cook showed that a certain problem called the "satisfiability problem" is *NP*-complete. This problem asks this question: Given an arbitrary Boolean expression— for example, (a OR b) AND (c OR d), where a, b, c, and d are Boolean variables—is there a set of truth values (TRUE and FALSE) for the terms in the expression such that the expression as a whole is TRUE? Cook showed that any problem in *NP* can be reduced to the satisfiability problem, which is also in *NP*.

The year after Cook's discovery, Richard Karp of the University of California, Berkeley, showed that some 20 other problems were *NP*-complete.[89] Since then the roster of *NP*-complete problems has greatly expanded, as Michael Garey and David Johnson of Bell Laboratories documented so thoroughly in their book *Computers and Intractability* (1979).[90]

So *P* is in *NP*. But *is P identical to NP*? That is, are there polynomial time algorithms for all *NP*-problems? This is the so-called *P* =? *NP problem*, arguably the most celebrated unsolved problem ("open problem" in mathematical jargon) in theoretical computer science: celebrated enough to have featured in an episode of the television series *Elementary* (an Americanized Sherlock Holmes series)! *If* for instance the satisfiability problem is shown to be in *P* then the answer to the *P* =? *NP* would be "yes." But no one has yet proved that *P* ≠ *NP*. If the latter is the case, as is widely believed, this would mean that there are problems (in *NP*) that are *inherently* intractable—inherently really difficult. And if they are *NP*-complete then all other problems reducible to them are also intractable.

So how does one show that a problem is in *NP*? What kind of reasoning is involved in this, the most abstract realm of formal computer science?

Consider, for example, the claim that the satisfiability problem (SAT) is in *NP*: Is there an assignment of truth values (TRUE, FALSE) to the variables in a Boolean expression that will make the expression TRUE?

Here, the barest gist of the argument is offered, taking great liberty with the proof presented in one of the most influential books on algorithms and complexity theory of the 1970s, authored by Alfred Aho of Bell Laboratories, John Hopcroft of Cornell University, and Jeffrey Ullman of Stanford University.[91]

Consider the set of all *satisfiable* Boolean expressions Σ. That is, Σ contains expressions whose values are always TRUE. A NDTM begins by "guessing" an assignment of values TRUE and FALSE to the distinct variables that appear in an input expression σ in Σ. (By "guessing" a correct assignment of values means to try out all possible assignment of values to the variables in parallel. Thus nondeterminism allows one to "parallelize" the problem situation.) The assigned values of the variables are then substituted for the corresponding variables in σ. The resulting expression is evaluated to verify its value as TRUE. As it happens, this evaluation can be done in polynomial time (as a function of the number of variables in σ).[92] Thus there is a NDTM that can solve SAT in polynomial time. Hence the claim that SAT is in *NP* is proved.

As for showing that SAT is *NP*-complete, this entails showing that every other problem Π in *NP* can be reduced in polynomial time to SAT. The proof is long and quite involved, but the way it goes is to consider a NDTM that solves Π in polynomial time, take an instance π of Π, and prove that there is a polynomial time algorithm for the NDTM that constructs a Boolean expression for π.[93]

NOTES

1. I. Asimov, 1989. "Foreword," pp. vii–viii in C. B. Boyer, *A History of Mathematics* (2nd ed.). Revised by U. C. Merzbach. New York: Wiley, p, vii.

2. C. A. R. Hoare, 1969. "An Axiomatic Approach to Computer Programming," *Communications of the ACM*, 12, 10, pp. 576–580, 583, esp. p. 576.

3. G. H. Hardy, [1940] 1969. *A Mathematician's Apology* (with a foreword by C. P. Snow). Cambridge: Cambridge University Press, pp. 150–151.

4. Knuth's writings in and about computer science is vast. For glimpses into his "Knuthian" philosophy, see, e.g., D. E. Knuth, 1996. *Selected Writings on Computer Science*. Stanford, CA: Center for the Study of Language and Information.

5. A. H. Eden, 2007. "Three Paradigms of Computer Science," *Mind and Machine*, 17, 2, pp. 135–167.

6. A classic discussion of the verification principle is A. J. Ayer, [1936] 1971. *Language, Truth and Logic*. Harmondsworth, UK: Penguin Books.

7. P. W. Bridgeman, 1927. *The Logic of Modern Physics*. New York: Macmillan.

8. P. Lucas and K. Walk, 1969. "On the Formal Description of PL/1," *Annual Review of Automatic Programming*, 6, 3, pp. 105–182.

9. P. E. Lucas, 1972. "On the Semantics of Programming Languages and Software Devices," pp. 41–58 in R. Rustin (ed.), *Formal Semantics of Programming Languages*. Englewood Cliffs, NJ: Prentice-Hall.

10. P. Wegner, 1972. "The Vienna Definition Language," *ACM Computing Surveys*, 4, 1, pp. 5–67.

11. Lucas and Walk, 1969, p. 105. Italics added.

12. D. E. Knuth, 1968. "The Semantics of Context Free Languages," *Mathematical Systems Theory*, 2, 2, pp. 127–145.

13. J. McCarthy, 1963. "Towards a Mathematical Science of Computation," pp. 21–28 in *Proceedings of the 1962 IFIP Congress*. Amsterdam: North-Holland; P. J. Landin, 1964. "The Mechanical Evaluation of Expressions," *Computer Journal*, 6, 4, pp. 308–320.

14. A personal note: I realized *how* nontrivial this task was when, in 1973–1974, working on a master's thesis on the design of a high-level microprogramming language, I constructed a VDL-based operational semantics of my language. With much effort I completed the construction but found the complexity of the resulting semantics far too obscure and cumbersome for my taste. Eventually, and with much regret, I chose to omit this lengthy semantic definition from my thesis!

15. This example is adapted from Lucas and Walk, 1969, pp. 154–155.

16. A. Birman, 1974. "On Proving Correctness of Microprograms," *IBM Journal of Research & Development*, 9, 5, pp. 250–266; G. B. Leeman, W. C. Carter, and A. Birman, 1974. "Some Techniques for Microprogram Validation," pp. 76–80 in *Information Processing 74* (Proceedings of the IFIP Congress). Amsterdam: North-Holland; G. B. Leeman, 1975. "Some Problems in Certifying Microprograms," *IEEE Transactions on Computers*, C-24, 5, pp. 545–553; W. H. Joyner, W. C. Carter, and G. B. Leeman, 1976. "Automated Proofs of Microprogram Correctness," pp. 51–55 in *Proceedings of the 9th Annual Microprogramming Workshop* (MICRO-9). New York: ACM/IEEE; W. C. Carter, W. H. Joyner, and D. Brand, 1978. "Microprogram Verification Considered Necessary," pp. 657–644 in *Proceedings of the National Computer Conference*. Arlington, VA: AFIPS Press.

17. R. Milner, 1971. "An Algebraic Definition of Simulation Between Programs," pp. 471–481 in *Proceedings 2nd International Joint Conference on Artificial Intelligence* (IJCAI-71). San Francisco: Morgan Kauffman.

18. D. S. Scott and C. Strachey, 1971. "Toward a Mathematical Semantics for Computer Languages." Tech. Monograph PRG-6, August. Oxford: Oxford University Computing Laboratory.

19. McCarthy, 1963.

20. J. E. Stoy, 1977. *Denotational Semantics: The Scott–Strachey Approach to Programming Language Theory*. Cambridge, MA: MIT Press.

21. This example is drawn from J. de Bakker, 1980. *The Mathematical Theory of Program Correctness*. London: Prentice-Hall International.

22. Hoare, 1969, p. 576.

23. M. A. Arbib and S. Alagic, 1979. "Proof Rules for **goto**s." *Acta Informatica*, 11, pp. 138–148.

24. S. Alagic and M. A. Arbib, 1978. *The Design of Well Structured and Correct Programs*. New York: Springer-Verlag.

25. For example, S. Owicki and D. Gries, 1976. "An Axiomatic Proof Technique for Parallel Programs," *Acta Informatica*, 6, pp. 319–340; R. Milner, 1980. *A Calculus of Communicating Systems*. Berlin: Springer-Verlag; S. Owicki and L. Lamport, 1982. "Proving Liveness Properties of Concurrent Programs," *ACM Transactions on Programming Languages and Systems*, 4, 3, pp. 455–495; G. R. Andrews and F. B. Schneider, 1983. "Concepts and Notatons for Parallel Programming," *ACM Computing Surveys*, 15, 1, pp. 3–44; N. Sounderajan, 1984. "A Proof Technique for Parallel Programs," *Theoretical Computer Science*, 31, 1&2, pp. 13–29; C. A. R. Hoare, 1986. *Communcating Sequential Processes*. Englewood Cliffs, NJ: Prentice-Hall International; R. M. Chandy and J. Misra, 1988. *Parallel Program Design: A Foundation*. Reading, MA: Addison-Wesley.

26. Owicki and Gries, 1976.

27. J. E. Donahue, 1976. *Complementary Definitions of Programming Language Semantics*. New York: Springer-Verlag. A similar perspective was offered by E. A. Ashcroft and W. W. Wadge, 1982. "Rx for Semantics," *ACM Transactions on Programming Languages and Systems*, 4, 2, pp. 283–294.

28. J. de Bakker, 1980. *Mathematical Theory of Program Correctness*. London: Prentice-Hall International.

29. E. W. Dijkstra, 1976. *A Discipline of Programming*. Englewood Cliffs, NJ: Prentice-Hall.

30. S. Dasgupta, 1991. *Design Theory and Computer Science*. Cambridge: Cambridge University Press.

31. See, in particular, Alagic and Arbib, 1978; C. B. Jones, 1980. *Software Development: A Rigorous Approach*. Englewood Cliffs, NJ: Prentice-Hall International; D. Gries, 1981. *The Science of Programming*. New York: Springer-Verlag; K. R. Apt, 1981. "Ten Years of Hoare Logic," *ACM Transactions on Programming Languages and Systems*, 3, 4, pp. 431–483; R. C. Backhouse, 1986. *Program Construction and Verification*. Englewood Cliffs, NJ: Prentice-Hall International; C. B. Jones, 1986. *Systematic Software Development Using VDM*. Englewood Cliffs, NJ: Prentice-Hall International; C. A. R. Hoare, 1987. "An Overview of Some Formal Methods of Program Design," *Computer*, 20, 9, pp. 85–91.

32. S. Dasgupta, 1983. "On the Verification of Computer Architectures Using an Architecture Description Language," pp. 158–167 in *Proceedings of the 10th Annual International Symposium on Computer Architecture*, New York: IEEE Computer Society Press; S. Dasgupta, 1984. *The Design and Description of Computer Archietctures*. New York: Wiley.

33. D. A. Patterson, 1976. "STRUM: A Structured Microprogram Development System for Correct Firmware," *IEEE Transaction on Computers*, C-25, 10, pp. 974–985; S. Dasgupta and A. Wagner, 1984. "The Use of Hoare Logic in the Verification of Horizontal Microprograms," *International Journal of Computer & Information Sciences*, 13, 6, pp. 461–490; S. Dasgupta, P. A. Wilsey, and J. Heinanen, 1986. "Axiomatic Specifications in Firmware Development Systems," *IEEE Software*, 3, 4, pp. 49–58; W. Damm, G. Doehman, K. Merkel, and M. Sichelschmidt, 1986. "The AADL/S* Approach to Firmware Design Verification," *IEEE Software*, 3, 4, pp. 27–37.

34. R. E. Shostak, 1983. "Formal Verification of Circuit Designs," pp. 13–30 in T. Uehara and M. R. Barbacci (eds.), *Computer Hardware Description Languages and Their Applications* (CHDL-83). Amsterdam: North-Holland.

35. T. S. Kuhn, [1962] 2012. *The Structure of Scientific Revolutions* (4th ed.). Chicago: University of Chicago Press.

36. S* was described in some detail by its designer in S. Dasgupta, 1980. "Some Aspects of High Level Microprogramming." *ACM Computing Surveys*, 12, 3, pp. 295–324.

37. This logic is fully and rigorously described in W. Damm, 1988. "A Microprogramming Logic," *IEEE Transactions on Software Engineering*, 14, 5, pp. 559–574.

38. W. Damm, 1985. "Design and Specification of Microprogrammed Computer Architectures," pp. 3–10 in *Proceedings of the 18th Annual Microprogramming Workshop* (MICRO-18). Los Alamitos, CA: IEEE Computer Society Press; W. Damm and G. Doehman, 1985. "Verification of Microprogrammed Computer Architectures in the AADL/S* System: A Case Study," pp. 61–73 in *Proceedings of the 18th Annual Microprogramming Workshop* (MICRO-18). Los Alamitos, CA: IEEE Computer Society Press.

39. The classic introductory text on switching theory from the 1960s was E. J. McCluskey, 1965. *Introduction to the Theory of Switching Circuits*. New York: McGraw-Hill. An advanced monograph from the same period was J. Hartmanis and R. E. Stearns, 1966. *Algebraic Structure Theory of Sequential Machines*. Englewood Cliffs, NJ: Prentice-Hall. The state of the art of switching theory and logic design at the beginning of the 1970s was comprehensively and elegantly captured in Z. Kohavi, 1970. *Switching and Finite Automata Theory*. New York: McGraw-Hill.

40. G. G. Langdon, Jr., *Logic Design: A Review of Theory and Practice*. New York: Academic Press, pp. 105–107.

41. See S. Dasgupta, 1992. "Computer Design and Description Languages," pp. 91–155 in M. C. Yovits (ed.), *Advances in Computers, Volume 21*. New York: Academic Press. See also *Computer*, 18, 2, 1985, a special issue of the journal on CHDLs edited by Mario Barbacci.

42. M. Shahdad, R. Lipsett, E. Marschner et al., 1985. "VHSIC Hardware Description Language," *Computer*, 18, 2, pp. 94–104.

43. C. A. Mead and L. Conway, 1980. *Introduction to VLSI Systems*. Reading, MA: Addison-Wesley.

44. Ibid.; W. W. Lattin, J. A. Bayliss, D. L. Budde et al., 1981. "A 32b VLSI Micro-mainframe Computer System," *Proceedings of the 1981 IEEE International Solid State Circuits Conference*, pp. 110–111.

45. L. Snyder, 1984. "Supercomputers and VLSI: The Effect of Large Scale Integration on Computer Architectures," pp. 1–33 in M. C. Yovits (ed.), *Advances in Computers, Volume 23*. New York: Academic Press.

46. A truly scintilliating tutorial on the hierarchical structure of IC systems from the level of device physics to logical subsystems published the same year as Mead and Conway's book was Wesley Clark, 1980. "From Electron Mobility to Logical Structure: A View of Integrated Circuits," *ACM Computing Surveys*, 12, 3, pp. 325–356.

47. J. A. Rowson, 1980. "Understanding Hierarchical Design." Tech. Report (PhD dissertation). California Institute of Technology, Pasadena, CA.

48. H. B. Curry and R. Feys, 1958. *Combinatory Logic, Volume 1.* Amsterdam: North-Holland.

49. V. Pitchumani and E. P. Stabler, 1984. "An Inductive Assertion Method for Register Transfer Level Design Verification," *IEEE Transactions on Computers,* C-32, 12, pp. 1073–1080.

50. R. E. Shostak, 1983. "Formal Verification of Circuit Designs," pp. 13–30 in T. Uehara and M. R. Barbacci (eds.), *Computer Hardware Description Languages and the Applications* (CHDL-83, 6th International Symposium). Amsterdam: North-Holland.

51. M. J. C. Gordon, 1979. *The Denotational Description of Programming Languages.* Berlin: Springer-Verlag.

52. M. J. C. Gordon, 1986. "Why Higher-Order Logic is a Good Formalism for Specifying and Verifying Hardware," pp. 153–177 in G. J. Milne and P. A. Subhramanayam (eds.), *Formal Aspects of VLSI Design.* Amsterdam: North-Holland; A. Camilleri, M. J. C. Gordon and T. Melham, 1987. "Hardware Verification Using Higher-Order Logic," pp. 43–67 in D. Borrione (ed.), *From HDL Descriptions to Guaranteed Circuit Designs.* Amsterdam: North-Holland; M. J. C. Gordon, 1988. "HOL—A Proof Generating System for Higher Order Logic," pp. 73–128 in G. Birthwhistle and P. A. Subrahamanyam (eds.), *VLSI Specification, Verification and Synthesis.* Boston, MA: Kluwer Academic Publishers; J. J. Joyce, 1988. "Formal Verification and Implementation of a Microprocessor," pp. 129–157 in Birtwhistle and Subrahamanayam, 1988.

53. This is taken from Joyce, 1988.

54. C. A. R. Hoare, 1986. *The Mathematics of Programming.* Oxford: Clarendon.

55. C. Alexander, 1964. *Notes on the Synthesis of Form.* Cambridge, MA: Harvard University Press.

56. P. Brinch Hansen, 1977. *The Architecture of Concurrent Programs.* Englewood Cliffs, NJ: Prentice-Hall.

57. L. March, 1977. "A Boolean Description of a Class of Built Forms," pp. 41–73 in L. March (ed.), *The Architecture of Form.* Cambridge: Cambridge University Press; J. P. Steadman, 1977. "Graph-Theoretic Representation of Architectural Arrangements," pp. 94–115 in March (ed.), 1977.

58. W. J. Mitchell, 1990. *The Logic of Architecture.* Cambridge, MA: MIT Press.

59. D. G. Gries, 1986. *The Science of Programming.* New York: Springer-Verlag.

60. R. De Millo, R. J. Lipton, and A. J. Perlis, 1979. "Social Processes and Proofs of Theorems and Programs," *Communications of the ACM,* 22, 5, pp. 271–280.

61. Kuhn, 1962.

62. B. Latour and S. Woolgar, [1979] 1986. *Laboratory Life: The Construction of Scientific Facts* (2nd ed.) Princeton, NJ: Princeton University Press.

63. De Millo et al., 1979, p. 271.

64. Ibid.

65. Ibid.

66. I. Lakatos, 1976. *Proofs and Refutations.* Cambridge: Cambridge University Press.

67. De Millo et al., 1979, p. 274.

68. Ibid., p. 272.

69. C. A. R. Hoare, 1971. "Proof of a Program: FIND," *Communications of the ACM,* 14, 1, pp. 39–45.

70. De Millo et al., 1979, p. 278.

71. F. DeRemer and H. Kron, 1975. "Programming-in-the-Large Versus Programming-in-the-Small," *ACM SIGPLAN Notices,* 10, 6, pp. 114–121.

72. De Millo et al., 1979, p. 277.

73. Among the formalists missing from the ACM Forum was C. A. R. Hoare who, however, reviewed the De Millo et al. paper in the *ACM Computing Reviews*, 20, 1, August, 1979.

74. W. Haken, K. Appel, and J. Koch, 1977. "Every Planar Map is Four Colorable," *Illinois Journal of Mathematics*, 21, 84, pp. 429–467. Also K. Appel, 1984. "The Use of the Computer in the Proof of the Four-Color Theorem," *Proceedings of the American Philosophical Society*, 178, 1, pp. 35–39.

75. W. L. Scherlis and D. S. Scott, 1983. "First Steps Towards Inferential Programming," pp. 199–212 in *Information Processing 83* (Proceedings of the IFIP Congress). Amsterdam: North-Holland.

76. See, e.g., de Bakker, 1980; Gries, 1981; Hoare, 1986.

77. R. S. Boyer and J. S. Moore, 1981. *The Correctness Problem in Computer Science.* New York: Academic Press.

78. J. H. Fetzer, 1988. "Program Verification: The Very Idea," *Communications of the ACM*, 37, 9, pp. 1048–1063.

79. Ibid., p. 1049.

80. K. R. Popper, 1965. *Conjectures and Refutations: The Growth of Scientific Knowledge.* New York: Harper & Row; K. R. Popper, 1968. *The Logic of Scientific Discovery.* New York: Harper & Row.

81. As was the case with the De Millo et al. paper, Fetzer's article spawned considerable, even violent, reactions, mostly from the formalists' camp. For example, the ACM Forum and Technical Correspondence section of the March 1989 issue of the *Communications of the ACM* was dedicated to letters on the Fetzer thesis.

82. K. R. Popper, 1972. *Objective Knowledge.* Oxford: Clarendon.

83. M. Minsky, 1967. *Computation: Finite and Infinite Machines.* Englewood Cliffs, NJ: Prentice-Hall; J. E. Hopcroft and J. D. Ullman, 1969. *Formal Languages and Their Relation to Automata.* Reading, MA: Addison-Wesley.

84. J. Hartmanis and R. E. Stearns, 1965. "On the Computational Complexity of Algorithms," *Transactions of the American Mathematical Society*, 177, pp. 285–306.

85. The historical circumstances in which Alan Turing created the abstract machine named after him and the general structure of this machine are described in S. Dasgupta, 2014. *It Began with Babbage: The Genesis of Computer Science.* New York: Oxford University Press, chapter 4.

86. M. O. Rabin and D. Scott, 1959. "Finite Automata and Their Decision Problems," *IBM Journal*, 3, 2, pp. 114–125.

87. A. V. Aho, J. E. Hopcroft, and J. D. Ullma, 1974. *The Design and Analysis of Computer Algorithms.* Reading, MA: Addison-Wesley, pp. 364–365.

88. S. Cook, 1971. "The Complexity of Theorem Proving Procedures," *Proceedings, 3rd ACM Symposium on Theory of Computing*, Shaker Heights, Ohio, pp. 151–158.

89. R. M. Karp, 1972. "Reducibility Among Combinatorial Problems," pp. 85–104 in R. E. Miller and J. W. Thatcher (eds.), *Complexity of Computer Computations.* New York: Plenum.

90. M. R. Garey and D. S. Johnson, 1979. *Computers and Intractability: A Guide to the Theory of NP-Completeness.* San Francisco: W. H. Freeman.

91. Aho, Hopcroft, and Ullman, 1974, p. 376.

92. Ibid.

93. Ibid., pp. 379–383.

6

A SYMBOLIC SCIENCE OF INTELLIGENCE

6.1   THE SYMBOL SYSTEM HYPOTHESIS

Human Problem Solving (1972) by Allen Newell and Herbert Simon of Carnegie-Mellon University, a tome of over 900 pages, was the *summa* of some 17 years of research by Newell, Simon, and their numerous associates (most notably Cliff Shaw, a highly gifted programmer at Rand Corporation) into "how humans think."[1]

"How humans think" of course belonged historically to the psychologists' turf. But what Newell and Simon meant by their project of "understanding . . . how humans think" was very different from how psychologists envisioned the problem before these two men invaded their milieu in 1958 with a paper on human problem solving in the prestigious *Psychological Review*.[2] Indeed, professional psychologists must have looked at them askance. Neither was formally trained in psychology. Newell was originally trained as a mathematician, Simon as a political scientist. They both disdained disciplinary boundaries. Their *curricula vitae* proclaimed loudly their intellectual heterodoxy. At the time *Human Problem Solving* was published, Newell's research interests straddled artificial intelligence, computer architecture, and (as we will see) what came to be called cognitive science. Simon's multidisciplinary creativity—his reputation as a "Renaissance man"—encompassing administrative theory, economics, sociology, cognitive psychology, computer science, and the philosophy of science—was of near-mythical status by the early 1970s.[3] Yet, for one prominent historian of psychology it would seem that what Newell and Simon did had nothing to do with the discipline: the third edition of Georgetown University psychologist Daniel N. Robinson's *An Intellectual History of Psychology* (1995) makes no mention of Newell or Simon.[4] Perhaps this was because, as Newell and Simon explained, their study of thinking adopted a pointedly information processing perspective.[5]

Information processing: Thus entered the computer into this conversation. But, Newell and Simon hastened to clarify, they were not suggesting a metaphor of humans as computers. Rather, they would propose an information processing system (IPS) that would serve to describe and explain how humans "process task-oriented symbolic information."[6] In other words, human problem solving, in their view, is an instance of representing information as *symbols* and processing them. The computer was not intended to be a metaphor; but they had grounded their theory in computer science because computer science afforded "possible problem solving mechanisms" that would suggest the kinds of mechanisms are actually used by humans.[7]

In fact the part of computer science Newell and Simon were alluding to was artificial intelligence (AI)—"devoted to getting computers (or other devices) to perform tasks requiring intelligence."[8]

This was the widely accepted working definition of AI in the early 1970s when *Human Problem Solving* was published. For instance, James Slagle of Johns Hopkins University began his *Artificial Intelligence: The Heuristic Programming Approach* (1971) with the announcement that his book "shows that machines are doing things often called intelligent when done by human beings."[9] This working definition seemed to remain stable over the next decade. Thus Avron Barr and Edward Feigenbaum of Stanford University, as editors of the three-volume *Handbook of Artificial Intelligence* (1983), began their introduction by noting that AI was the branch of computer science concerned with the design of systems that manifest behavior one normally associates with human intelligence.[10]

So, in a certain sense, for Newell and Simon, it would seem that AI was *a means to a particular end*: providing possible mechanisms as models to explain human thinking. AI and the Newell–Simon style of psychology appeared to have different aims: the former to build "intelligent" computational artifacts, the latter to gain understanding of human intelligence. One in the realm of the artificial, the other in the realm of the natural.

The same year *Human Problems Solving* appeared Marvin Minsky and Seymour Papert of MIT produced *Artificial Intelligence: Progress Report* (1972), an account of researches conducted at MIT's Artificial Intelligence Laboratory.[11] The Laboratory's goal, they stated, was "understanding the principles of intelligence." And under the latter they included artificial and natural intelligence.[12]

In fact, like that of Minsky and Papert, the Newell–Simon vision encompassed both. For them, computing and human thinking shared something fundamental: They were both symbol processing systems. They articulated this sense of unity between human and artificial intelligence in their 1976 Turing Award lecture by proposing a qualitative principle:

The Physical Symbol System Hypothesis. A physical symbol system (PSS) has the necessary and sufficient means for general intelligent action.[13]

6.2 WHAT *IS* INTELLIGENCE, THEN?

This still begged the question: What, for Newell and Simon, for Slagle, for Minsky and Papert, for Barr and Feigenbaum, constituted intelligence? What was it, they believed, *to be* intelligent?

A common way of answering this was by way of examples. Minsky and Papert, for instance, cited as examples of AI problems they and their collaborators were investigating circa 1972:

Robotics, computer vision, mechanical manipulation, learning models, induction, analogy, and so on.

A hodgepodge of technologies, techniques, and reasoning strategies. As for natural intelligence, they gave as examples of the kinds of things that interested them, models of commonsense thinking and natural language understanding.[14]

For James Slagle, instances of intelligent action and behavior included answering questions by deduction from facts, game playing, proving theorems, and balancing assembly lines.[15]

Barr and Feigenbaum mentioned "understanding language, learning, reasoning, solving problems."[16] As for Newell and Simon, their case studies in intelligent behavior in *Human Problem Solving* entailed three specific tasks: solving cryptarithmetic problems, proving propositions in logic, and playing chess.

There is something unsatisfactory—something unscientific—about characterizing intelligence by way of examples. The scientist—natural or artificial—seeks unity amid diversity. Surely there has to be something common to these phenomena that renders them all as manifestations of intelligent behavior.

Slagle appealed to his dictionary and accepted what it had to say: that intelligence entails (a) the ability of an agent adapting to a novel situation by appropriate adjustments to behavior; and/or (b) having the capacity to detect relationships between given facts that facilitate actions in response to goals.[17]

Newell and Simon, writing their 900+-page text in 1972, seemed reluctant to commit themselves to an explicit definition. Rather, we had to tease out a sense of what they thought what intelligence is. What we find is that, like Slagle, *adaptive behavior* in response to goals or problems was their defining feature,[18] a view they reiterated more explicitly in their Turing Award lecture of 1976, relating *goal-directed* adaptive behavior to intelligence.[19]

6.3   SO WHAT IS A PHYSICAL SYMBOL SYSTEM?

So let us return to the PSS hypothesis in the light of this identification of intelligence with goal-oriented (purposive or teleological) adaptive behavior. According to the PSS hypothesis a PSS has all that is necessary and sufficient for such adaptive—hence intelligent—behavior. It seems then that the entire burden of intelligent action was assigned by Newell and Simon to PSSs. But what *is* such a system?

In fact, what Newell and Simon called "physical symbol system" in 1976 they had termed *information processing system* (IPS) in 1972—a system comprising memory, processor, and "effectors" and "receptors."[20]

They then elaborate:

(1) There is a set of elements called *symbols*.

(2) A *symbol structure* consists of . . . symbols connected by a set of *relations*.

(3) A *memory* is a component of an IPS capable of storing and retaining symbol structures.

(4) An *information process* is a process that has symbol structures for . . . its inputs and outputs.

(5) A *processor* . . . consists of (a) a fixed set of *elementary information processes* (eip's); (b) a *short-term memory* (STM) that holds the input and output symbol structures of the eip's; (c) an *interpreter* that determines the sequence of eip's to be executed by the IPS as a function of the symbol structures in STM.

As for a symbol structure, it either designates (or refers to) some other object stored in an IPS memory that can be affected by information processes—they are data structures—or is a program.[21]

It is quite possible that a psychologist, unfamiliar with Newell and Simon's previous writings, reading this in 1972, would be quite disconcerted as to its bearing on human cognition. A computer scientist, on the other hand, would experience a shock of recognition, for what Newell and Simon described as an IPS (in 1972) and as a PSS (in 1976) was a specification of the architecture of a computer at (roughly) the ISP level of abstraction (Chapter 3, Section 3.5). We should not be too surprised by this: Recall that Allen Newell coauthored *Computer Structures: Readings and Examples* in which the ISP level was specified in some detail.

And just as a computer architecture is a liminal artifact—intrinsically abstract yet bound causally to physical processes in physical machines—so also an IPS is a liminal artifact. The point of the word *physical* in "physical symbol system" was to remind the reader of this liminal nature of IPS or PSS: the fact that such systems obey physical laws and that they are not confined to human symbol systems.[22]

6.4   ARTIFACTS AND HUMANS: TWO CLASSES OF PHYSICAL SYMBOL SYSTEMS

What grounds were there for Newell and Simon to believe in the PSS hypothesis? They were unequivocal on this: The PSS hypothesis was an empirical one; the evidence in its support was—had to be—empirical.

As theorizers both Newell and Simon were grounded in the empirical world. They titled their Turing Award lecture "Computer Science as *Empirical* Inquiry" (my italics). The evidence in support of (or, if one was a Popperian, that refuted) the PSS hypothesis lay in experiments. In their 1976 lecture they summarized the empirical evidence gleaned over the previous 20 or so years. Some of this evidence addressed the *suffi-ciency* aspect of the PSS hypothesis; others focused on the *necessity* condition.

The sufficiency condition could be tested by constructing symbol processing artifacts that manifested intelligent behavior. As for the necessity condition, the most obvious source of evidence lay in humans—"Man, the intelligent system best known to us."[23] Here, one was required to demonstrate that "his cognitive activities can be explained as the workings of a physical symbol system."[24]

In these two domains—the artifactual and the human (natural)—they proclaimed, lay the Janusian faces of the science of intelligence: The search for sufficiency evidence was the province of artificial intelligence; the quest for necessity evidence lay in the domain of cognitive psychology.[25]

There was, of course, a huge difference in methodology. In AI the method entailed identification of a task domain that demanded intelligence, then constructing a pro-gram to perform tasks in that domain.[26] In fact, this had been the methodological agenda of AI from its very beginnings in 1956. Newell and Simon pointed out that AI had evolved from relatively "easy" tasks—"puzzles and games, . . . problems of sched-uling and allocating resources, simple induction tasks"—to arguably more difficult tasks such as natural language understanding, visual scene analysis, design, speech un-derstanding, and so on.[27] In contrast, in cognitive psychology, the methodology would be that of "standard" experimental psychology but with a vital twist: by designing and performing experiments on human subjects requiring intelligence and demonstrating that such intelligent behavior was explainable in terms of symbol processing theories.

6.5 THE HEURISTIC SEARCH HYPOTHESIS

But how would a PSS go about its task of symbol processing? The answer lay in an-other qualitative law, a companion to the PSS hypothesis. This too was presented in a compact manner by Newell and Simon in their Turing Award lecture, but, like the PSS hypothesis, it was an encapsulation of many years of research by them and others located elsewhere:

Heuristic Search (HS) *Hypothesis*: A PSS exercises its intelligence in problem solving by HS—that is, by generating and progressively modifying symbol structures until it produces a solution structure.[28]

HS was a practical necessity engendered by the limited computational (in the case of machines) and cognitive (in the case of humans) resources. Herbert Simon's theory of *bounded rationality* originally formulated in the context of administrative decision making in the mid-1940s, later expanded to economic decision making (for which he received a Nobel Prize in 1979) and thence to problem solving in general,[29] applied as much to computational artifacts as to human beings.

In deploying heuristic search, Newell and Simon argued, lay the second common ground between human and machine intelligence. Their book *Human Problem Solving* was a paen to the primacy of heuristic search in problem solving.

But by then, HS as a distinct programming *strategy* was soundly and deeply ensconced in the AI community's consciousness. The distinction from the algorithmic mentality was so stark it would not be remiss to speak of HS as a distinctive design style for the construction of computational artifacts. It was not, however, accepted by all and sundry in the AI community of the time as the sole means by which intelligence is exercised. There were (as we will later see) other schools of thought. When James Slagle titled his 1971 textbook *Artificial Intelligence: The Heuristic Programming Approach*,[30] he was implying (and would state this explicitly in the text) the existence of other approaches. But by the beginning of the 1970s HS and programming had developed into a powerful body of knowledge and techniques. Slagle's book offered a plethora of examples of problem solving using HS—in playing board and card games, solving problems in geometry and calculus, proving theorems, balancing assembly lines, and so on.

But for Newell and Simon, at the time *Human Problem Solving* was published, the "jewel in the crown" of their PSS/HS theory of intelligent behavior was undoubtedly the "General Problem Solver" (GPS), which was conceived around 1960[31] and evolved and was applied through the '60s, by the Newell–Shaw–Simon team along with George Ernst, Newell's student at Carnegie-Mellon University. This work culminated in a monograph by Ernst and Newell titled *GPS: A Case Study in Generality and Problem Solving* (1969).[32]

6.6   GPS'S LEGACY: WEAK METHODS

As the name suggests, GPS was a task-independent problem-solving program—the first of its kind—that was designed to solve problems in a variety of task domains. Its most significant contribution to the science of intelligence was its recognition and implementation of a class of search heuristics Newell and Simon in 1976 called *weak methods*. They are "weak" in that they are heuristics one must fall back on when knowledge about the task environment is scarce. Or, more interestingly, they are heuristics that in some fashion are derived from experience and common sense logic.

One such weak method we have already encountered is the heuristic called *divide and conquer*. The idea is to decompose a task into simpler subtasks that (we expect) are simpler to solve. It was one of the foundations of the structured programming methodology invented by Edsger Dijkstra, Nicklaus Wirth, and others (Chapter 2, Section 2.6).

But GPS's major legacy to heuristic computing was a weak method Newell et al. called *means–ends analysis* (MEA). This can be stated informally as follows:

MEA: The problem is characterized by an initial state and a goal state. At the start of the problem-solving process this initial state becomes the current state. In the course of the computation any state reached is the current state. Given a current state and a goal state, determine the difference between the two. Then attempt to reduce this difference by applying a relevant operator. If the application of the operator requires a precondition not satisfied by the current state, then reduce the difference between the current state and the precondition by applying, recursively, one or more operators.

In fact, MEA can be more compactly formulated as a recursive algorithm:

---

means-ends-analysis (current-state, goal-state):
        compute difference between current-state, goal-state;
        apply operator to reduce difference to produce (new) current-state;
        if current-state = goal-state **then return** 'done'
                else *means-ends-analysis* (current-state, goal-state)
Initially, *means-ends-analysis* (initial-state, goal-state) will be activated

---

We consider here as a very simple example of the MEA in action, the "Towers of Hanoi" problem that James Slagle discussed briefly and that became, in the 1970s, a textbook exemplar of a problem to which MEA could be applied.

Three disks of different sizes, L(arge), M(edium), and S(mall), are stacked upon a "tower" **A** with the largest disk at the bottom, the smallest on top. There are two other towers, **B** and **C**, which are initially unoccupied. The task is to move the disks from **A** to **C** such that the disks are arranged on **C** in the same order as they are initially on **A**. But the disks can be moved among the towers under the conditions that (i) only one disk can be moved at a time, and (ii) a larger disk cannot be stacked on top of a smaller disk.

Let us use the following notation: If on a tower **T** two or more disks α, β, γ, . . . , are stacked such that the largest is at the bottom, the next largest is just above it, and so on, then we represent this situation as $T(\alpha > \beta > \gamma > \ldots)$. If a tower **T** has no disks on it, we represent this as **T**( ).

So the initial state Si and goal state Sg are, respectively,

$$\text{Si: } \mathbf{A}(L > M > S), \mathbf{B}(\ ), \mathbf{C}(\ ),$$
$$\text{Sg: } \mathbf{A}(\ ), \mathbf{B}(\ ), \mathbf{C}(L > M > S).$$

The MEA heuristic when applied here will work as follows:

To go from Si to Sg, L must first be moved from **A** to **C**. A precondition for this move to occur is that there should not be any disk on top of L on **A** and **C** must be empty. That is, the following *subgoal* state must be achieved:

$$\text{S1: } \mathbf{A}(L), \mathbf{B}(M > S), \mathbf{C}(\ ).$$

A precondition to attaining S1 is that M should be free to be moved from **A** to **B** and **B** must be empty. That is, the following state must be attained:

$$\text{S2: } \mathbf{A}(L > M), \mathbf{B}(\ ), \mathbf{C}(S).$$

This is achieved by moving S from **A** to **C**. Now M can be moved, producing the state

$$S3: A(L), B(M), C(S).$$

The subgoal now is to transform $S3$ to $S1$. Disk S is moved from **C** to **B**. $S1$ is achieved. So the precondition for moving L from **A** to **C** is satisfied, and so L can now be moved. The result is the state

$$S4: A(\ ), B(M > S), C(L).$$

When Sg is compared with $S4$, M must be moved from **B** to **C**. The precondition for this is that M must be free to be moved from **B** to **C**. A possible state for this is

$$S5: A(S), B(M), C(L).$$

This is achieved by moving S from **B** to **A**. The precondition for M to be moved from **B** to **C** is satisfied. M is now moved, leading to the state

$$S6: A(S), B(\ ), C(L > M).$$

When Sg is compared with $S6$, the former state is now reachable by moving S from **A** to **C**.

That this was a *search* through a *problem state space* is depicted in Figure 6.1. There is an overall Towers of Hanoi state space comprising all possible configurations of the three disks on the three towers. The actual state sequence found by using MEA is shown as the sequence $Si \rightarrow S2 \rightarrow S3 \rightarrow S1 \rightarrow S4 \rightarrow S5 \rightarrow S6 \rightarrow Sg$ meandering through a subspace of this state space.

So what was so "heuristic" about this solution? The point is that unlike real algorithms the knowledge of the task (problem) domain is not integrated into the

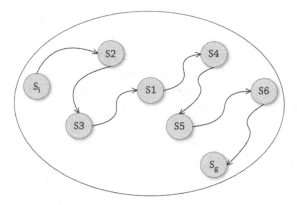

FIGURE 6.1 State space search for the Towers of Hanoi problem.

MEA strategy itself. Rather, the knowledge is in the *task environment*—the state space depicted in Figure 6.1—in which the MEA is let loose. Moreover, as important, unlike real algorithms there is no guarantee that in a given task domain MEA will terminate successfully. For example, given a pair of intermediate states S', S'' in the course of the search process, where the goal is to move from S' to S'', there may be more than one operator that might be applied to make this transition. The choice of which one to use may be the choice between success and failure. In the case of the Towers of Hanoi problem the task is simple enough that the differences between states can be identified and the "move" operator applied in an optimal fashion to lead directly through a successful state sequence to the goal state.

6.7   "SOMETHING WICKED THIS WAY COMES"

Quite coincidentally, a particular metaproblem—the problem of problems—was the subject of two very original—and, over time, widely referenced—papers published in 1973. Design theorist Horst Rittel and urban-planning theorist Melvin Weber, both of the University of California, Berkeley, wrote in the journal *Policy Sciences* about planning problems as inherently *wicked* problems.[33] And, writing in the journal *Artificial Intelligence*, Herbert Simon addressed the nature of *ill-structured* problems.[34]

Now, we have already encountered a metaproblem issue: Computational complexity theory distinguishes between *P*- and *NP*-classes of problems wherein the former comprises relatively easy problems and the latter contains intractable or really difficult problems. Moreover, the theory of *NP*-completeness shows the connectedness of certain kinds of *NP*-problems (Chapter 5, Section 5.16).

But the contrast Herbert Simon was interested in was between what he called *well-structured* and ill-structured problems. He identified several features of well-structuredness, most notably the following:[35]

(1) There is a definite criterion for testing any proposed solution to a well-structured problem and a mechanical process for applying the criterion.

(2) There is at least one problem state space in which the initial problem state, the goal state, and all other states reachable in the course of attempting to solve the problem can be represented. (See Section 6.6 and Figure 6.1.)

(3) Any knowledge the problem solver acquires about the problem can be represented in one or more problem spaces.

(4) If the actual problem involves acting on the external world then the state changes and the effects of the operators on states must cohere with the laws governing the external world.

(5) The problem-solving process requires only practical amounts of computation, and the information available to the process is available with practical amounts of search.

A problem is ill-structured if one or more of these attributes are violated. Indeed, the extent of ill-structuredness will be determined by the extent to which these attributes are violated.

Horst Rittel and Melvin Weber had their own term for well-structured problems. They are *benign* (or "tame"). For benign problems "the mission is clear. It is clear, in turn, whether or not the problems have been solved."[36]

But the metaproblem issue that Rittel and Weber were engaged with involved problems of a very different breed. Consider, in particular, planning problems: for example, determining the location and trajectory of a new freeway through a town; or preparing a proposal to expand the academic and research goals of a university. These are societal problems. And they are anything but benign. "Planning problems," Ritten and Weber state baldly, "are inherently wicked."[37]

This word wicked was more than an evocative metaphor for Rittel and Weber. In their hands it became a technical term. And they identified several distinguishing characteristics of wicked problems of which the most telling were the following:[38]

(a) There is no definite formulation of a wicked problem. In fact, formulating a wicked problem is itself the problem.

(b) Wicked problems have no terminating criteria. There is always the possibility of improving on the solution of such a problem.

(c) Consequently, solutions to wicked problems are potentially unbounded in variety. There are no means for ascertaining that all possible solutions have been identified.

(d) Solutions to wicked problems are neither true nor false. They are just good or bad.

(e) There is no testing for a solution to a wicked problem, for any proposed solution will provoke "waves of consequences" over, possibly, an extended period of time.

Having done seminal work in administrative decision theory,[39] Simon would have been thoroughly familiar with the kinds of planning problems Rittel and Weber were addressing. As a co-creator of the information processing approach to cognitive psychology he was also familiar with University of Michigan psychologist Walter Reitman's concept of "ill-defined problems" discussed in 1964–1965.[40] Drawing on Reitman's idea, Simon proposed the concept of the ill-structured problem that bore close similarity to the wicked-problem concept. And just as for Rittel and Weber wicked problems were problems that were not benign, so also for Simon; problems were ill-structured when they violated the significant features of well-structured problems.

But there was an important distinction between their respective concerns. Rittel and Weber published their paper in *Policy Sciences*: They were addressing readers of this journal, presumably policy and planning theorists and practitioners. Their goal was to raise their readers' consciousness about a fundamental *limit* to planning problems; and the wicked problems posed—as the title of their paper said—certain "Dilemmas in a General Theory of Planning." The largest dilemma was that there was no adequate theory "that might dispel wickedness."[41]

This was where Simon parted ways from Rittel and Weber: Simon's paper appeared in the journal *Artificial Intelligence,* and his agenda was quite *constructive*: To explore how problem-solving systems (that is, heuristic-based AI programs) could be adapted "in the domains usually regarded as ill-structured."[42]

Analyzing two types of problems with which he was thoroughly familiar, proving theorems in logic and playing chess (both discussed at length in *Human Problem Solving*), Simon made the point that whether these problems were well-structured or ill-structured was not a cut-and-dried matter. The answer was "It depends."

In the case of theorem proving, Simon's "well-structured conditions" (1) and (2) will be met under certain restrictions: Availability of an automatic proof checker, for instance, would satisfy condition (1), and limiting the theorems to be proved to (what logicians call) "well-formed formulas" (that is, legitimate formulas in that logic) might satisfy condition (2).

But what about condition (3)? If the theorem prover (human or machine) restricted the knowledge used to strictly the knowledge of the axiomatic (logical) system itself (its axioms, definitions, previously proved theorems, and rules of deduction) then theorem proving would be a well-structured problem. But what if the theorem prover appeals to kinds of knowledge that transcend the strictly logical? As, indeed, humans do. For example, they may use the proof of one theorem as an analogical guide to prove another theorem. If such kinds of "extralogical" knowledge—heuristics, in fact—are not defined in advance, if they are creatively introduced "on the fly" during problem solving, then theorem proving becomes ill-structured.

Likewise, chess playing would satisfy conditions (1) and (2). And if playing chess entails looking ahead at just one's own next move and its immediate consequences and no further, then condition (5) would be satisfied. But if one were to "look ahead" several moves, one's own and the opponent's, condition (5) would be violated, in which case chess playing is an ill-structured problem. As Simon put it, chess playing is well-structured "in the small" and ill-structured "in the large."[43]

6.8   A MONSIEUR JOURDAIN–LIKE IGNORANCE!

Of course, there are problems that are *inherently* ill-structured or wicked. The kind of activity we normally call *creative* is almost invariably ill-structured.

Designing artifacts that are to satisfy novel requirements or requirements that are ill-understood, diffuse, or amorphous is one such activity. Not all design, of course, manifests such characteristics. Design can be humdrum or "routine," in the words of David Brown of Worcester Polytechic and B. Chandrasekaran of Ohio State University, writing in 1986 on their AI experiments with mechanical design.[44] But design can be enormously creative when one has no initial grasp as to what shape the artifact will eventually take. A classic case was the design by the Scottish engineer Robert

Stephenson and his associates of the famous Britannia Bridge in the 1840s, a wrought-iron railway bridge spanning the Menai Straits in Wales.[45] This was a textbook case of an ill-structured engineering design problem.[46]

As for scientific discovery, like design, it has its well-structured elements. To some extent what Thomas Kuhn called "normal science" certainly has elements that deal with well-structured problems.[47] But of course, the most valuable kind of scientific practice, producing outcomes that find their way into monographs, textbooks, and reference works, is dominated by problems that began as ill-structured ones. Likewise, other arenas of creative work, like painting, sculpting, composing, and writing poetry, plays, and fiction, cope with problems and topics that range from the well-structured to the ill-structured.

But what Simon was interested in were the following questions: How does, or can, computers solve ill-structured problems? What are the implications of the well-structured–ill-structured dichotomy for AI?

Simon published his paper on the nature of ill-structured problems in 1973. By then AI had made quite impressive progress, and to some extent his questions were rhetorical: He already had at least partial answers. But, as he pointed out, like the French dramatist Molière's character Monsieur Jourdain, who one day discovered that he had been speaking prose all his life without knowing it, so also AI researchers had been concerned with, perhaps without realizing, and challenged by problems that were mostly ill-structured.[48]

But AI researchers were not alone in harboring such Monsieur Jourdain–like ignorance. All those computer scientists and practitioners engaged in the design of computational artifacts—computer architectures, computer languages, operating systems, and so on—were also, perhaps without realizing it, grappling with ill-structuredness.

And this is one of the reasons why this paper by Simon was so significant, not only in AI but in other domains of computer science. It brought the concepts of well-structuredness and ill-structuredness into the center of the computer scientists' consciousness—at least to those who cared to read the paper carefully.

Moreover, now that we had a proper sense of the nature of this dichotomy, and we realized that in fact this dichotomy was not cast in stone—that the boundary between them was "vague and fluid"[49]—the scope and domain of AI could be further expanded to cope better with the ill-structuredness of problems. It enhanced and further enlarged the range of research topics toward this end: hence the need to study such realms as semantics, knowledge representation, and understanding and responding to natural language instructions.[50]

6.9 MICROWORLDS: "INCUBATORS FOR KNOWLEDGE"

It is in the nature of scientific inquiry to isolate a given phenomenon, establish an artificial boundary between it and its normal environment, and examine the phenomenon as if it is a world of its own, its only interaction with the rest of the world being limited to what the boundary allows.

This is what happens in laboratory sciences, whether physical, biological, or behavioral. A cognitive scientist, to take an example, studying the mentality of chimps creates a physical environment—the boundary—that insulates the subject chimp from all other influences save those the investigator uses to stimulate the chimp's behavior. The assumption is that the chimp's behavior in its laboratory setting will reflect its natural behavior—behavior "in the wild."

Or consider the construction of opinion polls: Here an entire population is encapsulated in a small "sample" that, it is assumed, statistically represents the population. This sample becomes the closed world of the pollster's study.

It is also what happens in the analytical sciences. For example, a structural engineer analyzing the behavior of a structure such as a bridge will isolate its principal "members," idealize a member as a "free body," identify the forces acting on it, and thereby compute its behavior (bending, buckling, etc.) as if that structural element is a tiny world of its own.

Some researchers in AI also adopted this idea to cope with the complexities of intelligent behavior—especially of machines—but took it to an extreme. At MIT in 1972, Marvin Minsky and Seymour Papert, who led its Artificial Intelligence Laboratory, advanced their concept of *microworlds*.[51]

A microworld is precisely that: an idealized, tiny universe, a kind of fairyland comprising entities so stripped of complexity that it bears little resemblance to the real world.[52] Fairyland, perhaps, but one endowed with more than magical dust. Minsky and Papert (and their students) believed that investigating such microworlds would yield "suggestive and predictive powers" that could be fruitfully embedded in larger, more complex, real-world systems.[53]

So a microworld à la Minsky and Papert is a microcosm of a larger world that can be explored computationally to illuminate larger questions concerning cognition and intelligence such as those related to learning, understanding, reasoning, perceiving, planning, and designing. Seymour Papert in his book *Mindstorms* (1980) called microworlds "incubators for knowledge."[54]

6.10   THE BLOCKS MICROWORLD

The most well-recognized microworld designed at the MIT AI Laboratory was the *blocks world*. Indeed, the very mention of microworld conjures up an image of the blocks world—and "image" is apposite here, for the blocks world by its very nature is a visual world.

At MIT several versions of the blocks world were devised, depending on their precise purposes. Patrick Winston's project was to design programs that would illuminate how agents learn structural concepts from examples, a version of what philosophers call the "problem of induction."[55] Toward this objective Winston created a microworld of simple, three-dimensional structures made solely of objects such as bricks, cubes, and wedges. For instance, his program would learn the concept of an *arch* by being "exposed" to block structures and being informed that some were a-kind-of arch, and others were not-a-kind-of arch. In other words, learning obtained from examples and counterexamples.

Another member of the Minsky–Papert community, Gerald Jay Sussman had a microworld comprising a tabletop and solid blocks of different sizes and shapes. Sussman's program, called HACKER, was, like Winston's program, an interactive system that served as an experimental apparatus to study skill acquisition. Sussman's problem was to shed light on this question: "What is the process by which a person develops a skill through practice?"[56] This formidable and wide-ranging problem (imagine the varieties of skill humans acquire and practice, both in everyday life and as vocations) was pared down to its simplest by deploying a blocks-type microworld and the "skill" was HACKER's ability to move blocks around and to learn to place blocks on one another even when there were obstacles to doing so. Learning to cope with obstacles was a part of the skill-acquisition process.

Arguably, the most celebrated blocks-type microworld was created by Terry Winograd, also of the MIT AI Laboratory. Winograd's project was the following: How do we understand a natural language, such as English? His interactive program, named SHRDLU, was a simulated robot that manipulated objects on a tabletop. The objects were more varied than HACKER's objects, comprising solid blocks, pyramids, and boxes open at one end of different sizes and colors. Given a scenario of a configuration of such objects variously disposed on the tabletop—for example, a blue block on top of a red one or a green pyramid inside a red box—SHRDLU would manipulate these objects in response to commands given to it in English.[57]

6.11 SEYMOUR PAPERT'S TURTLEWORLD

The microworlds created by the likes of Winston, Winograd, and Sussman were intended to simplify the problem state space (to use the Newell–Simon term) to the barest minimum in which particular types of intelligent behavior could be experimentally studied. They each pared reality down to its sparest representation sufficient for the purpose at hand.

Their co-mentor at MIT's AI Laboratory, Seymour Papert had a different purpose in mind for the microworld idea. Trained by the eminent Swiss development psychologist Jean Piaget (following his mathematical training), Papert was deeply interested in how children learn concepts and how they perceive things. We get a glimpse of this interest in *Artificial Intelligence Progress Report* in which Minsky and Papert describe children's development of such concepts as quantity and geometry.[58]

Papert went beyond this. He wanted to effect something of a revolution in education by using the computer as a pedagogical tool (in a time when personal computers were still a laboratory curiosity in the most advanced research centers). His book *Mindstorms* (1980) was subtitled "Children, Computers and Powerful Ideas,"[59] and the "powerful ideas" were mathematical and physical ideas children could learn by interacting with computers.

Papert created microworlds out of computers. That is, when children interacted with the computer they were *inhabiting* a microworld. The first and most widely known was the LOGO *turtle* microworld—composed of a small object on the computer screen called the turtle. Depending on its implementation, a turtle could look like an actual turtle, but it could be something else, a triangle for instance.

The turtle could be moved around and made to draw lines by a child "talking" to it by typing in commands on a keyboard. For instance, the command "FORWARD 25" would cause the turtle to move in the direction it was facing and draw a line 25 units long—move 25 "turtle steps." The command "RIGHT 90" would cause the turtle to turn 90 deg. The command set constituted LOGO, the turtle geometry language, and by engaging in such "turtle talk" a child could make the turtle anything involving lines, even curves. Papert documented a conversation between two children as they explored turtle geometry, producing geometric shapes of different kinds, even flower-like patterns, speaking to each other of what could be done with the turtle and how to go about it. In effect they learned elements of geometry in a *constructive* manner: Rather than learn *what* something is, they came to know that something in terms of *how* it is created. This was a case of *procedural knowledge*.

It was the child then who inhabited the turtle microworld. It was, as Papert described it, a completely closed world defined by the turtle, its possible moves, and the LOGO language it could "understand." Yet, despite its simplicity, Papert pointed out in a lecture delivered in 1984, it was a world rich in possibilities, for the child to explore and experiment with, the world of Euclidean geometry. It was, he wrote, a *safe* world in the sense that a child could try all kinds of things without fearing that he or she would get into trouble, would not feel stupid, would never be embarrassed by doing something wrong.[60]

The point of the turtle microworld was that it afforded a platform for "powerful kinds" of mathematical learning—and was also "fun."[61]

The turtle microworld was extended to enable children to learn elementary physical principles: "here we do for Newton what we did for Euclid."[62] The particle of Newtonian mechanics—having no dimensions but possessing such physical properties as position, mass, and velocity—became a turtle. "Turtle-talk" was enriched so that children could inhabit a Newtonian microworld. For example Newton's first law of motion was recast as the "turtle law of motion":

Every turtle remains in its state of rest until compelled by a turtle command to change that state.[63]

Additional Newtonian microworlds were created, such as a "velocity turtle" and an "acceleration turtle."[64]

The reader may have realized that turtle microworlds à la Papert are not especially of import to the AI project *narrowly conceived*—that is, a project dedicated to creating "intelligent" artifacts. Turtle geometry did not do geometry on its own but only under the direction of its user. But if one takes AI as *broadly conceived*—dedicated not only to the construction of intelligent artifacts but also to constructing computational theories, models, systems, and apparati that help understand *natural* intelligence— then an argument can be made that the computer-based turtle geometry *afforded* an experimental testbed for Papert to test his Piaget-inspired theory of learning. Papert's microworlds were thus of relevance to the study of natural intelligence.

6.12   WHAT SHRDLU COULD DO AND HOW IT DID WHAT IT COULD DO

If there were "microworld skeptics" within the AI community (or the larger computing community)—after all, what were these microworlds but (as Minsky and Papert said) fairylands (or toylands)—they may well have been dramatically quietened by Terry Winograd's language-understanding system SHRDLU.[65] This was the product of a doctoral dissertation completed at MIT in 1971 that was first published as an entire issue of the journal *Cognitive Psychology* and then as a book in 1972.

SHRDLU was a simulated robot. It inhabited a microworld comprising variously colored objects: solid blocks of different shapes (cubic, brick-like) and sizes, pyramids, and open boxes situated in some configuration on a tabletop. It would engage in a dialogue with a speaker (Winograd himself) in which it would accept commands or questions or declarations concerning a given scenario of objects on the table from the speaker in English and respond to them, including (simulated) actions that changed the configuration of the objects.

The scene itself would be represented in SHRDLU's database, along with other pieces of factual knowledge about this microworld. For example, stated informally in English, if a section of the tabletop contained a particular configuration of objects, this would be represented by assertions of the following sort:

B1 is *Block*

B2 is *Block*

B3 is *Block*

B4 is *Pyramid*

B5 is *Box*

B6 is *Pyramid*

ColorOf B1 is *Red*

ColorOf B2 is *Green*

ColorOf B3 is *Green*

ColorOf B4 is *Red*

B1 *Supports* B2

TopOf B2 is *Clear*

B3 *Supports* B4

ColorOf B6 is *Blue*

B5 *Contains* B6

So this records several facts, such as that a red block (B1) is supporting a green block (B2), that there is nothing on B2, that B2 has nothing on top of it, that a box (B5) contains a blue pyramid (B6).

Consider now an interaction between a person (PERSON) and SHRDLU. The following is a paraphrase of fragments of an actual dialogue between the two as documented by Winograd.

A command from PERSON to pick up a red block causes SHRDLU to "survey" the tabletop and determine that the only red block in the scene (B1) has a green block (B2) upon it. So it must first remove B2. Because B2 has nothing atop it (as recorded in SHRDLU's database) it can be moved after it finds space on the tabletop where B2 can be placed. Thus SHRDLU engages in planning. It plans by executing a MEA (Section 6.6). A goal–subgoal hierarchical plan is established along the following lines:

Grasp B1
    GetRidOf B2
        PutOn B2 on TABLE1
            Put B2 at *Location* (xxx)

Thus the means to achieve the goal of grasping B1 is to get rid of B2 (a subgoal); this subgoal is effected by putting B2 on TABLE1; to achieve this latter subgoal, an empty space on TABLE1 is sought, and, when found, B2 is put at this location. Each of these goals–subgoals, *Grasp, GetRidOf, PutOn,* and *Put,* is in fact a program or procedure.

Now if SHRDLU is instructed to grasp "the pyramid," it responds by stating that it does not understand which pyramid is being referred to as there are two of them on the tabletop (and its database). In so responding SHRDLU appeals to its knowledge that a phrase beginning with the definite article refers to a specific object. PERSON asks what the box contains. SHRDLU answers that it is a blue pyramid. And now when PERSON asks what "the pyramid" is supported by, SHRDLU responds correctly: It is the box (B5). Previously, it did not understand which pyramid PERSON had referred to, but now it assumes that PERSON is referring to the one just mentioned. It utilizes its knowledge that a box supports anything it contains.

As we've just seen, SHRDLU can answer questions. When asked whether the table can pick up blocks it referred to its knowledge that only animate entities can pick up things and so it responds in the negative. When asked if a pyramid can be supported by a block SHRDLU finds an actual example on the tabletop of a pyramid (B4) supported by a block (B3), and so it infers that this is indeed possible. And when PERSON declares ownership of all nonred blocks but not any object that supports a pyramid—a fact about possession of an object previously unknown to it—SHRDLU stores this new fact in its database. Now when PERSON asks whether he owns the box, SHRDLU appeals to its knowledge that the box (B5) supports the blue pyramid (B6) and the new fact it has just recorded about possession and answers in the negative.

There was, of course, much else involving the robot's linguistic prowess, its knowledge of the blocks world, and its reasoning capacity. But this tiny fragment, extracted and paraphrased from the lengthy actual dialogue between Winograd and SHRDLU, will give the reader a sense of the nature and scope of the "thinking" and "understanding" SHRDLU was capable of.

Underpinning these capabilities were some significant innovations in the system, markers of Winograd's creativity and SHRDLU'S originality as an artificial language-understanding system—the SHRDLU project's undoubted theoretical contributions to the science of intelligence, both natural and artificial. (It is worth noting that *Understanding Natural Language* was first published as an entire issue of a psychology journal, clearly indicating that the journal's editor(s) and the paper's reviewers felt this work to be significant in the realm of *human* intelligence.)

Arguably, its most prominent contribution was its widespread adoption of the *principle of proceduralism*. Indeed, one of his publications on SHRDLU was titled "A Procedural Model of Language Understanding."[66] The general principle is the use of procedures (in the strict computational sense) to describe or define actions, concepts, states of affairs, and so on. In other words the *semantics* of such entities were specified universally as procedures.

We have previously encountered this idea of procedural semantics, though by way of a slightly different term: as operational semantics—of programming and microprogramming languages (Chapter 5, Section 5.5). And, of course, because computational processes are procedures, we should not be too surprised that an AI system espouses proceduralism. Its significance in the case of SHRDLU lay in Winograd's espousal of proceduralism as a kind of foundational theory of language understanding.

For example, SHRDLU's blocks world contained a concept called CLEARTOP. Its semantics was defined procedurally along the following lines:

CLEARTOP (X):
L : **If** X supports an object Y
   then get-rid-of (Y);
     goto L
   else X is cleartop

The action GRASP is described as a procedure along the following lines:

GRASP (X):
 if X is not manipulable **then** fail
   else if X is already being grasped **then** succeed
     else if another object Y is currently grasped
       then get-rid-of (Y);

Move to the top-center of X;

Assert (X is grasped)

If SHRDLU is commanded to find a green cube that supports a pyramid, that particular *state* of the tabletop scene would also be procedurally described:

L: Find a block X;
>   **if** X is not green **then goto** *L*
>> **else**
>>> **if** X is not equidimensional **then goto** *L*
>>>> **else** Find a pyramid Y;
>>>>> **if** X does not support Y **then goto** *L*
>>>>>> **else** succeed

For convenience and readability these procedures have been described in a Pascal-like notation. Winograd implemented SHRDLU by using LISP, a language invented by John McCarthy between 1956 and 1958 and widely used for AI programming,[67] and a version of the more recently developed PLANNER, designed and implemented by Winograd's then–fellow graduate student Carl Hewitt.[68] PLANNER played an important role in in implementing SHRDLU: It could not only express statements, commands, and questions but it could also perform deductions of the kind SHRDLU had to make as it went about its tasks. PLANNER was thus both a language and a language processor: It could accept expressions as inputs and evaluate them. As we have seen, SHRDLU was able to *plan* sequences of goals and subgoals by way of MEA. This was facilitated by the PLANNER system: The PLANNER programmer would specify his or her program as a set of statements—called theorems in PLANNER jargon—about how to achieve goals. As a simple example, consider the PLANNER theorem (stated in English):

To show that someone is mortal show that he or she is human.

The phrase "someone is mortal" is called a "pattern." In solving this problem, if a goal is generated,

G : ARISTOTLE IS MORTAL.

Then if the system can prove the theorem

G* : ARISTOTLE IS HUMAN,

it can deduce Aristotle's mortality: **G*** then becomes a subgoal to achieve the goal **G**.

If proceduralism was an overarching principle for handling meaning and knowledge representation in SHRDLU, another significant feature was its espousal of *interactionism*. The SHRDLU system in fact consists of a dozen functional subsystems that interact intimately during its operation. The most notable interaction was between the subsystems named GRAMMAR and SEMANTICS, responsible for syntax analysis and semantic interpretation, respectively. Winograd pointed out that at that time (1968–1972) current linguistic theories tended to focus on syntax, ignoring semantics.[69] (Perhaps Noam Chomsky's influence still held sway.) In contrast, Winograd was interested in the problem of how the syntactic structure of language (its form) conveys semantics (meaning).[70]

He found a theory of grammar called systemic grammar, invented in 1967–1968 by the English linguist M. A. K. (Michael) Halliday of University College London, which suited his purpose.[71] Rather than separate the analysis of syntax from semantics, systemic grammar attended to the manner in which language was organized into units, each having a particular role in communicating meaning.[72]

The nature of the interaction among syntax analysis, semantic interpretation, and deduction characterizing SHRDLU can be illustrated with a simple example. Consider PERSON commanding SHRDLU to place the red pyramid in the box.

The subsystem GRAMMAR would recognize the phrase describing the red pyramid as a syntactically feasible unit (a noun phrase). SEMANTICS would then check whether this was a meaningful phrase. SHRDLU's knowledge about the blocks world and the scene at hand, held in a subsystem called BLOCKS, would suggest that there was indeed a red pyramid in the scene. GRAMMAR would then have to confirm that the remainder of the command, IN THE BOX, was of a particular kind of phrase. A plan to grasp the pyramid and place it in the box would be generated by SEMANTICS, with the help of BLOCKS and GRAMMAR.

6.13   SHRDLU'S ORIGINALITY, WINOGRAD'S CREATIVITY

At the time of its publication SHRDLU was considered a remarkable—and exciting—achievement in the realm of AI. In particular, its ability to conduct with the speaker a realistic conversation involving references across sentences captured the imagination of both AI researchers and people outside the AI community. As also did its ability to carry out realistic tasks, albeit in its limited blocks microworld. Writing 8 years later in a position paper titled "What Does It Mean to Understand Language?" (1980)[73]—published, incidentally, not in an AI journal (or even a computing journal) but in the journal *Cognitive Science* (on which more later)—Winograd, summarizing its scientific contributions to the topic of natural language understanding systems, mentioned in particular its embrace of proceduralism; its focus on how language effects action—"The meaning of a sentence was represented not as a fact about the world but as a command for the program to do something"; and its emphasis on the cognitive context in the understanding process. (For instance, reference to an object in a command or a question was understood by appealing to what might have been said in preceding sentences.)[74]

There is another noteworthy aspect of SHRDLU. Writing in 1973, Winograd pointed out that the SHRDLU project attempted to demonstrate the kinds of details necessary to create a detailed working model of a language.[75]

SHRDLU was indeed a working model: a fully operational computational artifact that embodied a theory of natural language understanding in a very limited environment. And as such it stood far apart from the methodology of language understanding pursued by linguists and psychologists, whether conducted experimentally (in a laboratory) or theoretically. Winograd, of course, recognized this: The conventional wisdom in language studies, he stated, attempted to make psychological sense of "isolated" language components; the SHRDLU project held a holistic or systemic stance wherein the relationships between different knowledge types and procedures entailed in language use could be explored in a unified manner.[76]

So what can we say about the originality of SHRDLU as a computational artifact that was intended to manifest intelligence of a certain sort? Correspondingly, what can we say about its artificer, Terry Winograd's creativity?

To answer this, let us appeal to some ideas proposed by creativity researchers. Recall from Chapter 4 (Section 4.8) the notion of *historical* originality (H-originality): An artifact is deemed *H-original* (and its artificer, *H-creative*) if a functionally identical artifact has ever been previously built.[77] Thus SHRDLU was undoubtedly H-original; Winograd was correspondingly H-creative.

Another related concept is the notion of *consequential* originality (C-originality): An artifact is deemed *C-original* (and its artificer, *C-creative*) if it significantly influences the future in some sense; that is, the artifact has consequences.[78] This is more difficult to demonstrate, but one way to do this is to identify the richness of the knowledge generated by the act of that artifact's creation and embedded in the artifact. This has been called "epistemic complexity."[79] If not the artifact itself, this knowledge becomes available to future researchers. The extent of an artifact's epistemic complexity is a marker of C-originality. Given the kinds of knowledge that were embedded in SHRDLU—the efficacy of proceduralism, interactionsism, its overall systemic quality, its ability to conduct realistic conversations and carry out realistic tasks, among others—one could certainly claim that SHRDLU was unequivocally C-original; and by the same token Winograd was C-creative.

6.14   FINDING FAULTS WITH SHRDLU

All artifacts are specialized: They can do only so much and no more. Inevitably, SHRDLU had its critics.

Winograd himself was ready to point out its deficiencies—there were, he noted, many details that were "inadequate" and "whole areas" of language that were missing.[80] Many aspects of language understanding and language use were not dealt with.[81] Writing in 1980, with the benefit of a historical perspective, he dispassionately dissected SHRDLU's achievements and limitations. He noted that some of the features of the system were quite ad hoc, clearly an unsatisfactory situation. For instance, in obeying a command to pick up a big red pyramid, the robot would identify a pyramid in the scene as big if its dimensions exceeded a certain numerical value. But, of course, big is a qualitative and relative term, and this relativity was something SHRDLU could not handle. More generally, the meaning of a word depended on the context of the situation in which the word was used.

As another limitation, Winograd noted that "natural reasoning" was not restricted to making logical deductions. Indeed, deductive logic constitutes a small part of a human being's reasoning apparatus. They generalize, for instance, based on analogies and experience. Reasoning in AI systems such as SHRDLU would need to extend its range of logic beyond deductive logic.

But Winograd stood by his methodology: Any further advances in understanding how agents (humans or machines) understand language would do well to follow the systemic approach embedded in SHRDLU. Single, isolated semantic rules and heuristic principles would not do.[82]

Perhaps the most consequential lesson SHRDLU taught was a Janus-faced one. On the one hand, the assumption underlying the microworld philosophy was that the insights gleaned by tackling a microworld could be "scaled up" to real-world situations. SHRDLU'S limitations taught that this was not possible. The flaws that Winograd himself (and others) identified demanded other theories, other techniques, other models. In particular, some AI researchers realized that more comprehensive or universal frameworks—worldviews—were needed if they were to advance the science of intelligence appropriate for the real world.

On the other hand, SHRDLU's very success in its tiny blocks world suggested that perhaps rather than seeking universal human-like intelligent artifacts, the AI research agenda should focus on limited, highly specific domains for the pursuit of AI. Perhaps, certain AI researchers thought that the search for GPS-like problem solvers should be abandoned: one should seek artifacts that exhibited *expertise* in limited domains.

6.15 FROM SCHEMAS TO FRAMES (AND SCRIPTS): A BORROWING FROM PSYCHOLOGY

Consider the following scenario. Driving into a new city and realizing he is hungry, the driver searches for a restaurant. Eventually he comes upon a place that announces itself as offering Greek–Lebanese food.

The very word "Greek" may cause a certain set of impressions to surface on the person's conscious mind:

Images of a very blue Mediterranean Sea, of white, sundrenched villas, the Acropolis. The names of Plato and Aristotle, *The Republic*, Pythagoras' theorem. Raphael's fresco *The School of Athens*.

On the other hand, if the person has previously experienced Middle Eastern food, an entirely different set of impressions may surface in his consciousness:

Images of kabobs and chicken shwarma, grapeleaves, hummus, babaganouche, tabuli salad, pita bread. Memory of their tastes. Image of scooping up chunks of hummus with pieces of pita bread.

Regardless of whether the person has eaten Middle Eastern food, the sign that this is a restaurant may also precipitate in the person's conscious mind a certain *narrative*:

There will be an array of small tables with chairs arranged around them. He will be ushered to a table. He will be shown a menu by a server. There will be silverware and napkins laid out on the tables. The server will take his order, and soon after he will get his food and drink.

The point is, the sign at the front of the restaurant proclaiming it to be a Greek–Lebanese restaurant is a piece of information. But each of the separate sets of thoughts or images or imagined tastes this information may give rise to is an integrated structure that characterizes a particular archetypal concept or idea: the first, a certain image of Greece, the second of Middle Eastern food, the third of a restaurant. In fact, these concepts may constitute fragments of yet-larger structures that form the person's personal knowledge of Greece, the Middle East, Mediterranean culture, and restaurants. As knowledge they are composed of pieces of information; but the power of knowledge is much greater than the power of isolated information. For (a) that person's knowledge of Middle Eastern food or of restaurants will give rise to certain *expectations* in him about the experience of eating (and enjoying) such food and being in a restaurant; and (b) it facilitates certain kinds of *inferences* the person may make about this particular restaurant. (For instance, he may infer that the food at this restaurant will not be spicy or that this restaurant will not play rap or R &B music in the background.)

Cognitive psychologists have a particular term for integrated symbolic structures representing one's knowledge of complex concepts such as restaurants, Greece, or Middle Eastern cuisine. They call such structures *schemas* (or *schemata*, also the plural of schema).

The term, which is due to British psychologist Henry Head, was made prominent by Head's student Frederick Bartlett of the University of Cambridge in his book *Remembering* (1932). Bartlett used schema to mean "an active organization of past reactions or of past experiences" that will participate "in any well-adapted organic response."[83]

So for Bartlett, a schema was a mental structure storing some past experience that is activated in responding to a new situation. For developmental psychologist Jean Piaget (with whom Seymour Papert had trained), in describing the cognitive development process in children, a schema (or "scheme" as he preferred to call it) is a kind of general structure for some actions that can be applied to situations that vary under environmental conditions; moreover, a new experience or entity is *assimilated* into an existing schema, thus enlarging and enriching the latter.[84]

In 1975, reflecting on the state of AI, Marvin Minsky came to the disconsolate conclusion that most current theories were built on components that were "too minute, local and unstructured" to adequately model thinking.[85] A number of organizing principles had been proposed, including the problem state space of Newell and Simon and the microworld concept he and Papert had articulated. In the realm of language understanding, Roger Schank of Yale University had proposed a formalism he called "conceptual structures" to represent the meaning utterances in an unambiguous way.[86]

That same year, Schank's colleague at Yale, Kenneth Colby, had proposed organized representations of "belief systems."[87] These were all along the right direction, Minsky concluded, but they were not sufficient; something larger, more general was needed that could tie these distinct strands together into a unified coherent theory—or at least the "pretence" of such a theory.[88]

The theory he had in mind was founded on the schema concept. Minsky borrowed from Frederick Bartlett but translated into the language of computer science. He described the Bartlettian schema as a data structure and called such data structures *frames*. A frame would represent "stereotyped" situations like "a certain kind of living room or going to a children's birthday party."[89] Thus, faced with a new situation, one would retrieve from memory a frame—a "remembered framework"—that would then be adapted to the new situation (or vice versa).[90]

Bartlett's influence is clear in this statement, and Minsky freely and frequently paid tribute to the British psychologist. But he went slightly further by also relating his "frame idea" to the Kuhnian paradigm.[91]

Indeed, anyone familiar with Thomas Kuhn's formulation of the scientific paradigm concept would, without much delay, find a similarity between the schema and the paradigm concepts. So Minsky's debt here was to both a psychological and a historical–philosophical theory. But where Minsky believed he had gone beyond both Bartlett and Kuhn was in his proposal that frames could be linked together to form *frame systems*.[92]

Envisioning frames as data structures afforded Minsky the power of computational thinking, something people like Bartlett, Piaget, and Kuhn, radically creative though they were, would never have embraced. Complex data structures—and, indeed, program structures—such as "linked lists" (the one unifying structure in the programming language LISP) could be envisioned as networks. So also a frame or a frame system could be envisioned as "a network of nodes and relations."[93] Such a network would be hierarchically structured.[94]

Thus was born what came to be called frame theory. Consistent with the philosophy expounded in his coauthored monograph *Artificial Intelligence Progress Report* (1972), [95] Minsky insisted that his frame theory of knowledge representation made no distinction between humans and machines—an abiding unity manifested by the pioneers of AI, notably Minsky and Papert, Newell and Simon, and their early students.

Much of Minsky's 1975 essay—and it really was a meditative essay rather than a hard-edged research paper—was given to demonstrating how his theory could deal with situations in visual imaging and language understanding. How, for example, the perspective appearances of a block could be explained in terms of frames, or how the changing perception of a room as one moved around in it could be represented. But here I give a sense of what a frame-like schema for a common or everyday situation might "look" like.

Consider the Mont Blanc fountain pen I'm writing these words with. I possess a hierarchical schema–frame system for my beliefs and knowledge concerning this kind of object. At the highest, most general level, this schema represents the archetypal nature of all fountain pens: their form or shape, their main components (top, body, nib, clip), how they are used, what to do when the ink runs out (unscrew the nib part, remove the cartridge, insert a new cartridge), what action to take if the ink does not flow smoothly (shake the pen).

That part of the schema that describes what to do when the ink runs out may be the "default" description because most times this is what one should do. But it is possible that I may have an alternative to this procedure in my fountain-pen schema, as the ink can also be held (as was common in the past) in long tubes siphoned in from ink bottles. If I have this alternative element in my fountain-pen schema then on encountering such a pen I won't be unduly surprised.

At a lower level there are subschemas corresponding to my experiences with a number of different brands of fountain pens I happen to own and use. The subschemas are *instantiations* of top-level representation for they represent the actual forms, colors, and other characteristics of the nibs (fine or medium) of each pen. These subschemas also contain representations of the idiosyncrasies of the different pens that cause me to change, however slightly and subtly, my writing mode and my expectations if and when I change pens.

Another aspect is worth noting. When I'm using a particular fountain pen, as I am now, it becomes, so to speak, an extension of myself. Thus the schema representing my *self-knowledge* is linked to my fountain-pen schema whenever I am using one of my pens. In Minsky's terms, they form a temporary frame system.

Minsky's frame concept was not the only schema-like principle to enter the AI domain. Almost contemporaneous with Minsky's publication, Roger Schank and Robert Abelson of Yale University presented at the 1975 International Joint Conference in Artificial Intelligence, held in Tblisi, Georgian SSR, their version of schemas they called *scripts*.[96] Two years later they elaborated their script theory into an influential book, *Scripts, Plans, Goals and Understanding* (1977).[97]

Like frames, scripts were knowledge-representation structures. Although in broad agreement with the frames philosophy, Schank and Abelson found it too general for actual application in AI. The script concept was a kind of specialization of the frame idea—a structure for describing "stereotyped" sequences of events.[98]

Roger Schank's larger project was to model natural language understanding but on a larger canvas (so to speak) than the microworld domain. As he explained in 1973 (he was then at Stanford University), if it is the case (as it undoubtedly is) that humans can understand natural language it should be possible to have the computer imitate humans.[99] This is an unequivocal assertion of what AI is about: In contrast to Minsky and Papert and Newell and Simon, who saw no boundary between human thinking and machine thinking, Schank's assertion suggested that the goal of AI is to have the computer *imitate* humans.

Toward this end, in 1973, he proposed a small set of "conceptual categories" (including "action" and "thing"). A natural language utterance (e.g., "John hit the large ball") could be represented as a "conceptual structure" comprising concepts (as instances of conceptual categories) and their relationships (called "dependencies"). Thus, "John hit the large ball" would generate a conceptual structure with a particular thing concept

("John"), an action concept ("hit"), a thing modifier concept ("large"), and a general thing concept ("ball") with dependencies representing how these concepts are related.

Script theory built on this earlier work to model the sequence of events constituting stories. A story comprises a script with interesting variations. Schank and Abelson mention, as instances of scripts, "a restaurant script, a birthday party script, a football game script, a classroom script, and so on."[100] Their description of a restaurant script is not unlike the Greek–Lebanese restaurant schema that introduced this section.

In fact, later commentators did not make much of a distinction between frames and scripts: They belonged to the same family of knowledge-representation schemes so that many writers would write "frames and scripts" conjunctively.[101] But in 1975, when both ideas were announced, there was one significant difference. Minsky's frame theory was *unfalsifiable*, so general and diffuse was its formulation. If one was a Popperian, one would label frame theory (following Karl Popper[102]) as a metaphysical proposition rather than as a scientific theory. Frame theory, in its first manifestation was, in fact, a bold prescription that invited further refinement that could claim scientific credibility.

Script theory as Schank and Abelson presented it had, in fact, been tested empirically: A program (called SAM—script-applying mechanism) that used scripts to make inferences in certain situations had been implemented. The input to SAM were conceptual dependency structures extracted by a "preprocessor" from English-language stories. SAM would then identify the relevant script, fill in details as new inputs were received, and make inferences.[103] The resulting structure could then be asked questions about the narrative and give answers or could be used to make summaries and paraphrases of the initial story.

6.16 "A VERITABLE JUNGLE OF OPINIONS"

The period circa 1975–1990 was enormously fertile in the matter of knowledge representation. Which is not to say that it was a time of harmony or of a general sense of shared clarity within the knowledge-representation research community. A marker of the prevailing discord was the "Special Issue on Knowledge Representation" (1980) published by ACM's Special Interest Group on Artificial Intelligence (SIGART), edited by Ronald Brachman, then of Bolt, Beranek and Newman (BBN), and Brian Smith, a doctoral student at MIT.[104] This issue published the results of a comprehensive questionnaire on knowledge representation sent to over eight research groups. The questions spanned over a wide range of topics, but what was remarkable was the extent of disagreement on even fundamental issues—prompting Allen Newell, writing in 1982, to comment that the outcome of the questionnaire was "a veritable jungle of opinions."[105]

Actually, no one within the AI community disagreed on the centrality of knowledge in the construction of intelligent artifacts. Nor on the necessity of knowledge representation in modeling or theorizing about intelligent agents. The PSS hypothesis

(Section 6.1) proclaimed as much. In his doctoral dissertation completed at MIT in 1982, Brian Smith encapsulated these assumptions in what he called:

The Knowledge-Representation (KR) Hypothesis: Any mechanically embodied intelligent process will comprise structural ingredients that (a) we as external observers naturally take to represent a propositional account of the knowledge that the overall process exhibits; and (b) independent of such external semantical attribution, play a formal but causal and essential role in engendering the behavior that manifests that knowledge.[106]

The fundamental bone of contention amid the "jungle of opinions" was the *nature* of KR itself; or, in terms of the PSS hypothesis, the form of symbol structures representing knowledge. The issue was complicated by the fact that knowledge is an active agent that, as the KR hypothesis postulated, plays a central role in producing behavior. So any KR scheme must allow for *reasoning* to be effected, thereby not only engendering behavior but also begetting, explicitly, knowledge that is implicit in the knowledge structure—just as an axiomatic system enables the generation of theorems that are implicit in the axioms and definitions; and as the formal rules of a grammar enable sentences to be generated that are implicit in the rules of the grammar.

There was also the matter of the *is–ought* dichotomy. When the object of KR is the construction of intelligent artifacts this question arises: What *ought* to be the "right" KR to facilitate efficiently intelligent action in the artifact? When the concern is to understand natural intelligence, the question is this: What *is* that KR that produces observed natural intelligence? A theory of KR will be prescriptive in the former situation but descriptive in the latter.

6.17   REPRESENTATION STYLES

By the mid-1980s certain styles of KR had clearly emerged, each with its own community of adherents, though, as in all creative endeavors, there were researchers who experimented by mixing styles.

A style represents choice. As Herbert Simon pointed out in 1975, a style comes into existence whenever there are more ways than one of doing something. When there is only one way of doing something—if, in the case of design, form inexorably follows function—the question of style does not arise.[107] Moreover, a style is a purposeful (or goal-oriented) invention: People invent styles. We do not associate style with natural processes.

A KR style, then, describes or characterizes the general features of some invented representative scheme. So we can think of a representation style as encompassing a family of particular representation schemes. We can even imagine a style as being a schema.

So the principle KR styles that, by the mid-1980s, were vying with one another (by way of their proponents) for dominance were based on *schemas, semantic networks, productions,* and *logic.* Let us consider the development of each of these in turn.

6.18   UNPACKING FRAME-BASED THINKING

We have already encountered schemas in the persona of frames and scripts. Recall that Minsky's 1975 formulation of frames was quite informal, indeed at time quite vague. But this very informality gave ample space for AI researchers to interpret Minsky's basic idea more precisely in the years that followed. One direction this interpretation took was the development of concrete frame-based representation schemes.

A notable and influential example of this was Knowledge Representation Language (KRL) designed by Daniel Bobrow of Xerox Palo Alto Research Center and Terry Winograd, by then at Stanford University, and published, significantly, not in a computing or AI journal but in the very first issue of *Cognitive Science* in 1977.[108]

Bobrow and Winograd began with some "aphorisms" that captured their main "intuitions" and that clearly revealed the Minskyan influence on their thinking:

Knowledge should be organized around conceptual entities with associated descriptors and procedures.

A description must be able to represent partial knowledge about an entity, and accommodate multiple descriptors that can describe the associated entity from different viewpoints.

Reasoning is dominated by a process of recognition in which new objects and events are compared with stored sets of expected prototypes and in which specialized reasoning strategies are keyed to these prototypes.[109]

The reader may have realized, from earlier discussions of schemas and their frame and script avatars, that this representation style is concept oriented—a more ambitious version of object orientation (Chapter 2, Section 2.14). For instance, a restaurant schema–frame–script describes the stereotypical concept of the restaurant experience. In addition, schema-style KRs are primarily representations of *declarative knowledge* rather than of the procedural knowledge (characterizing SHRDLU, for instance). A structure describing the restaurant experience concept with its many aspects *declares that* this is what a restaurant experience *is*. Bobrow and Winograd built the KRL around this organizational principle of declarative knowledge.

But was frame theory—or the "frame movement" that Minsky's paper seemed to have spawned—something fundamentally new? Writing in 1979, Patrick Hayes clearly thought otherwise. Calling his paper "The Logic of Frames" (1979), Hayes examined the basic ideas put forth by Minsky and later researchers, most notably Bobrow and Winograd, and argued, with supporting evidence, that (with one notable exception) there was nothing in frame-based representation that could not be expressed in logic.[110]

For example, suppose a concept C is represented by a frame with slots that are related to C by the relations $R_1, R_2, \ldots, R_n$. A frame-based scene-understanding system will then conclude that a given object in the scene is an instance of C if its features bear the relationship to C as specified by $R_1, R_2, \ldots, R_n$. In the notation of first-order logic this would correspond to the following assertion:

$$\forall x\, C\,(x) \equiv \exists y_1, \ldots, y_n.\ R_1\,(x, y_1)\ \&\ R_2\,(x, y_2)\ \&\ \ldots\ \&\ R_n\,(x, y_n)$$

Or, in English, for all entities x, x is an instance of the concept C if there exists elements $y1, \ldots, yn$ such that each yi relates to x according to the relation Ri.

Hayes went on to argue that most of the frame idea was "just a new syntax" for expressions in logic. But, he concluded, there was one notable exception, which he called "reflexive reasoning."

By this he meant the ability of one to reflect about oneself; to make the self itself an object of analysis. This ability to be self-aware, especially in regard to the process of reasoning, is integral to commonsense thinking. It entails examining an argument or a proof previously constructed and inferring *its* properties. Reflexive reasoning, Hayes stated, entails "descriptions of the self."[111]

In their discussion of KRL, Bobrow and Winograd talked about the desirability in a system such as KRL to represent "knowledge about knowledge"—a metalevel KR.[112] Presumably Hayes was referring to this, approvingly, as an example of reflexive reasoning. This awareness, he concluded, was "one of the few positive suggestions which the 'frames' movement has produced."[113]

6.19   KNOWLEDGE BY ASSOCIATION (AKA SEMANTIC NETWORKS)

The idea of *associative memory*—remembering something or someone by association with other things, persons, or events—has a rich history in psychology, reaching back to times long before the emergence of psychology as an experimental science in the mid-19th century. Nor was it the sole prerogative of the psychologist: Poets and novelists have exploited the phenomenon, perhaps most spectacularly by the French writer Marcel Proust in his monumental novel *À la recherché du temps perdu* (1913–1927), originally translated into English as *Remembrance of Things Past*. But here I give a vastly lesser-known instance of the poet's use of associative memory.

In 1942, an English poet, Robert Nichols, offered a detailed analysis of the genesis of a poem he had composed, titled "Sunrise Poem."[114] Nichols recalled how the first lines of his poem had their origins in his solitary contemplation from a ship's deck of the sun rising on a tranquil sea. The observation filled him, he wrote, with "an immense and pure emotion." This emotion in turn prompted him to see in the pattern of light reflected on the water an Arabic script of a kind he had seen (he did not remember where or when) in a 16th-century holy book. Thus began the composition of the poem.

The opening lines of the poem read:

> *The sun, a serene and ancient poet,*
> *Stoops and writes on the sunrise sea*
> *In softly undulant cypher of gold*
> *Words of Arabian character.*

Poetry making is a knowledge-rich process—as indeed, all other acts of creation are. The poet's repository of knowledge is not restricted to facts, hypotheses, laws, theories—declarative knowledge; it contains autobiographical knowledge and stored impressions of visual scenes, images, sounds. Perhaps some of this may be held in external memory—notebooks, journals, commonplace books, and so on. But large parts are held in one's own (internal) memory. Robert Nichols's self-analysis on the experience of writing "Sunrise Poem" reveals something of the nature of how he drew associatively on his own memory:

The sea pattern of reflected light caused him to imagine in the next instant that the script was written on the water by the sun. The sun itself was the poet. No sooner did this notion take shape but he was reminded of a postcard he had often bought at the British Museum of a seated poet, possibly Persian, wearing turban, caftan, and slippers. Simultaneously, the first line came to him.

Consider the first word, "stoops," of the second line. It originated in the poet's observation of the sun rising rapidly above the horizon. But for the sun to write upon the sea it would have to bend forward and down. This image, combined with the notion of "ancientness" in the first line led by association to the image of William Blake's painting *The Ancient of Days* (1794) in which the long-bearded old man kneels forward and "sets a compass upon the deep." The poet's recollection of this painting prompts him to think of, and select, the word stoop.

Like schemas, associative memory as a psychological concept has had a substantial influence on KR research in AI. Like schemas, it was renamed (as computer scientists are wont to do, in all their brashness) by AI researchers as *semantic networks*.[115]

Computer scientists understand the term network to mean a particular kind of structure comprising entities called nodes that are connected to other nodes by entities called links that signify relationships between the entities denoted by the connected nodes. Computer scientists also take some comfort in the fact that there is a natural mapping between networks and the mathematical structure called "directed graphs" (an aspect of graph theory): Nodes correspond to a graph's "vertices" and the links to a graph's directed "edges." This mathematical mapping is also valuable because it emphasizes the graphical or diagrammatic aspect of a network.

As an illustration of what a semantic network might diagrammatically look like, consider the knowledge the early computer pioneers would have possessed about computers circa 1950. Figure 6.2 depicts a tiny fragment of this knowledge in the form of a semantic network.

In Figure 6.2, the elliptical (or oval) nodes denote concepts expressed as English terms (such as "digital computer" or "memory") and particular objects are shown as rectangular nodes (such as "EDSAC," a computer built in 1949).

The directed links are of a number of types. The ones shown in Figure 6.2 are as follows:

Kind-of (K) link: indicating that one concept is a-kind-of another concept. For example, "digital computer is a-kind-of 'artifact.'"

Instance-of (I) link: indicating that a particular object is an-instance-of a concept. For example "EDSAC" is an-instance-of "digital computer."

Component-of (C) link: indicating that an object or a concept is a-component-of another object or concept respectively. For example, "EDSAC memory" is a-component-of "EDSAC."

Relationships tagging links conform to certain *rules of inferences* that enable appropriate conclusions to be drawn. For instance, the rule of inference—IF *A* is-an-instance-of *B* and *B* is-a-kind-of *C* THEN INFER *A* is-an-instance of *C*—which is a kind of transitive rule, allows one to infer, for instance, that the Analytical Engine (Charles Babbage's great 19th-century idea[116]) is an-instance-of digital computer and that EDSAC memory is an-instance-of memory.

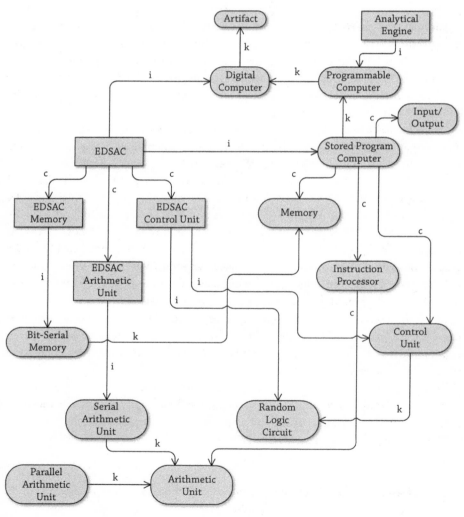

FIGURE 6.2 A partial semantic network for digital computers circa 1950.

In cognitive psychology, the assumption was that access to a particular entity in memory (a concept, a piece of factual knowledge, etc.) is by way of association—for example, the concept of "ancientness" led, by association, to Blake's painting in Robert Nichols's mind. But how is this association effected?

In 1975, Allan Collins (then) of BBN and Elizabeth Loftus of the University of Washington proposed what they called a *spreading-activation theory*.[117] The main idea of spreading activation is that given a semantic network, activating a node activates in turn other adjacent nodes, and they in turn cause activation of *their* adjacent nodes, and so on. So activation spreads through the network though decreasing in strength over time and distance from the originating node. A knowledge item is thus activated to the degree it is related to the source of the activation.

In their now-classic paper, Collins and Loftus provided experimental data as supporting evidence for their proposal. As a mechanism for activating and retrieving elements from long-term memory modeled as a semantic network, spreading activation was subsequently explored by several cognitive scientists, most notably John Anderson of Carnegie-Mellon University, as explicated in his book *The Architecture of Cognition* (1983).[118]

For AI researchers, however, whose perpetual burden is the feasibility of ideas, theories, and models as means of constructing working intelligent artifacts, the issue was to explore the semantic network as a computational structure and elicit its adequacy for representing real-world knowledge. In other words, to develop an empirically feasible theory of semantic networks.

Writing in 1975, William Woods of BBN pointed out that where semantic networks stood apart from other knowledge-representation styles was that it offered a single, unified means to represent both factual (declarative) knowledge *and* (by means of links) their associations.[119] But he issued this warning:

> One should keep in mind that the assumption that such a representation is possible is merely an item of faith, an unproven hypothesis used as the basis of the methodology. *It is entirely conceivable that no such single representation is possible.*[120]

His paper did not dispel this fear. His project, in this paper, was clarification—at great length and detail—of some fundamental matters. First, the very notion of semantics and its varied interpretations by linguists, philosophers, and computer scientists (especially programming language theorists) and how it must be perceived in AI research. Second, what philosophers of a certain persuasion might call the *problematique* of semantic network representation: the problems that must be resolved for semantic networks to make the move from "an item of faith" to, at the very least, a promising and testable hypothesis.

He noted that, from the perspective of AI, the general *desiderata* of this representation style were twofold. First, any interpretation a person may impose on a sentence must be representable.[121] Woods called this the problem of "logical adequacy."

Second, there must exist a procedure or algorithm to convert a given sentence into a representation along with algorithms that could be used to draw inferences from such representations.[122]

What Woods demanded, then, was nothing less than a knowledge-representation *system* comprising a representation language and its "preprocessor" that translated natural language utterances into this language and a "postprocessor" that could execute inferences from such representations.

Having said this, however, Woods devoted his attention to the problem of "logical adequacy." Or rather, his paper was a critical meditation, by way of many examples of different kinds of expressions, of the "logical *inadequacy* of the current (naïve) assumptions of what semantic networks do and can do."[123] He had no reservations of the importance of knowledge-representation issues and of semantic networks as a style in furthering our understanding of both human and machine intelligence. He could only hope that his paper would help stimulate research on semantic networks "in a productive direction."[124] Indeed, insofar as Woods's paper became a manifesto on "what must be done" in the cause of KR, it would prove to be seminal in the development of semantic network research.

He noted, for example, that philosophers interested in the problem of knowledge had passed over the meanings of atomic propositions—the "truth conditions of basic propositions."[125] The philosophers had focused on specifying the meanings of complex expressions assuming that the meanings of atomic propositions were known. AI researchers could not afford this luxury. If they wished to construct computational models of knowledge they must deal with the representation of meanings of atomic propositions as well as how they are assembled into larger expressions.

Drawing on philosophical jargon, Woods dwelt, as an example, on the distinction between "extensional" and "intensional" meanings of a predicate such as the word "red."[126] Extensionally, this predicate would refer to the set of all red things. Intensionally, red would refer to the abstract concept of *what it is to be red*: the quality (or, as philosophers would call it, "qualia") of redness. It is not enough for AI researchers to characterize a word such as red (or, for that matter, a term such as bit-serial ultrasonic memory in Figure 6.2) extensionally to mean a (possibly infinite) set of entities, but also to have an intensional (and finite) representation. To take another example, extensionally, the predicate "cognitive" would refer to a set such as "thinking," "language understanding," "problem solving," "planning," "designing," and so on. An intensional meaning would characterize the concept "cognitive" in some finite way. And one of the questions Woods posed was this: how to represent such a concept using semantic network notation.

As another example, Woods considered the sentence

I need a wrench.

Here, the speaker is not asserting the existence of a *particular* wrench. Rather the speaker is asserting the need for *something* (or anything) that is a wrench. So a representation of this sentence would demand an (intensional) representation of the concept of wrench as the object of the need.

Exploring and extending the "logical adequacy" of semantic networks entailed enlarging the notation itself. The reader will get a sense of these changes with the following examples.

Visualizing knowledge is an obviously attractive feature of semantic networks. A proposition such as "EDSAC is a digital computer" or "parallel arithmetic unit is a kind of arithmetic unit" (Fig. 6.2) can be nicely represented in graphical form. But, of course, propositions or sentences can be vastly more complex. Logicians have evolved a variety of formalisms to help depict such complexities, such as the universal quantifier $\forall$ denoting "for all" ["for all X such that P(X) is true"], the existential quantifier $\exists$ denoting "there exists" ["there exists Y such that Q(Y) is true"] in predicate logic, and the "temporal" operators $\square$ denoting "always" and $\diamond$ denoting "sometimes" [as in "X is always greater than 3 and sometimes less than 10," represented as "$\square$ (X > 3) and $\diamond$ (X < 10)"] in temporal logic.

In 1976, Lenhart Schubert (then) of the University of Alberta explored how the notation of semantic networks could be extended to increase its expressive adequacy to cope with such statements. Schubert proposed notational formalisms for such kinds of propositions as, for example, "9 is *necessarily* greater than 8" (a statement expressible in "modal" logic), or a "possible-world" scenario such as "If there were a major planet beyond Pluto, the number of major planets *would be* 10."[127]

As another example—this time stimulated by Woods's concern with intensional concepts but going far beyond—Anthony Maida, then of Brown University and Stuart Shapiro of the State University of New York (SUNY) Buffalo explored, in 1982, a certain kind of semantic network called a *propositional semantic network* that satisfied the following properties:

(1) Each node will represent a unique concept.
(2) Each concept is represented by a unique node.
(3) The links represent nonconceptual binary links between nodes.
(4) The knowledge represented about each concept is represented by the entire network connected to the node representing the concept.[128]

So in their formalism, in a propositional network every assertion is considered a concept and represented by a node, rather than by a linked relationship between nodes. Maida and Shapiro claimed that their formalism offered "promising approaches to several knowledge representation problems which often lead to confusion."[129] They give, as an example, a problem that has a philosophical history reaching back to the 19th-century German mathematician and logician Gottlieb Frege. In its modern, AI-contextual form it was posed by John McCarthy of Stanford University in 1979 as the "telephone number problem."[130] In their version of the problem, Maida and Shapiro considered this sentence sequence:

Pat knows Mike's telephone number. Ed dials Mike's phone number. Tony correctly believes that Mike's phone number is the same as Mary's phone number.[131]

On the basis of their formalism, Maida and Shapiro presented a semantic network representing this little narrative. Given this network, one could make certain inferences such as that, in dialing Mike's phone number, Ed is also dialing Mary's phone number; but not that Tony knows either Mike's or Mary's number.

6.20   DISAFFECTION WITH SEMANTIC NETS

These studies illustrate the general temper of research on the semantic network style. If Quillian's work in the late 1960s sounded the clarion call for a network-structural framework for associative thinking,[132] the work of such people as Woods, Schubert, and Maida and Schapiro—there were, of course, others [133]—demonstrated the exploratory, formal nature of the research conducted in the 1970s and 1980s. An abiding theme was the comparison of the semantic net as a KR formalism and its effectiveness as a calculus (for retrieving knowledge and making inferences) with other representation styles, most notably first-order predicate calculus, but also such "competitors" as frames and scripts.

These issues were brought to a head in 1990 by Lenhart Schubert (by then) of the University of Rochester. The title of his paper, "Semantic Nets Are in the Eye of the Beholder," virtually said it all.[134] Schubert voiced his disaffection quite bluntly:

I want to argue that the term "semantic nets" has become virtually meaningless, at least in its broadest sense . . . .It no longer signifies an objectively distinguishable species in the K[knowledge] R[representation] taxonomy. *All* KR schemes I have lately encountered, which aspire to cope with a large, general propositional knowledge base, qualify as semantic nets, appropriately viewed.[135]

Schubert then argued that semantic networks failed to be distinctive as a KR style in both its representational capacity (as a graph-theoretic information structure) and its computational capacity.

So should the term "semantic network" be expunged from the "AI lexicon"? Schubert's disaffection did not go that far. He suggested that perhaps the term should be reserved to refer to a specific kind of representation, proposed originally to explain associative mental processing that use, for example, spreading activation (Section 6.18) or other similar propagation processes that enable inferencing or understanding.[136] Alternatively, Schubert suggested, the term might narrowly apply to graphical representations that were appropriate for depicting hierarchical taxonomic structures.[137]

So this would confine semantic nets to hierarchical tree structures in which the links between nodes are of the "component-of" (C) or "kind-of" (K) types depicted in Figure 6.2.

6.21   RULES AS REPRESENTATIONS

In any science of the artificial—including computer science—implementation of a theory or (equivalently) a design plays the same role as a laboratory experiment does in the natural sciences. Implementation of a theory (of an artifact or class of artifacts) in the form of an operational (working) artifact serves to test—falsify or corroborate—the theory.

And insofar as implementation of KR styles are means of testing their feasibility in real-world settings then arguably *production rules* were the most tested of the knowledge representation styles that emerged in the 1970s.

The origins of production rules actually reach back to the logician Emil Post's 1943 concept of productions for the formal manipulation and transformation of symbol structures.[138] The concept of "Post productions" may also have led to the idea of defining a language's grammar in terms of "rewriting rules" proposed by linguist Noam Chomsky of MIT in 1957[139] and that, in the context of the syntax of programming languages, Robert Floyd, then of the Carnegie Institute of Technology (later Carnegie-Mellon University), called "productions" in 1961.[140] Production rules entered the AI lexicon as a formalism for modeling cognitive behavior in the early 1960s through the work of Allen Newell and was amply manifested in the book *Human Problem Solving* (1972) coauthored by Newell and Herbert Simon.[141] As a KR style production rules were thus an integral part of the Newell–Simon style of AI.

Yet, ironically, production rules received their strongest endorsement when some AI researchers, as early as the mid-1960s, came to believe that the agenda Newell and Simon had so meticulously constructed for a *general* theory of problem solving along the lines of GPS (Section 6.6) had not materialized. Domain-independent heuristics ("weak methods") such as MEA alone were insufficient to solve the kind of problems AI aspired to. Rather, it was domain-specific knowledge—and lots of it—pertaining to real-world domains (such as engineering design, computational artifact design, medical diagnosis, scientific problem solving) that needed to be represented, manipulated, and controlled to elicit intelligent behavior. This mentality was not new: It was precisely what Newell and Simon had embarked on in developing Logic Theorist in 1956.

And in this renewed mentality, production rules constituted the most favored mode of knowledge representation.

The very general architecture of a production-rule-based *system* as envisioned was of the form shown in Figure 6.3. Here, knowledge in the form of an organized body

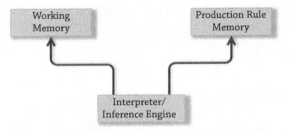

FIGURE 6.3 Organization of a production system.

of production rules (or simply *rules*) would be held in a *rule memory*. A rule is of the form

IF *condition* THEN *action*,

which specifies that if the "current state" of a cognitive (or computational) process—in thinking, planning, designing, problem solving, and so on—is such that *condition* is satisfied, then a plausible step is to perform the associated *action*.

The current state of a cognitive (or computational) process would be held in a *working memory*. This is, in fact, the problem state space of the Newell–Simon model. Thus it would depict the trajectory of a hypothetical search through this space in the course of a computation (as shown in Figure 6.1).

In addition, an *inference engine* will decide, based on the current state of working memory, which rule(s) to select, based on a match between the *condition* parts of the rules in rule memory and the current state of the working memory. A selected rule will "fire"—that is, execute the specified action—causing the state of working memory to change. This process is effected by the inference engine, which can be envisioned as a finite-state machine that cycles through three states, *match, select,* and *execute:*

CYCLE FOREVER:
    match: identify all the rules in rule memory such that their *condition* parts
              satisfy the current state of working memory; collect these rules into
              a "conflict set";
    select: select a preferred rule from the conflict set to fire;
    execute: fire the selected rule.
END CYCLE

The selection from members of the conflict-resolution set will be determined by some strategy inherent in the inference engine. This strategy may itself constitute a rule but at a higher level of abstraction than the rules contained in rule memory. It would be thus a metarule.

The consolidation of production rules as a KR style was stimulated and driven by the domain-specific "knowledge-based" *expert systems* (as the term emerged) that proliferated through the '70s and '80s—though this movement really began earlier, most notably with DENDRAL, a system developed at Stanford University through the second half of the '60s and the early '70s. DENDRAL analyzed mass-spectrographic and other chemical data and inferred plausible structures of unknown compounds.[142] By the 1980s, rule-based KRs for expert systems were being widely explored.

As a sampler of the kind of rules being identified, consider the following examples.

IF a current goal is volumetric analysis THEN separate the solid and analysis parts.
IF a current goal is fluid analysis THEN first compute water saturation.

In 1985, David Barstow showed how such rules as these, relevant to the domain of oil-well logging, were used in the design of programs for computing geological characteristics from oil-well measurements.[143]

IF signal (CTL) = 'high' at time t THEN signal (OUT) = signal (IN) at time t + 1.

This example expresses the behavior of an unidirectional pass transistor with ports IN, OUT, and CTL and was used as a rule in the Palladio VLSI design environment created in 1983 at the California Institute of Technology:[144]

IF the goal is to make the postcondition *Rout = Rin1 + Rin2*
THEN generate the microoperation sequence *Aout ← Ail + Air*; *Rout ← Aout*
and generate a new goal: *Ail = Rin1 ∧ Air = Rin2*.

This rule, based on the automatic microcode synthesis project conducted by Robert Mueller and Joseph Varghese of Colorado State University in 1985, specified a piece of knowledge relevant to the synthesis task.[145]

IF current state has a room-A *and* right-wall-of (room-A) is clear
THEN conjoin right-wall-of (room-A) with a room-C such that top = wall-of (room-C) is clear *and* right-wall-of (room-C) is clear *and* left-wall-of (room-C) is clear.
IF current context is to satisfy room configuration goal *and* site context goal
THEN achieve room configuration goal first.

These two examples, proposed in 1987 by architectural theorists John Gero and Richard Coyne of the University of Sidney, show rules relevant to building design. The second rule is a metarule that may be used to resolve conflict between two coexisting goals.[146]

Rules can be somewhat more complex than these examples. For instance, in 1983, Donald Thomas and his collaborators at Carnegie-Mellon University published a paper on the automatic synthesis of computer data paths that (among many others) described the following rule:

IF
current goal is to maximize parallelism
and there is a link from constant C1 to some Bus2 or Multiplexer2
and there is a link from Bus1/Multiplexer1 to LEFT-IN (addero
and there is a link from Bus2/Multiplexer2 to RIGHT-IN (addero
THEN
Delete link from OUT (PC) to Bus!/Multiplexer1
and delete link from C1 to Bus2/Multiplexer2
and delete link to IN(PC)
and create PC as counter.

One of the most well-known production-rule-based expert systems was MYCIN, developed in the mid–late 1970s by a team of researchers at Stanford University led by Bruce Buchanan, a computer scientist, and Edward Shortcliffe, a medical scientist.[147] Its primary goal was to ascertain how computational symbol reasoning could be used in medical diagnosis and therapy planning. It is also of great interest as a case study in experimental AI—of the efficacy of production rules as a KR style in a robust problem-solving domain—and also because of a remark made by Bruce Buchanan that MYCIN's ability to reason in the manner of a medical expert "was a valuation of the Turing test."[148]

MYCIN was implemented to diagnose meningitis. Its knowledge base comprised some 500 rules pertaining to the analysis and diagnosis of the disease. In the experiment, the knowledge base was first manually refined so that it was correctly diagnosing some 100 "training cases" of meningitis. This was, then, the "training phase" of the experiment. The actual tests then followed. The program was supplied with 10 different test cases of meningitis selected randomly from medical records. These cases were submitted to eight different medical personnel at the Stanford Medical Center—including faculty members specializing in infectious diseases, an infection disease research fellow, and a medical student. Their recommendations for therapy were collected along with the actual therapy given by the physician who had treated the patients. MYCIN's own recommendations were determined and included (without being identified) in the mix, and the total set of therapy recommendations were submitted to eight "recognized experts" in meningitis. They were asked to evaluate for each patient the 10 therapy recommendations. Their evaluations were binary: Either the recommended therapy was "appropriate" or it was not. The outcome was that MYCIN's recommendations were "at the top of the list" among a collection of recommendations from Stanford's infectious disease specialists.[149]

Hence Buchanan's remark pertaining to the Turing test. As for analyzing MYCIN's performance, the most striking part was that the analysis produced several hypotheses about rule-based expert reasoning itself, beyond the domain of medical diagnosis. In fact, these domain-independent hypotheses were implemented in another program called EMYCIN ("E" for "essential") that had the same system architecture as MYCIN but without any domain-specific knowledge base.[150] EMYCIN was itself implemented and experimented with across "many diagnostic and classification problems."[151]

This situation seemed to be a repeat of the historical situation Newell and Simon had created: abstracting from the Logic Theorist experience to GPS.

6.22   REPRESENTING INTUITIVE KNOWLEDGE (AKA NAÏVE PHYSICS)

To repeat, implementations (in the form of robust working programs, say) serve as possibly stringent tests of the validity of a theory, a model, or a design of some intelligent agent. This is why expert systems seem so impressive. But for Patrick Hayes, the trend toward hasty implementations appeared unseemly, a case of "more haste, less speed." More attention should be paid to constructing the theory or model itself. His ire was also directed toward the toy problems that so dominated AI research: artificial, contrived puzzle-like situations such as the toy block microworlds of the Minsky–Papert persuasion.

In a paper tantalizingly titled "The Naïve Physics Manifesto" (1979) Hayes issued a call for a more realistic AI, one that engaged with "non-toy" problems: the physical world of everyday experiences of shape, space, movement, time, matter, and so on.[152]

So this was the world of "naïve physics" or (as it is also sometimes called) "folk physics." This is the world of experiential, commonsense knowledge and reasoning wherein humans develop mental models or internalized theories of everyday physical phenomena and act on the basis of such privately held mental models. The knowledge here is the intuitive physics that precedes scientific physics, the intuitive technology that precedes scientific technology.

Hayes's formalization of naïve physics had certain *desiderata*:[153]

(a) It must be *thorough*; that is, it must cover the "whole range of physical phenomena."
(b) It must aspire to *fidelity* to the details of actual physical phenomena.
(c) It must acknowledge the *density* of facts: "The ratio of facts to concepts needs to be fairly high." Concepts are associated with a plethora of facts, and this must be reflected in the formalism.
(d) There must be a *uniform* representational framework for this formalism.

There is a fundamental sense in which what Hayes was advocating went against the scientific ethos. In science, as in mathematics, the goal is reductionism. In mathematics one begins with a small set of axioms and definitions—the smaller, the better. Natural science aspires to "universal" laws that "cover" entire ranges of seemingly diverse facts or observations. In mathematics and science Occam's principle reigns.

But in the realm of the everyday world of commonsense reasoning, such reductionism would be "astonishing"—even though, Hayes noted, AI researchers aspire such reductionism for their theories and models. Such "low-density representation" is anathema to his goals for a naïve physics.

Like Minsky's 1975 paper on frame theory, Hayes's "manifesto" was a call for action rather than a fully worked out technical document. Like Minsky's paper it was sprinkled with informal examples of situations that naïve physics must attend to. One of his examples, for instance, dealt with substances: *stuff* such as iron, water, sand, and so on. These can exist in different kinds of physical states: solid, liquid, paste, slime, powder, and so on. But there is a certain "default" state for each kind of stuff: solid for iron, liquid for water, powder for sand, and so on. Certain kinds of stuff change state if heated or cooled enough. Some substances "decompose" into "useless" substances (e.g., iron rusting), others "mature" into something useful (e.g., wine). Then there are the more obvious physical properties stuff possesses: rigidity, color, hardness, mal-leability, and so on. Our intuition about substances, to be formalized as knowledge, would have to take into account all such features.

Six years later, Hayes offered "The Second Naïve Physics Manifesto" (1985)[154] as a revision—or correction—of the earlier manifesto because, he confessed, the former had been overoptimistic in its reach, had underestimated the difficulties, and was "in-appropriately naïve" in some respects.[155] Still, he was encouraged by the work that had been evoked by his 1979 paper. Perhaps some of the earlier "passion" had subsided;[156] the time was appropriate for a more tempered discussion.

To illustrate in more concrete terms his vision for naïve physics than he had pursued in 1979, Hayes (somewhat ironically, given his earlier eschewal of "toy problems") chose the microworld of toy blocks as a case study.

To represent the blocks world one must begin with the concept *block*. Associated with this concept are several *states* resulting from moving blocks around a tabletop. One example of a state is the relation "on" such that one can say that a given block is on an-other block or on the table. This can be expressed by the following *axiom* (in words):[157]

(1) If s is a state and b is a block, then there is some block c such that in state s either b is on c or b is on the table.

Another concept needed for the blocks world is *state change*. This could be represented by actions that cause states to change. Hayes's examples were the actions "pickup" and "putdown" a block, respectively. Thus, for example, the result of picking up a block can be expressed by the following axioms:

(2) If a block b is picked up then that b is held in one's hand.
(3) If a block b is held then for all other entities x, b cannot be on x.
   Two possible effects of putting down are described by the following axioms:
(4) If b is held in state s and c is clear in that state then the effect of putting down b on c is that b is on c.
(5) If there are no objects that are held in state s, then the effect of putting down is to leave the state s unchanged.

Given these and other axioms, certain inferences can (or cannot) be made:

If we know that block *A* is on another block *C* on the table and *B* is also on the table, then we can infer from axiom (2) that after being picked up *A* will be held. However, we cannot infer that *B* is clear, for the axioms tell us nothing about *B*'s status.[158]

This latter situation is an example of the celebrated and much-discussed *frame problem* (not to be confused with Minskyian frame theory of KR), first identified by John McCarthy and Patrick Hayes in 1969:[159] the problem of differentiating those elements of a state that are unaffected by some action; the problem of characterizing those aspects of the "background" ("frame") of a situation, so to speak, that remain invariant following a "foreground" action.

This small example illustrates what Hayes had in mind in formulating a naïve physics view of the everyday physical world. We can *now* recognize that what Hayes was demanding of AI came to be called by computer scientists in the 1990s (appropriating a philosophical term) an *ontology* of the physical world: the structure of representable reality of the everyday world.

6.23   ONE AI OR MORE?

The very idea of AI—the idea that mental phenomena could be understood through computational means—of course had its skeptics, not only from outside the AI community but, paradoxically, a few from within. Perhaps the most well-known opponents were two philosophers, Hubert Dreyfus[160] and John Searle.[161] But, rather famously, there was also Joseph Weizenbaum,[162] one of the first generation of AI researchers. Whereas Dreyfus and Searle rejected the AI agenda on logical and empirical grounds, Weizenbaum, an apostate from within the AI community, came to reject the idea of AI on moral grounds.

But this chapter ends with a remark on another controversy. This did not concern the idea of AI—the idea itself was accepted—but rather what *form* this idea should take.

A community took gradual shape, beginning in the mid-1970s but more insistently by the mid-1980s, which rejected the direction AI had taken over the past two decades: a direction framed around the symbol system and heuristic search hypotheses (Sections 6.1, 6.5) judiciously blended with the formalism of logic: AI-as-symbol processing.

In 1985, philosopher John Haugeland of the University of Pittsburgh had labeled this direction "good old fashioned artificial intelligence" (GOFAI).[163] The "old fashioned" part must have stung the adherents of the symbol processing school. In Kuhnian terms, GOFAI became, in the eyes of the new young Turks, the old paradigm that had not met the hopes, aspirations, and expectations its practitioners had held ever since its emergence in the 1950s with Logic Theorist. Indeed (again in Kuhnian terms), it was so established that it merited paradigm status as far as AI as a science was concerned. It was what the most influential textbooks on AI, authored by its leading exponents, taught up to the mid-1980s.[164] It was what was encapsulated in the three-volume *Handbook of Artificial Intelligence* (1981) edited by Avron Barr and Edward Feignebaum.[165]

The challenge to this paradigm arose phoenix-like, in the mid-1980s, from the ashes of an earlier perspective reaching back to 1943. The challengers—perhaps they thought themselves to be revolutionaries—took their inspiration from biology, the architecture of the brain to be more precise; and thus the new philosophy came to be called *connectionism* because of the conviction of its protagonists that the proper path to enlightenment in the matter of mind, cognition, and intelligence lay in theories and models that imitated the neuronal connections in the brain.

But there was a larger metasense in which the word connectionism seems apposite— not just in the AI context but related to other realms of computer science. This more general connectionism lay in the fact that computer scientists of different persuasions had started making interesting biological connections relevant to their respective specializations. All of which leads us to the next part of this story.

NOTES

1. A. Newell and H. A. Simon, 1972. *Human Problem Solving*. Englewood Cliffs, NJ: Prentice-Hall, p. 1.

2. A. Newell, C. J. Shaw, and H. A. Simon, 1958. "Elements of a Theory of Human Problem Solving," *Psychological Review*, 65, pp. 151–166.

3. For an analysis of Simon's multidisciplinarity see S. Dasgupta, 2003. "Multidisciplinarity: The Case of Herbert A. Simon," *Cognitive Science*, 27, pp. 683–707.

4. D. N. Robinson, 1995. *An Intellectual History of Psychology* (3rd ed.). Madison: University of Wisconsin Press.

5. Newell and Simon, 1972, p. 5.

6. Ibid.

7. Ibid., p.6.

8. Ibid.

9. J. R. Slagle, 1971. *Artificial Intelligence: The Heuristic Programming Approach*. New York: McGraw-Hill, p. v.

10. A. Barr and E. A. Feigenbaum (eds.), 1981. *The Handbook of Artificial Intelligence, Volume 1*. Stanford, CA: HeurisTech Press/Los Altos, CA: Kaufman, p. 3.

11. M. L. Minsky and S. Papert, 1972. *Artificial Intelligence: Progress Report*. Artificial Intelligence Laboratory. Cambridge, MA:MIT.

12. Ibid. There are no page numbers.

13. A. Newell and H. A. Simon, 1986. "Computer Science as Empirical Inquiry: Symbols and Search," *Communications of the ACM*, 19, 3, pp. 113–126. Reprinted, pp. 287–318 in Anon, 1987. *ACM Turing Award Lectures: The First Twenty Years 1966–1985*. New York: ACM Press/Reading, MA: Addison-Wesley, p. 293. All page references are to the reprint.

14. Minsky and Papert, 1972.

15. Slagle, 1971, p. v.

16. Barr and Feigenbaum, 1981, p. 3.

17. Slagle, 1971, pp. 1–2.

18. Newell and Simon, 1972, p. 53.

19. Newell and Simon, 1976, p. 290.

20. Newell and Simon 1972, p. 20.

21. Newell and Simon 1972, pp. 20–21.

22. Newell and Simon 1976, p. 282.

23. Ibid., p. 297.

24. Ibid.

25. Ibid.

26. Ibid.

27. Ibid., p. 298.

28. Ibid., p. 300.

29. Simon's papers on bounded rationality are numerous. For a concise discussion see H. A. Simon, 1987. "Bounded Rationality," pp. 266–268 in J. Eatwell, M. Milgate, and P. Newman (eds.), *The New Palgrave: A Dictionary of Economics*. London: Macmillan. See also H. A. Simon, 1996. *The Sciences of the Artificial* (3rd ed.). Cambridge, MA: MIT Press. The process by which Simon expanded the scope of bounded rationality from the social sciences through cognitive psychology to artificial intelligence is traced in S. Dasgupta, 2003. "Innovation in the Social Sciences: Herbert Simon and the Birth of a Research Tradition," pp. 458–470 in L. V. Shavinina (ed.), *The International Handbook on Innovation*. Amsterdam: Elsevier Science.

30. Slagle, 1971.

31. A. Newell, C. J. Shaw, and H. A. Simon, 1960. "A Variety of Intelligent Learning in a General Problem Solver," pp. 153–189 in M. C. Yovits and S. Cameron (eds.), *Self-Organizing Systems*. New York: Pergamon.

32. G. W. Ernst and A. Newell, 1969. *GPS: A Case Study in Generality and Problem Solving*. New York: Academic Press.

33. H. W. J. Rittel and M. M Weber, 1973. "Dilemmas in a General Theory of Planning," *Policy Sciences*, 4, pp. 155–169.

34. H. A. Simon, 1973. "The Structure of Ill-Structured Problems," *Artificial Intelligence*, 4, pp. 181–200. Reprinted, pp. 145–166 in N. Cross (ed.), 1984. *Developments in Design Methodology*. Chichester, UK: Wiley. All page references are to the reprint.

35. Simon 1973, pp. 146–147.

36. Rittel and Weber, 1973, p. 160.

37. Ibid.

38. Ibid., pp. 161–167.

39. See H. A. Simon, 1976. *Administrative Behavior* (3rd ed.) New York: Free Press; H. A. Simon, 1957. "Rationality in Administrative Decision Making," pp. 196–206 in H. A. Simon, *Models of Man*. New York: Wiley.

40. W. R. Reitman, 1964. "Heuristic Decision Procedures, Open Constraints, and the Structure of Ill-Defined Problems," pp. 282–315 in M. W. Shelley and G. L. Bryan (ed.), *Human Judgements and Optimality*. New York: Wiley; W. R. Reitman, 1965. *Cognition and Thought*. New York: Wiley.

41. Rittel and Weber, 1973, p. 169.

42. Simon, 1973, p. 146.

43. Ibid., p. 150.

44. D. C. Brown and B. Chandrasekaran, 1986. "Knowledge and Control for a Mechanical Design Expert System," *Computer*, 19, 7, pp. 92–110.

45. N. Rosenberg and W. G. Vincenti, 1978. *The Britannia Bridge: The Generation and Diffusion of Technological Knowledge*. Cambridge, MA: MIT Press.

46. The ill-structuredness of this design problem, the gradual evolution of the form of this bridge, the nature of the problem state space, and the manner of its search are described in S. Dasgupta, 1996. *Technology and Creativity*. New York: Oxford University Press, pp. 78–86, 110–115.

47. T. S. Kuhn, 2012. *The Structure of Scientific Revolutions* (4th ed.). Chicago: University of Chicago Press.

48. Simon, 1973, p. 164.

49. Ibid., p. 165.

50. Ibid., p. 164.

51. Minsky and Papert, 1972, sections 5.1.3, 5.2.

52. Ibid., section 5.1.3.

53. Ibid.

54. S. Papert, 1980. *Mindstorms: Children, Computers and Powerful Ideas*. New York: Basic Books, p. 120.

55. P. H. Winston, 1975. "Learning Structural Descriptions From Examples," pp. 157–210 in P. H. Winston (ed.), *The Psychology of Computer Vision*. New York: McGraw-Hill.

56. G. J. Sussman, 1973. *A Computer Model of Skill Acquisition*. New York: American Elsevier, p. 1.

57. T. Winograd, 1972. *Understanding Natural Language*. New York: Academic Press.

58. Minsky and Papert, 1972.

59. Papert, 1980.

60. S. Papert, 1984. "Microworlds: Transforming Education," ITT Key Issues Conference, Annenberg School of Communication, University of Southern California, Los Angeles, CA.

61. Papert, 1980, p. 13.

62. Ibid., p. 122.

63. Ibid., p. 127.

64. Ibid., p. 128.

65. Winograd, 1972.

66. T. Winograd, 1973. "A Procedural Model of Language Understanding," pp. 152–186 in R. C. Schank and K. M. Colby (eds.), *Computer Models of Thought and Language*. San Francisco: W. H. Freeman.

67. See J. McCarthy, 1981. "History of LISP," pp. 173–185 in R. L. Wexelblat (ed.), *History of Programming Languages*. New York: Academic Press. For its place in the larger history of computer science, see S. Dasgupta, 2014. *It Began With Babbage: The Genesis of Computer Science*. New York: Oxford University Press, pp. 236–238.

68. C. Hewitt, 1969. "PLANNER: A Language for Proving Theorems in Robots," pp. 295–301 in *Proceedings of the International Joint Conference on Artificial Intelligence* (IJCAI). Bedford, MA: Mitre Corporation.

69. Winograd, 1972, p. 16.

70. Ibid.

71. M. A. K. Halliday, 1967–1968. "Notes on Transitivity and Theme in English," Parts 1–3. *Journal of Linguistics*, 3(1), pp. 37–81; 3(2), pp. 199–244; 4(2), pp. 179–215.

72. Winograd, 1972, p. 17.

73. T. Winograd, 1980. "What Does It Mean to Understand Language?" *Cognitive Science*, 4, pp. 209–241.

74. Winograd, 1980, p. 213.

75. Winograd, 1973, p. 183.

76. Ibid., p. 183.

77. M. A. Boden, 1991. *The Creative Mind.* New York: Basic Books.

78. S. Dasgupta, 2011. "Contesting (Simonton's) Blind Variation, Selective Retention Theory of Creativity," *Creativity Research Journal*, 23, 2, pp. 166–182.

79. S. Dasgupta, 1997. "Technology and Complexity," *Philosophica*, 59, 1, pp. 113–139.

80. Winograd, 1973, p. 183.

81. Ibid., pp. 183–184.

82. Ibid., pp. 185–186.

83. F. C. Bartlett, 1932. *Remembering.* Cambridge: Cambridge University Press, p. 210.

84. J. Piaget, 1955. *The Child's Construction of Reality.* New York: Routledge & Kegan Paul. See also J. Piaget, 1976. *The Child and Reality.* Harmondsworth, UK: Penguin.

85. M. Minsky, 1975. "A Framework for Representing Knowledge," pp. 211–277 in Winston (ed.), 1975. (This paper originally appeared in 1974 as an MIT AI Laboratory Memo.)

86. R. C. Schank, 1973. "Identification of Conceptualization Underlying Natural Language," pp. 187–247 in Schank and Colby, 1973.

87. K. M. Colby, 1973. "Simulations of Belief Systems," pp. 251–282 in Schank and Colby, 1973.

88. Minsky, 1975, p. 212.

89. Ibid.

90. Ibid.

91. Ibid., p. 213.

92. Ibid., p. 212.

93. Ibid., p. 212.

94. Ibid.

95. Minsky and Papert, 1972.

96. R. C. Schank and R. P. Abelson, 1975. "Scripts, Plans, and Knowledge," pp. 151–157 in *Proceedings of the International Joint Conference on Artificial Intelligence, Volume 1.* San Francisco: Morgan Kaufmann.

97. R. C. Schank and R. P. Abelson, 1977. *Scripts, Plans, Goals and Understanding.* Hillsdale, NJ: Lawrence Erlbaum.

98. Schank and Abelson, 1975, p. 151.

99. Schank, 1973, pp. 187–188.

100. Schank and Abelson, 1973, p. 152.

101. See, e.g., Barr and Feigenbaum, 1981, pp. 216–222.

102. K. R. Popper, 1968. *The Logic of Scientific Discovery.* New York: Harper & Row.

103. Schank and Abelson, 1975, p. 153.

104. R. J. Brachman and B. C. Smith (eds.), 1980. "Special Issue on Knowledge Representation." *SIGART Newsletter*, 70, February.

105. A. Newell, 1982. "The Knowledge Level," *Artificial Intelligence*, 18, pp. 87–127, esp. p. 89.

106. B. C. Smith, 1982. "Prologue," in "Reflections and Semantics in a Procedural Language," PhD dissertation. Tech. Report MIT/CCS/TR-272. Massachusetts Institute of Technology. The "Prologue" is reprinted, pp. 31–40 in R. J. Brachman and H. J. Levesque (eds.), 1985. *Readings in*

Knowledge Representation. San Mateo, CA: Morgan Kaufmann. The knowledge-representation hypothesis is on p. 33 of the reprint.

107. H. A. Simon, 1975. "Style in Design," pp. 287–309 in C. M. Eastman (ed.), *Spatial Synthesis in Computer-Aided Building Design*. New York: Wiley.

108. D. G. Bobrow and T. Winograd, 1977. "An Overview of KRL: A Knowledge Representation Language," *Cognitive Science*, 1, 1, pp. 3–46. Reprinted, pp. 264–285 in Brachman and Levesque, 1985. All page references are to the reprint.

109. Bobrow and Winograd, 1977, p. 265.

110. P. J. Hayes, 1979. "The Logic of Frames," pp. 46–61 in D. Metzing (ed.), *Frame Conceptions and Text Understanding*. Berlin: de Gruyter. Reprinted, pp. 288–295 in Brachman and Levesque, 1985. Page references are to the reprint.

111. Hayes, 1979, p. 294.

112. Bobrow and Winograd, 1985, p. 272.

113. Hayes, 1979, p. 294.

114. R. B. M. Nichols, 1942. "The Birth of a Poem," pp. 104–126 in R. E. M. Harding (ed.), *An Anatomy of Inspiration* (2nd ed.). Cambridge: W. Heffer.

115. Other terms continued to be used contemporaneously, such as "associative network" and "conceptual network."

116. Dasgupta, 2014, chapter 2.

117. A. M. Collins and E. F. Loftus, 1975. "A Spreading Activation Theory of Semantic Processing," *Psychological Review*, 82, pp. 407–428.

118. J. R. Anderson, 1983. *The Architecture of Cognition*. Cambridge, MA: Harvard University Press, chapter 3.

119. W. A. Woods, 1975. "What's in a Link? Foundations for Semantic Networks," pp. 35–82 in D. G. Bobrow and A. M. Collins (eds.), *Representation and Understanding: Studies in Cognitive Science*. New York: Academic Press. Reprinted, pp. 218–241 in Brachman and Levesque, 1985. All page references are to the reprint.

120. Woods, 1975, p. 222. Italics added.

121. Ibid., p. 223.

122. Ibid.

123. Ibid., p. 240. Italics added.

124. Ibid.

125. Ibid., pp. 220–221.

126. Ibid., p. 225.

127. L. K. Schubert, 1976. "Extending the Expressive Power of Semantic Networks," *Artificial Intelligence*, 7, pp. 163–198.

128. A. S. Maida and S. C. Shapiro, 1982. "Intensional Concepts in Propositional Semantic Networks," *Cognitive Science*, 6, 4, pp. 291–330. Reprinted, pp. 170–189 in Brachman and Levesque, 1985. Page references are to the reprint.

129. Maida and Shapiro, 1982, p. 170.

130. J. McCarthy, 1979. "First-Order Theories of Individual Concepts and Propositions," pp. 129–147 in J. E. Hayes, D. Michie, and L. I. Mikulich (eds.), *Machine Intelligence* 9. Chichester, UK: Ellis Horwood.

131. Maida and Shapiro, 1982, p. 177.

132. For a systematic and scholarly account of a general move from associative thinking to "structuralism" as it was understood circa 1980, the reader is referred to the book *From*

Associations to Structure (Amsterdam: North-Holland, 1980) by Kellogg V. Wilson of the University of Alberta.

133. An important collection of papers was assembled in N. V. Findler (ed.), 1979. *Associative Networks: Representation and Use of Knowledge by Computers.* New York: Academic Press.

134. L. K. Schubert, 1990. "Semantic Nets Are in the Eye of the Beholder." Tech. Report 346, Department of Computer Science, University of Rochester, Rochester, New York.

135. Ibid., p. 2.

136. Ibid., p. 11.

137. Ibid.

138. E. Post, 1943. "Formal Reductions of the General Combinatorial Decision Problem," *American Journal of Mathematics*, 65, pp. 197–268.

139. N. Chomsky, 1957. *Syntactic Structures.* The Hague: Mouton.

140. R. W. Floyd, 1961. "A Descriptive Language for Symbol Manipulation," *Journal of the ACM*, 8, 4, pp. 579–584.

141. Newell and Simon, 1972.

142. E. A. Feigenbaum, B. G. Buchanan, and J. Lederberg, 1971. "On Generality and Problem Solving: A Case Study Using the DENDRAL Program," pp. 165–190 in B. Meltzer and D. Michie (eds.), *Machine Intelligence 6.* Edinburgh: Edinburgh University Press; B. G. Buchanan and E. A. Feigenbaum, 1978. "DENDRAL and Meta-DENDRAL: Their Application Dimension." *Artificial Intelligence,* 11, pp. 5–24.

143. D. R. Barstow, 1985. "Domain-Specific Automatic Programming," *IEEE Transactions on Software Engineering*, SE-11, 11, pp. 1321–1336.

144. H. Brown, C. Tong, and G. Foyster, 1983. "Palladio: An Exploratory Environment for Circuit Design," *Computer*, 16, 12, pp. 41–56.

145. R. A. Mueller and J. Varghese, 1995. "Knowledge-Based Code Selection in Retargetable Microcode Synthesis," *IEEE Design and Test*, 2, 3, pp. 44–55.

146. J. S. Gero and R. D. Coyne, 1987. "Knowledge-Based Planning as a Design Paradigm," pp. 339–373 in H. Yoshika and E. A. Waterman (eds.), *Design Theory for CAD.* Amsterdam: Elsevier Science.

147. R. Davis, B. G. Buchanan, and E. A. Shortcliffe, 1977. "Production Rules as Representation for a Knowledge-Based Consultative Program," *Artificial Intelligence*, 8, 1, pp. 15–45.

148. B. G. Buchanan, 1988. "AI as an Experimental Science," pp. 209–250 in J. H. Feltzer (ed.), *Aspects of Artificial Intelligence.* Dordecht, The Netherlands: Kluwer.

149. Ibid., p. 213.

150. B. G. Buchanan and E. H. Shortcliffe, 1984. *Rule-Based Expert Systems: The MYCIN Experiments of the Stanford Heuristic Programming Project.* Reading, MA: Addison-Wesley.

151. Buchanan, 1988, p. 214.

152. P. Hayes, 1979. "The Naïve Physics Manifesto," pp. 240–270 in D. Michie (ed.), *Expert Systems in the Micro-Electronic Age.* Edinburgh: Edinburgh University Press. Reprinted, pp. 171–205 in M. Boden (ed.), 1990. *The Philosophy of Artificial Intelligence.* Oxford: Oxford University Press. All page references are to the reprint. The quote is from p. 171.

153. Ibid., p. 172.

154. P. J. Hayes, 1985. "The Second Naïve Physics Manifesto," pp. 1–36 in J. R. Hobbs and R. C. Moore (eds.), *Formal Theories of the Commonsense World.* Norwood, NJ: Ablex. Reprinted, pp. 468–485 in Brachman and Levesque, 1985. All page references are to the reprint.

155. Ibid., p. 468.

156. Ibid.

157. Hayes actually describes his axioms in the language of predicate logic.

158. Hayes, 1985, p. 471.

159. J. McCarthy and P. J. Hayes, 1969. "Some Philosophical Problems from the Standpoint of Artificial Intelligence," pp. 463–502 in D. Michie and B. Meltzer (eds.), *Machine Intelligence 4*. Edinburgh: Edinburgh University Press.

160. H. E. Dreyfus, 1979. *What Computers Can't Do* (Revised ed.). New York: Harper & Row.

161. J. R. Searle, 1980. "Minds, Brains and Programs," *Behavioral and Brain Sciences*, 3, pp. 417–424; J. R. Searle, 1985. *Minds, Brains and Science*. London: BBC Publications.

162. J. Weizenbaum, 1976. *Computer Power and Human Reason*. San Francisco, CA: W. H. Freeman.

163. J. Haugeland, 1985. *Artificial Intelligence: The Very Idea*. Cambridge, MA: MIT Press.

164. See, e.g., P. H. Winston, 1977. *Artificial Intelligence*. Reading, MA: Addison-Wesley; N. J. Nilsson, 1980. *Principles of Artificial Intelligence*. Palo Alto, CA: Tioga Publishing Company; E. Charniak and D. McDermott, 1985. *Introduction to Artificial Intelligence*. Reading, MA: Addison-Wesley; M. R. Gensereth and N. J. Nilsson, 1987. *Logical Foundations of Artificial Intelligence*. Los Altos, CA: Morgan Kaufmann.

165. Barr and Feigenbaum, 1981.

7

MAKING (BIO)LOGICAL CONNECTIONS

7.1   AN ODD COUPLE

At first blush, computing and biology seem an odd couple, yet they formed a liaison of sorts from the very first years of the electronic digital computer.[1] Following a seminal paper published in 1943 by neurophysiologist Warren McCulloch and mathematical logician Warren Pitts[2] on a mathematical model of neuronal activity, John von Neumann of the Institute of Advanced Study, Princeton, presented at a symposium in 1948 a paper that compared the behaviors of computer circuits and neuronal circuits in the brain. The resulting publication was the fountainhead of what came to be called *cellular automata* in the 1960s.[3] Von Neumann's insight was the parallel between the abstraction of biological neurons (nerve cells) as natural binary (on–off) switches and the abstraction of physical computer circuit elements (at the time, relays and vacuum tubes) as artificial binary switches. His ambition was to unify the two and construct a formal universal theory.

One remarkable aspect of von Neumann's program was inspired by the biology: His universal automata must be able to *self-reproduce*. So his neuron-like automata must be both computational and constructive.

In 1955, invited by Yale University to deliver the Silliman Lectures for 1956, von Neumann chose as his topic the relationship between the computer and the brain. He died before being able to deliver the lectures, but the unfinished manuscript was published by Yale University Press under the title *The Computer and the Brain* (1958).[4] Von Neumann's definitive writings on self-reproducing cellular automata, edited by his one-time collaborator Arthur Burks of the University of Michigan, was eventually published in 1966 as the book *Theory of Self-Reproducing Automata*.[5]

7.2 "A CONFEDERACY OF DUNCES"

A possible structure of a von Neumann–style cellular automaton is depicted in Figure 7.1. It comprises a (finite or infinite) configuration of cells in which a cell can be in one of a finite set of states. The state of a cell at any time t is determined by its own state and those of its immediate neighbors in the preceding point of time $t - 1$, according to a state transition rule.[6] At any given time, the state of the whole system is defined by the set of individual cell states. Such a system along with an assigned initial system state (at time $t = 0$) becomes a fully defined cellular automaton. So we can imagine a cellular automaton as a *cellular space.*

Cellular automata studies as new branch of automata theory entered the portals of the newly established computer science in the 1960s. Arthur Burks's edited volume *Essays on Cellular Automata* (1970)[7] and the monograph *Cellular Automata* (1968) by Edgar Codd[8]—whom we have earlier encountered as the inventor of relational database theory (Chapter 2, Section 2.16)—based on his doctoral dissertation at the University of Michigan, captured the technical spirit of cellular automata research in the 1960s. The intent was always to explore the computational and self-constructive capacity of cellular automata. The point was that though the computational capacity of the individual cells could be quite primitive, the power lay in the *collective* behavior of the interconnected ensemble of cells. A cellular automaton was envisioned (with apologies to the novelist John Kennedy O'Toole) as a "confederacy of dunces" out of

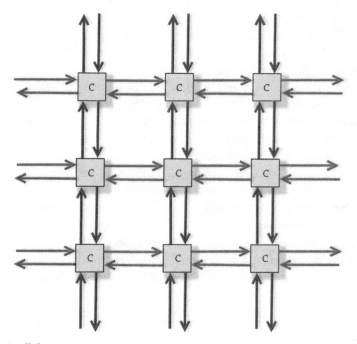

FIGURE 7.1 A cellular space.

which a certain genius would emerge. It was this characteristic of the cellular space that most intrigued researchers.

7.3 INVENTING, DESIGNING, EXPLORING CELLULAR SPACE

But for designers of practical computational artifacts, the cellular space became a terra incognita ripe for exploration in the 1970s, in directions quite different from the abstract automata-theoretic path. In particular, the prospect of a highly regular space of identical processing elements as a source of parallel processing (Chapter 4) to enhance either the computational or communicative throughput of computational artifacts was seductive, to say the least.

The emergence of VLSI technology (Chapter 3, Section 3.7) was hugely influential in this growth of interest in the design of physical (rather than abstract) cellular spaces. As we have noted (in Chapter 3), the density of circuit components in IC chips being fabricated in the '70s demanded, for the sake of complexity management, highly regular structures for laying out circuit elements on a planar chip. The quest for harnessing potential parallelism in such physical cellular spaces followed the track laid out by the technological imperative.

The biological connection in this exploration may sometimes appear highly tenuous or none at all. Yet the very idea of computing in cellular space reaches back through cellular automata to neuronal structures of the brain. And for the more reflective researchers and designers of practical cellular systems there was an awareness of this legacy.[9]

7.4 NETWORKING IN CELLULAR SPACE

Driven by the needs of parallel processing, cellular spaces of a certain kind became the subject of intense scrutiny in the 1970s and remained of great interest in the following decade. The concern was with the design of efficient *interconnection networks* for multiprocessor and multicomputer systems: a concern pertaining to computer architecture design (Chapter 3).

In very general terms, given a set of devices (e.g., processing units) that need to communicate with one another, an interconnection network (IN) is an arrangement of switching elements (or, simply, switches) that facilitate communication paths between pairs of such devices. The connections may be between processors or between processors and memory modules (Figure 7.2). Such was the interest in the design of INs prompted by the needs of parallel processing that, by the end of the 1980s, authors of textbooks on parallel processing were able to present a rich taxonomy of INs.[10] Our interest here is a subclass of INs that manifested cellular architectures.

In all INs the basic building block is the *switch*. In cellular INs, the switch *is* the cell. A two-input, two-output (2 × 2) switch can be set to one of four states; by setting a

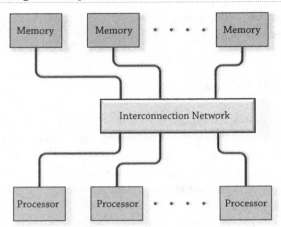

FIGURE 7.2 Processor–memory interconnection architecture.

switch to one of these four states, four pairs of input–output (I/O) combinations can be set.

A major concern of IN designers was the number of switches needed to implement an IN architecture. Here, we look at a particular IN called the Ω network designed by Duncan Lawrie of the University of Illinois circa 1975.[11] This is an instance of a "multistage" IN—that is, a network consisting of two or more "stages" (pictorially, columns), each of one or more switches. In such a network it is usually possible to establish a transmission path between arbitrary pairs of devices connected to the IN.

To understand the cellular structure of the Ω network and its behavior, we need to understand the structure and behavior of its component stages, all of which are identical. Each stage happens to be a single-stage network developed and studied by Harold Stone of the University of Massachusetts circa 1971 and called the perfect shuffle (or "shuffle exchange").[12] In this structure a processor (say) I is connected to another processor J such that J's address is obtained from I's address by a circular left shift by one position of the binary representation of I.

The perfect shuffle constituted a single stage in Duncan Lawrie's Ω network. Figure 7.3 shows an 8 × 8 Ω network consisting of three such stages, each stage containing four 2 × 2, four-state switches. Here we witness a case of networking in cellular space. More generally, for some $N = 2^\gamma$, where γ is an integer, an $N \times N$ Ω network will consist of $\gamma = \log_2 N$ identical perfect shuffles consisting of $N/2$ switches. Such a network thus will require $O(N \log_2 N)$ switches.

Data can be transmitted through the Ω network from a particular input processor to a particular output processor by setting the switches to specific states according to a simple algorithm that draws on the addresses of the input and output processors. For example, suppose we wish to send data from processor $I = 4$ to processor $J = 7$. Figure 7.3 shows the settings of the switches (indicated by dotted lines) required to bring about the required transfer.

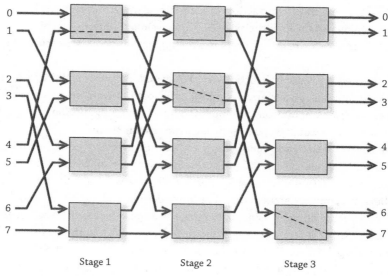

FIGURE 7.3 An 8 × 8 Omega network.

In the Ω network there is, in fact, exactly one path between any I/O pair of processors. So, for a given mapping from one set of inputs to one set of outputs the algorithm will always provide a unique set of paths to achieve the mapping. However, there may be situations in which a particular pair of unique paths may conflict with one another because either they require two different states of the same switch or they require the same path connecting the stages. For example, it can be verified that a mapping from input processors (0, 4) to output processors (4, 7) will engender such a conflict. In his 1975 paper, Duncan Lawrie presented and proved several theorems that established certain classes of feasible mappings for the Ω network.

7.5 SYSTOLIC PULSATIONS IN CELLULAR SPACE

The brain was certainly a major biological inspiration to computer scientists. But, as we will see, it was not the only biological influence. A particularly imaginative linkage due to H. T. Kung of Carnegie-Mellon University, drew upon an altogether different biological analogy.

Kung was interested in exploiting VLSI to build special-purpose machines on a single chip. These could then be attached to a general-purpose host computer. As befitting VLSI's technological imperative, these special-purpose processors would possess highly regular structures.

Circa 1978–1979, Kung conceived the idea of the *systolic array*, a cellular network of primitive processing *cells* that would perform a particular computation in a regular, synchronous fashion.[13] In using the word systolic Kung was appealing to the action of the heart, its periodic contraction causing blood to be pumped out by and from the heart and pulsed through the arteries. The physiological term for this contraction is "systole." (In measuring blood pressure, the upper number is a measure of systolic pressure.)

Kung likened his artifact's action to the systolic action of the heart at each "beat": Each cell in a systolic array would "pump" data in from an adjacent cell, perform some simple computation, and "pump" data out to a neighboring cell. In effect, a regular flow of data would be pulsed through the cellular space.

In the course of the early 1980s Kung and his collaborators published a series of papers describing several special-purpose single-chip systolic arrays for a variety of applications.[14] Very quickly, the idea of systolic arrays caught the imagination of researchers in computer architecture, hardware design, and algorithm design, and a whole new research domain—a subparadigm, so to speak—came into being. Through the 1980s, the first conferences on systolic arrays were held, and the first textbooks on the topic were published. The originality of Kung's conception was thus highly *consequential*: It was C-original; and Kung was C-creative (Chapter 6, Section 6.13).

One characteristic of a special-purpose computational artifact (whether this is a Turing machine or a physical artifact) is that its architecture design and its application design are tightly coupled. Thus, in the case of a systolic system, algorithm and architecture are entirely "made for each other," they are symbiotic. If the architecture is cellular so must be the algorithm, prompting Kung to use the suggestive term *cellular algorithm* to describe such systems.[15] Elsewhere, he and his collaborators preferred the term *systolic algorithm*.[16] Regardless of these terms, the significant point is that this constituted a different class of algorithms, that would share the following properties.[17]

(1) Systolic algorithms will be implemented using only a few types of quite simple cells.
(2) Their data and control flow will be simple and regular.
(3) Architecturally, systolic algorithms will exploit pipelining and multiprocessing styles of parallel processing (Chapter 4, Section 4.13) with the vital caveat that the stages of the pipeline or the processing elements of the multiprocessor would comprise simple cells.
(4) The parallelism manifested in systolic arrays would be "fine grained"; that is, the computational units participating in parallel activities would be very simple, very primitive.

H. T. Kung's originality—and that of his collaborators, collectively the first explorers of systolic computing—lay in no small measure in their ability to imagine a cellular space for this kind of fine-grained parallelism. They invented a new style of parallel thinking that others would emulate. Therein lay Kung's C-creativity.

Not unexpectedly, given the strong empirical practice of building real-world experimental systems at Carnegie-Mellon University, systolic computing research expanded into a project, beginning in 1984, to build a high-performance *programmable systolic array computer.* The resulting machine, called Warp, was still an "attached processor" connected to a general-purpose host. Only now, Warp could be programmed to implement a variety of applications including vision processing for robot vehicles, signal processing, magnetic resonance imaging (MRI) image processing, and scientific computations.[18] Warp was, in a manner of speaking, a general-purpose attached computer. Two identical prototypes, each consisting of a linear array of 10 identical cells, were constructed by General Electric and Honeywell, and they both became operational at Carnegie-Mellon University in 1986. A faster, "production-grade" version, called PC Warp, was constructed by General Electric and became operational in 1987.

Cellular computing had indeed traveled far from its abstract roots in von Neumann's vision.

7.6   DREAMS OF A MATERIAL "THINKING MACHINE"

Despite the appeal to the physiological systole and Kung's passing acknowledgment of cellular automaton, the biological connection never went further. Systolic arrays—whether special-purpose, algorithm-hardwired chips or the Warp-like programmable variety—were out and out computational artifacts driven by the possibilities of VLSI technology. But, in this same technological milieu, there were others who could not forget the brain. Some dreamed of a massively parallel machine architecture modeled on the neuronal architecture of the brain. Others reasoned that if neuronal architecture is the source of natural intelligence (which it undoubtedly is) then perhaps the pursuit of AI should profit from a biological analogy.

Of course, the discipline of AI had been explored to some extent along these lines: A semantic network was an exemplar (Chapter 6, Section 6.19). But this line of thinking had been thus far largely at the conceptual—or abstract—level.

In the first half of the 1980s, as a doctoral student at MIT's Artificial Intelligence Laboratory, Daniel Hillis conceived the idea of a material thinking machine that would blend these two broad biological connections. Hillis called his conception the *Connection Machine*.[19] Later in the '80s, as a cofounder of Thinking Machines Corporation, he would engage in building versions of this architectural idea.

There is a double entendre in the name of this machine: Literally, it might be construed to mean a machine made up by connecting a very large number of identical processors: a physical embodiment of a cellular automaton as he suggested by the title of one of his papers.[20] Analogically it was an idea that connected computer architecture to brain architecture: a "thinking machine," as he would audaciously call it.[21]

Hillis's idea originated in an empirical paradox. Comparing the computer with the brain, he noted that the switching speed of a transistor (the time to switch from one binary state to another—"off" to "on" or vice versa) circa mid-1980s was of the order of a few nanoseconds. The transistor was a million times faster than the milliseconds-long switching speed of neurons. Yet, why was it that the brain can execute a cognitive task (such as making sense of the scene in a painting) so much faster than even the fastest computer of the time?[22]

It was true that, according to the estimates of the time, the brain contained some 10^{10} neurons, an order of magnitude more than the number of circuit elements in computers of the time. Assuming that each neuron can switch states no more than a thousand times a second, the brain's capacity to switch states would be about 10^{13} per second. The computer, in contrast containing about 10^9 transistors each with a switching rate of 10^9 per second, would switch at the rate of 10^{18} per second—about 100,000 times greater than the brain's switching rate. Yet the brain's speed for cognitive processing was vastly superior to that of the computer.

Thus went Hillis's argument. His explanation of this paradox was that the computer's inefficiency lay in bad design. The "modern" computer's architecture was as old as the first electronic stored program computer conceived in the mid-1940s.[23] And for the next two decades processors were much faster and costlier than memory because the physical technology for the two were different. Memory and processor were kept physically and logically apart, and the more expensive processor was kept as busy as possible by such innovations of the late '50s and '60s as memory hierarchy, multiprogramming, and time sharing.

In the meantime, however, the physical technology had changed. With the advent and incredible progress in IC technology (leading to VLSI in the late '70s) processors and memories were being made from the same technology: They were all made of transistors.

The speed disparity still prevailed but was much less; the cost disparity still prevailed but was much less. Moreover, because a very large part of the "silicon real estate" was given to memory, though the processor could still be kept busy most of its operational life, much of memory would be idle for much of *its* operational life, because only a small segment of memory would read from or written into in any given slice of time. So in the operation of a computer, the actual proportion of transistors switching states would be much less than the theoretical maximum as previously calculated. The "modern" computer, circa 1985, was quite inefficient in terms of utilizing its switching capacity.

For Hillis, the resolution of this dilemma entailed something like a paradigm shift: Banish the classical, stodgy, yet still-prevailing architecture and replace it with a different architecture in keeping with the spirit of the (technological) times. Hillis was by no means the only person who was urging a kind of revolution in computer design. But his specific idea differed from others. He proposed a uniform machine architecture comprising a network of "hundreds of thousands or even millions of tiny processing cells."[24] Each cell would comprise a single processor and memory (P/M) complex, all the cells would have the capacity to be active concurrently. Each P/M cell, then, would perform its own "local" process. The computing power of the machine would lie in the interconnectedness of its P/M cells and in its potential parallel processing capacity rather than in its power of the individual P/M cells.

This, then, in very broad outline, was the concept underlying the Connection Machine. Its analogy with the neuronal architecture was quite evident. And, as we have noted, Hillis referred to it as a machine architecture based on cellular automata. Hillis was also mindful of its relation to systolic arrays. The Connection Machine defined a cellular space as did its illustrious von Neumann–inspired predecessor and its Kungian contemporary.

But the Connection Machine as an idea caught the imagination of the larger world in a way that the systolic array did not: For instance, in December 1986 *The New York Times Magazine* carried a lengthy article on Hillis and his vision. And in 1993 the film director Steven Spielberg placed the Connection Machine as the supercomputer in his film *Jurassic Park*.[25] Perhaps this was because of the ambition of its reach: its promise of using "hundreds of thousands or even millions" of processing elements—albeit low-performing ones—all working harmoniously and in parallel (but not synchronously in zombie fashion as in systolic arrays), and following a different architectural style may have seemed more exciting to many, even computer scientists, than the mighty but stodgy supercomputers of the time that were all, still, designed according to the von Neumann style. Moreover, it was conceived as a *thinking machine*: a machine for AI. It was no coincidence that while Hillis was imagining the Connection Machine into existence his doctoral dissertation advisor and mentor Marvin Minsky was articulating his idea of the mind as a "society" of essentially "mindless" agents. Minsky's book *The Society of Mind* (1985) was first published the same year as Hillis's *The Connection Machine* appeared.[26] And Hillis would acknowledge that "Most of the ideas in the book have their root, directly or indirectly in discussions with Marvin."[27] (Minsky was also a cofounder of Thinking Machines Corporation.[28]) The Connection Machine embodied this spirit of a community of relatively humble physical agents out of which *an intelligence emerges*.

This idea of intelligence as an *emergent* property of a society of intrinsically "unintelligent" agents was clearly in Hillis's mind. The Winter 1988 issue of *Daedalus* (Journal of the American Academy of Arts and Sciences), devoted to AI, included an essay by Hillis on intelligence as an emergent behavior of randomly connected neurons. One could then envision constructing a "thinking machine" by "simply hooking together" a large collection of artificial neurons. From the notion of emergence, one might expect that eventually, after reaching a certain "critical mass," such a network would eventually "spontaneously begin to think."[29]

So one can view the building of the Connection Machine as the instrument for such an experiment: an experimental apparatus that would facilitate the testing of a theory of intelligence as emergent behavior.

The Connection Machine would indeed be a "confederacy of dunces" from which an intelligence was expected to emerge (Section 7.2). But what kind of behavior must these "dunces" manifest? And what would be the nature of their confederacy? Hillis's problem was thus—despite his airy remark about "simply hooking together a large network of artificial neurons"—an engineering design and implementation problem.

Consider the problem of simulating a VLSI circuit consisting of several hundred thousand transistors. Such simulation is essential in the empirical verification of VLSI circuit designs. In an actual VLSI circuit the transistors would be active concurrently; so also in the simulated circuit. Each transistor in the circuit will be simulated by an individual processing element, but the processors must be connected in the same pattern as the physical circuit being simulated. This is not an AI problem, but its scale and the demands of massive parallelism in the circuit made its simulation an ideal problem for the Connection Machine.

Or consider the representation and behavior of a semantic network (Chapter 6, Section 6.19). Each concept in the network can be represented by an individual cell, and the connections between cells will represent the relationships between concepts in the semantic net. The pattern of connections between the cells must conform to the patterns of links between the concepts.

Extracting a piece of knowledge or deducing new knowledge efficiently from a semantic network entails concurrent spread of activation through relevant parts of the network (Chapter 6, Section 6.19). Such spreading activation may be modeled by finding the shortest paths between node (or concept) pairs in the semantic net—that is, by computing the shortest path between processor cells in the representation of the network. The relevant computational solution is to apply a "shortest-path algorithm" between vertices of a graph, the vertices denoting the nodes of the network, the edges of the graph denoting links. But even the most efficient shortest-path algorithms designed for sequential computation when applied to a graph containing hundreds of thousands of vertices, nodes, or concepts will be unacceptably slow. Spreading activation must occur concurrently: A concurrent shortest-path algorithm must be used.

These and other examples led Hillis to formulate two "motherhood" requirements for his Connection Machine:[30]

Requirement 1: There must be a very large number of processor cells in the machine that can be allocated as befitting the size of various computational problems.

Requirement 2: The patterns of connectivity between processor cells must be variable, hence programmable, according to the demands of individual computational problems.

Programmability thus set the Connection Machine apart from Kung's original vision of a systolic array in which connections were hardwired according to the nature of specific computational tasks—although, as we have seen (Section 7.5), the Warp was a programmable systolic computer. Of course, at the *physical* level the connections between cells in the Connection Machine would have to have a fixed topology—a fixed interconnection network. Connection variability would then be a virtual feature, an abstraction effected by software.

Hillis's vision entailed in essence the following:

The Connection Machine would be composed of a physically regular configuration of cells, each cell being a processor element along with a "local" memory (P/M complex). In the prototype design (called CM-1) there were 65,536 (64K) P/M cells with 4096 (4K) bits of memory per cell. So, in effect, the CM-1 could be viewed as a 32-Mbyte "intelligent memory" or (alternatively) as a 64K processor complex, each processor possessing its own memory.

The P/M cells would be connected by means of an appropriate interconnection network (of the kind discussed in Section 7.4).

The Connection Machine would also be connected to a general-purpose host computer by means of a bus.

The Connection Machine would also be connected by means of a high-bandwidth I/O channel to peripheral devices.

Activation of the P/M cells are "orchestrated" by a host computer. A single host machine command would cause several thousand P/M cells to be energized simultaneously.

The P/M cells are themselves incapable of initiating their own or their neighbors' activations. That must emanate from the host machine by way of a microprogrammed control unit.

We see, then, that although the Connection Machine by itself has the feature of what Michael Flynn would have called a MIMD machine (Chapter 3, Section 3.19), the host machine–Connection Machine complex is a variant of Flynn's SIMD category.

What Flynn's taxonomy could not depict is the Connection Machine's interconnection network topology. In CM-1, the chosen topology was the *Boolean n-cube* (also called the *hypercube*)[31] conceived, designed, and first implemented by Charles Seitz of the California Institute of Technology, circa 1985.[32] Figure 7.4 shows a Boolean 3-cube connecting eight P/M cells. Here, if the P/Ms are numbered in binary form, each P/M is connected to another P/M whose binary representation differs by just one binary digit.

7.8   THE CONNECTION MACHINE'S LISP

The LISP family of programming languages (dialects of the language invented in the mid-1950s by John McCarthy[33]) constitutes, collectively, the *lingua franca* of the AI community. Given Hillis's aspiration that the Connection Machine would be a thinking machine, it was inevitable that some version of LISP would be used to write Connection Machine programs in. The new variant was named *CmLisp*. Like many other high-level programming languages, CmLisp was an abstraction of the Connection Machine; yet an abstraction that revealed something of an underlying machine architecture, like a pentimento. Or, to take an alternative view and paraphrasing Hillis, the Connection Machine was the hardware embodiment of CmLisp.[34]

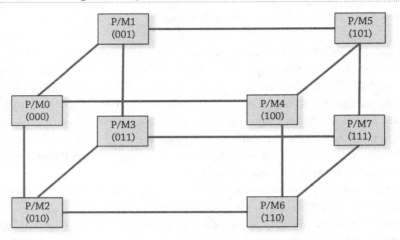

FIGURE 7.4 A Boolean 3-cube.

The principal data structure in CmLisp was called the *xector*: a set of abstract processors with values stored in each processor. The xector was thus a distributed data structure that mapped onto a set of P/M cells. Concurrent processing in CmLisp entailed concurrent operations by the abstract processors of xectors on their local data. For example, two xectors could be added by simultaneously adding the values in the corresponding xector memory elements and producing a new xector. A single add operation thus could effect the addition of possibly hundreds of thousands of data elements.[35] The language offered several types of concurrent operations that would create, modify, combine, or reduce xectors, and much as in a vector programming language such as APL, invented by Kenneth Iverson of IBM, circa 1962,[36] or Actus, invented by Ron Perrot of Queen's University, Belfast, circa 1979 (Chapter 4, Section 4.10), various vector operations were provided.

The first physical Connection Machine, CM-1, with 65,536 P/M cells—with a price tag of $5 million—was released by Thinking Machines in 1985. A year later an improved version, CM-2, with floating-point hardware and FORTRAN support, was released. Alas, by 1990, other massively parallel supercomputers with more conventional von Neumann-ish architectures had come into being. And although a successor to CM-2 called CM-5 was announced in 1991, by then other supercomputers were being preferred by users. And not necessarily for AI applications. In 1993 Daniel Hillis left Thinking Machines and the company filed for bankruptcy.[37] All that remained of Hillis's dream of a real thinking machine was what was documented in his book.

7.9   NEO-DARWINIAN OVERTURES

We have seen earlier that for some AI researchers (notably James Slagle, Herbert Simon, and Allen Newell), intelligence entailed adaptation to changing circumstances

(Chapter 6, Section 6.2). And for believers in cellular automata-like systems, intelligence was an emergent property of a plethora of rather primitive, perhaps "stupid" agents working in conjunction (Section 7.6).

Adaptation, however, has a biological significance quite independent of its role as a marker of intelligent behavior. The marriage, in the 1920s and 1930s, of the Darwinian theory of evolution by natural selection with the Mendelian science of genetics—a marriage famously called "The Modern Synthesis" by English biologist and humanist Julian Huxley[38]—came to be known more technically as *Neo-Darwinism*.[39] Adaptation in a biological sense (rather than a cognitive or psychological sense) is closely associated with neo-Darwinian evolution.

In *very* general terms, this process can be described as follows.[40]

(1) Within any population of organisms of a given species there is considerable random variation among individuals largely brought about by genetic factors—random in the sense that variations appeared irrespective of the organism's immediate environment.

(2) In a given environment, the likelihood of an individual surviving to adulthood and reproducing successfully will be influenced by the particular genotype of that individual.

(3) On the one hand, if the individual survives so as to reproduce then the offspring will inherit the genetic traits and increase the frequency of occurrence of these traits in the population. On the other hand, if the traits are such that the individual is unable to reach maturity and reproduce, then the traits will less likely perpetuate in the population. This process is called natural selection.

(4) However, there will be variations in the genotype of offspring as a result of certain transformations in the genes, most notably through gene mutation and recombination.

(5) There needs to be a large number of variations on which natural selection can work. In biological terms, organisms must be fecund in their reproductive capabilities. It is this fecundity in the offspring production that makes selection possible.

(6) By this process organisms are constantly tested against the environment, and those that are genetically endowed to survive and reproduce successfully may be said to be fit relative to the environment. (Hence the tautological expression "survival of the fittest.")

(7) If the environment changes, then some forms of the organism within the population may become genetically "fit" relative to the new environment while other forms may not be so fit and will die out. Thus organisms appear to constantly adapt to their environments.

AI researchers encountering natural selection for the first time may well experience a shock of recognition. For they would realize that, as a process, natural selection bore a similarity to a weak heuristic search strategy (Chapter 6, Section 6.6) Allen Newell and Herbert Simon called *generate-and-test*.[41]

This heuristic can be succinctly described as follows:

Generate-and-test:
    Given a *goal* G:
repeat
        generate a possible solution S
        if S satisfies G **then** return S; exit
    until no possible solution is found;
    return failure; exit

Both generate-and-test and natural selection undertake *search* through a space of possibilities—the problem space in the former case, a population space in the latter case. Both generate-and-test and natural selection are *gradualistic*—that is, evolutionary—processes. The crucial difference is that generate-and-test works with a prior goal whereas natural selection does not. The generation of alternative possible solutions in the former case is goal directed and relies on knowledge of prior candidate solutions; that is, by history. Generate-and-test thus does a teleological search whereas the production of variant offspring in natural selection is effected blindly. Finally, generate-and-test assumes that the goal against which solutions are tested remain fixed whereas the environment against which organisms are tested can change.

Can natural selection be imported into the artifactual domain? More specifically, what would a computational strategy modeled along the lines of natural selection look like?

From the early 1960s, John Holland of the University of Michigan had been exploring these kinds of questions. Holland's ambition was to construct a general theory of adaptive systems that embraced biological, economic, and technological domains. The culmination of his investigations, along with those of his students and collaborators, was a book, *Adaptation in Natural and Artificial Systems* (1975).[42]

7.10   GENETIC ALGORITHMS (OR DARWINIAN GENERATE-AND-TEST)

The term genetic algorithm was not due to Holland, though he came to be more or less regarded as its "father" by way of his 1975 book. The term apparently first appeared (according to Jeffrey Sampson of the University of Alberta and a former student of Holland[43]) in a 1967 PhD dissertation by another Holland student, J. D. Bagley.[44] Holland's preferred terms were reproductive plan or genetic adaptive plan for which by "plan" he meant "algorithm" or "procedure."

A genetic algorithm, in Holland's reproductive plan sense, would effect a heuristic search through some search space using the neo-Darwinian metaphor. Thus we might well view genetic algorithms as instances of a Darwinian generate-and-test heuristic. And, as can be imagined, the essence of a genetic algorithm would (a) entail reproduction of structures according to some "fitness" or "utility" criterion; and (b) deploy genetic operators, such as mutation, recombination, and crossover, to generate variation among "offspring" of the parent structures.

The genetic algorithm can be stated in different ways. One presentation is subsequently given. Here, the genetic term "chromosome" (a sequence of genes) is replaced by the generic term "structure."

---

(1) Create or choose initial current population of structures.
while *not* terminating condition **do**

    (2) Evaluate the current population of structures in the environment according to the fitness or utility criterion. The fitness or utility measures are used to construct a probability distribution on the population (representing the reproductive fitness of the structures in the population).

    (3) From the probability distribution computed in (2), select the required number of "parent" structures.

    (4) Apply genetic operators to the selected parents and produce offspring.

    (5) Replace the parent structures with offspring structures.

    (6) current population ← offspring population.

end while

Although much of the early studies of genetic algorithms were conducted by Holland, his students, and *their* students from the late '60s through the late '70s,[45] it soon began to attract much broader attention and by the end of the 1980s was an active research topic, marked by the publication of textbooks on the subject.[46] This interest was not confined to the realm of computer science and AI; it spread beyond.[47] Holland's invention of genetic algorithms must therefore be considered an act of C-creativity.

Not surprisingly, given the nature of genetic operators (such as mutation or crossover), *combinatorial optimization problems* became important sources of applications for genetic algorithms. An example is the shortest-path problem: to find the shortest path between two given cities (say) on a map. The initial population of structures will be an initial set of randomly selected paths between the start and end cities through a set of intermediate cities. The fitness of each such path is an inverse function of the sum of the distances between the cities in that path. Using mutation and crossover operators (applied to some segment of the paths) new candidate routes are generated and their fitness values computed. If an offspring's fitness is lower than those of its parents (that is, the distance between the start and end cities is longer in the offspring than in the parents), the parents replace the offspring. And so the cycle is repeated until the terminating criterion (say an upper bound on the number of iterations of the algorithm) is reached. At which point the shortest path (up to that terminating condition) will be present in the current population of paths.

7.11 DAVID MARR'S VISION

As we saw (in Chapter 2), a significant feature of the second age of computer science was the ubiquitous presence of abstraction and hierarchy in the computer scientist's consciousness. To think computationally meant (among other things) to "think abstraction," "think hierarchy."

To David Marr of MIT and trained formally in neurophysiology, this realization seemed to have occurred as a kind of epiphany after he encountered the computational paradigm—more precisely, with his realization that the phenomenon of vision did not just entail physiological or psychological investigations, but also it was an information processing phenomenon; and that understanding vision entailed understanding it at three distinct hierarchically related levels of abstraction that were dominated by computational considerations.

Marr's book *Vision* (1982) was published posthumously (he died at age 35 in 1980).[48] In it he laid out his theories of the visual process. What most interests us here is the extent to which the computational imagination permeated his theories.

Basically, he posited that vision entails *symbolic representation* of the external world in the nervous system in a manner that allows us to make visual sense of what is looked at. This was one part of his epiphany. To understand how vision works we must abstract up from the level of neurons to the level of symbols and analyze the problem as an information processing task.[49] This was not to deny the physiological level of understanding but rather to complement it, acknowledging that without analyzing the problem as an information processing problem "there can be no real understanding of the function of all those neurons."[50]

So Marr was reiterating the physical symbol system hypothesis (Chapter 6, Section 6.1), though he made no acknowledgement of Newell and Simon as the articulators of this hypothesis.

But there was another part to his epiphany: the realization that we must also theorize about the problem at three quite distinct levels of abstraction:

Computational level (as he called it). This is the most abstract level wherein we explain the visual system in terms of its goals and the logic by which the system realizes its goals. This is the visual system's *computational theory*.
Algorithmic level (as he called it). This is the second level wherein we attempt to explain how the system's computational theory might come about in terms of symbolic representations and algorithms.
Hardware level (his term). At this, the third and lowest level, we try to explain how the algorithmic theory is physically wired into the nervous system.

One cannot imagine a more explicitly computational perspective of a biological problem than this. Most likely Marr would have been unfamiliar with the multiple levels of abstraction of computer systems described by Gordon Bell and Allen Newell in their book *Computer Structures: Readings and Examples* (1972. See Chapter 3, Sections 3.4, 3.5). Closer to home, at the about the time Marr was writing *Vision* (between 1977 and 1979), Allen Newell was probably formulating a multilevel model of cognition that was strikingly similar to the Marr abstraction levels, except that what Marr termed the algorithmic level was Newell's physical symbol system level,[51] and Marr's computational level was Newell's knowledge level.[52]

Marr's separation of function from process was, of course, a well-established idea in computer science. It was isomorphic to the software methodologist's separation of specification design from specification implementation and the related concepts of abstract data types and information hiding (Chapter 2, Sections 2.9, 2.10)—concepts that had emerged well before Marr began writing *Vision*. So computer scientists may not have been terribly excited by Marr's ideas though, perhaps, they will have stirred the minds of vision researchers unfamiliar with computational thinking.

A curious feature of Marr's theory was his placing *goals* as the centerpiece of his computational level: He speaks of "the goal of the computation."[53]

This clearly smacks of teleology. Newell, in talking about his knowledge level of cognition, could refer to goals with impunity because Newell's knowledge-level agent is a purposive being who *consciously* articulates goals. But visual processing is a biological phenomenon. One does not begin with a conscious goal "to see" and then undertake the process of seeing. One just sees. To speak of the visual system as having a goal seems to ascribe teleology to a natural process. One might as well ascribe a goal to natural selection.

We can "explain away" this ascription by assuming that when Marr spoke of the goal of the computation he was really referring to the *function* of the computation. And function is not a teleological term.

Thus, for example, when Marr, citing the researches on the visual system of the housefly by Werner Reichardt and Tomaso Poggio of the Max Planck Institute in Tübingen, Germany, in the mid-1970s, mentioned that "the fly's visual apparatus controls its flight through a collection of five . . . very fast responding systems,"[54] he was really referring to the housefly's visual system's functions, not to conscious goals.

So what did Marr's "new computational approach to vision" look like?[55] Beginning with a function for vision—"to deliver a completely invariant shape description from an image," as he put it[56]—Marr postulated as his theory a "representational framework" consisting of three stages of representation, each of which would demand a computational and an algorithmic level of description:[57]

(a) A "primal sketch" whose purpose was to make explicit certain aspects of the two-dimensional (2)D image (such as intensity changes and image geometry).

(b) A "2½D sketch," which makes explicit additional properties of the image (such as surface orientation and surface reflectance).

(c) A "3D model representation" of the structure and organization of the viewed image.

Whether Marr's theory would be validated was a matter for future research in vision. From the viewpoint of the history of computer science, his claim to originality properly lay in his connecting the physiology of vision to computational thinking in terms of his three levels of abstraction.

7.12   DANCING TO THE SAME TUNE

About the same time John von Neumann was constructing his ideas about cellular automata, neuropsychologist Donald Hebb of McGill University published a book that would become enormously influential. In *The Organization of Behavior* (1949) Hebb proposed that the connectivity between the neurons in the brain continually changes—strengthens or weakens—through experience.[58] In particular, over time, a collection of weakly connected neurons organize themselves into assemblies that act in concert; repeated excitations of one neuron by another through their synaptic link strengthen their association, and so groups of neurons, if synchronously activated, learn to dance to the same tune, so to speak.

This was the idea of the *Hebbian cell assembly*, which would exercise much influence on psychology, neurophysiology, and (as we see) AI. Here was a "confederacy of dunces" to be sure (Section 7.2), but dunces that announced their confederacy by dancing to the same tune. More tantalizingly, the idea of a Hebbian cell assembly suggested that this confederacy of dancing dunces could *learn* to dance together. And if natural cell assemblies could learn then perhaps artificial cell assemblies could well follow Hebb's rule and be induced to dance together also.

As I write these words, the term *machine learning* has become something of a buzzword, like "paradigm": One hears it uttered in the most unlikely places. But the idea has a history about as old as digital electronic computing. Alan Turing speculated on computers that could be made to learn in his essay "Computing Machinery and Intelligence" (1950),[59] the philosophical manifesto for AI. But Turing did not propose by what means a computer could be made to learn. Now here was a possibility: a machine conceived as a network of *artificial neurons*—of the kind conceived by Warren McCulloch and Warren Pitts in 1943[60]—designed to obey some form of Hebb's rule of cell assembly and "trained" to dance together under appropriate stimulus. A simple version of Hebb's rules of cell assembly could be this:

If cell I and cell J are simultaneously excited, then increase the strength of their connection.

7.13 THE RISE AND FALL OF THE PERCEPTRON

One of the first attempts to achieve this kind of learning machine was by Albert Uttley who, in 1954, showed how a network of artificial neurons—later to be known as an *artificial neural network* (or more simply, as a *neural network*)—armed with Hebb-like modifiable connections, could learn to classify binary patterns into certain classes.[61]

Arguably, the most significant harbinger of the possibility of a Hebb-style learning machine was the *perceptron*, conceived in 1958 by Frank Rosenblatt of Cornell University.[62] Its basic architecture is depicted in Figure 7.5. Input "sensory units" (S-units) (such as may exist in the retina) are connected to a set of "association units" (A-units), which in turn is connected to a set of output "response units" (R-units).

The function of the perceptron was to activate appropriate R-units in response to particular patterns of input stimuli from the S-units. Signals from the latter would be transmitted to a set of A-unit cells. These stimuli may be either *excitatory* or *inhibitory* in their effect on the A-unit cells.

The strengths of the connections of S-units $S_1, S_2, \ldots, S_n$ to the A-units are indicated as *weights* $\Omega_1, \Omega_2, \ldots, \Omega_n$. Some weights are positive, indicating excitatory connections; some are negative, indicating inhibitory connections; and the weight values represent the strengths of excitatory or inhibitory connectivity. So there is a single layer of modifiable connections. An A-unit will "fire" (respond) if the net stimulus to that unit exceeds a certain threshold value.

The R-unit cells respond to signals from the A-units in the same manner as A-unit cells respond. However, whereas the connections from S-units to A-units are unidirectional, the connections between A-units and R-units are bidirectional. And the rule governing this feedback connection was along the following lines:

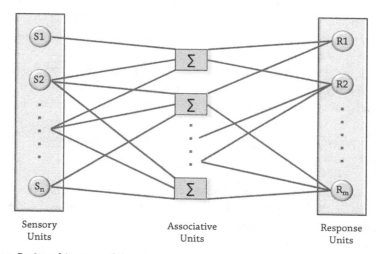

Sensory Units   Associative Units   Response Units

FIGURE 7.5 Basic architecture of the perceptron.

Each R-unit prohibits activity in any A-unit cell that does not transmit to it.

So both A-unit cells and R-unit cells have thresholds, and cells in both sets would fire when the sum total of the strengths (weights) of input connections for the cell exceeded the threshold. But only the strengths of the A-unit cells were modifiable, and it was this modifiability (or plasticity) that afforded the perceptron its *learning* capability. An elementary *learning rule* for the perceptron went something like this:

Assume the perceptron to be activated and that after a brief period one of the randomly determined R-units becomes active. Then (1) because of inhibitory feedback connections the A-units *not* connected to the active R-unit are suppressed. (2) Increase the active A-unit's strength.

Thereafter, if at a later time the same stimulus emanates again from the S-unit, the A-unit activated by it would show stronger activity, which in turn would strengthen the activation of the associated R-unit. That is, the relevant R-unit would probabilistically respond correctly.

Rosenblatt's original paper of 1958 devoted much space to a careful discussion of the behavior and learning characteristics of the perceptron, but viewed as an abstract automaton. He would further elaborate on the perceptron in a book, *Principles of Neurodynamics: Perceptrons and the Theory of Brain Mechanisms* (1961).[63] In 1960, he and his collaborators at the Cornell Aeronautical Laboratory built a physical perceptron that they called the Mark I, comprising an artificial retina as the set of S-units, 512 A-units, and 8 R-units, with each S-unit having up to 40 connections to an A-unit.[64]

In the '50s and '60's the fashionable term for this kind of system was *self-organizing*: a self-organized coupling of stimulus to response. Rosenblatt showed that the perceptron could learn random associations. But he also acknowledged that there were limits to this ability; for instance, if an excessive number of random stimuli were associated with a response the accuracy of the coupling of stimulus and response would decrease. He also noted the following:

The memory of the perceptron is distributed in the sense that any association may make use of a large proportion of the cells in the system, and the removal of a portion of an associative system . . . would begin to show up as a general deficit in *all* learned associations.[65]

And to psychologists and others interested in learning, who may want to know what the perceptron had accomplished, Rosenblatt answered:

. . . the fundamental phenomena of *learning, perceptual discrimination and generalization can be predicted entirely from six basic physical parameters* . . . [to wit] the number of excitatory connections per A-unit, . . . , the number of inhibitory connections per A-unit, . . . , the expected threshold of an A-unit, . . . , the proportion of R-units to which an A-unit is connected, . . . , the number of A-units in the system, and, . . ., the number of R-units in the system.[66]

Rosenblatt acknowledged that Hebb's work "has been a source of inspiration for much of what has been proposed here."[67] If Hebb had attempted to show "the plausibility of a bridge between biophysics and psychology," his (Rosenblatt's) theory "represents the first actual completion of such a bridge."[68] And his final sentence reads thus:

By the study of systems such as the perceptron, it is hoped that those fundamental laws of organization which are common to all information handling systems, *machines and men included* may eventually be understood.[69]

The perceptron created something of a "sensation" (in the words of James Anderson and Edward Rosenfeld[70]) within the motley interdisciplinary community who were interested in such matters as adaptation, learning, memory, and pattern recognition in brains and machines. This excitement induced much effort to be expended through the '60s to explore its possibilities, as evidenced by the conferences held in the early '70s on self-organizing systems.[71]

The results were disappointing. The problem was one that designers of computational artifacts were—and would increasingly be—familiar with: poor performance when the perceptron was "scaled up" to harder, larger, and more complex problems than the ones Rosenblatt had investigated. The tipping point came at the end of the '60s with the publication, by Marvin Minsky and Seymour Papert, of their book *Perceptrons* (1969).[72]

This was a mathematical monograph written for those, as the authors stated in the introduction, interested in computational theory, but not just theoretical computer scientists or automata theorists—Minsky had written 2 years earlier *Computation: Finite and Infinite Machines* (1967), an elegant book on automata theory[73]—but also "psychologists and biologists who would like to know how the brain computes thoughts and how the genetic program computes organisms."[74] It was, above all, a study of the functional *limits* of perceptron-like devices. The analysis and formalisms spread over some 200 pages were mathematically and conceptually dense, but the ultimate qualitative message was unequivocal and damning: There were significant types of basic processes pertaining to the brain, such as pattern recognition and perception, that perceptron-like devices could not compute. As a striking example, Minsky and Papert proved that a certain kind of perceptron could not determine whether all parts of a geometric figures are connected to another.

Perceptrons, in considerable part perhaps because of the prestige and influence of its authors, sounded the death knell to perceptron-oriented research. The prospect of connectionism as a way of studying brain-like machines or machine-like brains seemed doomed. In this sense *Perceptrons* was negatively consequential, and quite significantly so. The book was, in an ironic sort of way, C-original.

7.14   THE NEW CONNECTIONISM: PHOENIX RISING FROM THE ASHES

It would be more accurate to say that through the 1970s neural network research as a means of building adaptive, learning machines slipped into the subliminal level as far as biologically and cognitively conscious computer scientists were concerned. In particular, AI researchers, by and large, abandoned connectionism in favor of the symbolic approach à la Newell's and Simon's symbol system hypothesis (Chapter 6, Section 6.1). But others more interested in machine-like brains rather than in brain-like machines— an interdisciplinary lot composed of physicists (including Leon Cooper, a Nobel Laureate), neurophysiologists, electrical engineers, and applied mathematicians— were busily at work through the '70s theorizing and building neural network–inspired models of brain processes, such as associative memory, low-level feature detection, and stereoscopic vision.[75]

For computer scientists in general and AI researchers in particular, and others interested in adaptive, learning machines, a new connectionism arose in the 1980s, phoenix-like, from the ashes of the old perceptron. Indeed, we might well claim that the '80s marked the second age of neural network research in computer science.

What made this happen? Why this revival after a decade's hiatus? There is no single, simple answer to these questions.

In 1988, MIT Press issued an "expanded edition" of *Perceptrons* that added a new prologue and an afterword to the original 1969 edition.[76] In this edition Minsky and Papert offered their own explanations. They referred to a general idea that *Perceptrons'* "pessimism" had brought about the decline of connectionism following its publication. On the contrary, they believed (with the benefit of hindsight) that the book's effect had been beneficial after all: It induced a "necessary interlude," and prompted connectionist researchers to "take time off" and construct new ideas about knowledge representation. So for Minsky and Papert, writing in 1988, it was the lack of an understanding of knowledge representation (Chapter 6, Sections 6.16–6.22) that was the real reason for this abandonment of connectionism. Thus it was, they noted (as we have seen in Chapter 6), that the 1970s saw an insurgence of research into knowledge representation. Then, reflecting on the 1980s, they speculated on the battle between connectionism and symbolic AI. Perhaps the latter had "run out of steam." Or perhaps it was the exciting new promise of "massively parallel hardware." Or perhaps it reflected "a cultural turn toward holism."[77]

To repeat, the factors that resulted in the return of connectionism to the forefront of research on intelligent machines were several and intertwining. But two ideas stand out.

One was a notion that was not part of Rosenblatt's original perceptron: the effect of an internal or *hidden layer* of cells between input and output of a neural network. The perceptron of Figure 7.5 shows only a single layer of modifiable connections. This was the model Minsky and Papert fixated on in their critical (and ultimately damning) analysis of the perceptron's capabilities.

The second notable feature of the new connectionism was the invention of *backpropagation* as a learning algorithm. The idea itself apparently had origins reaching back to the early 1960s, but in the context of neural networks it was the creation of David Rumelhart and Ronald Williams of the University of California, San Diego (UCSD) and

Geoffrey Hinton of Carnegie-Mellon University in 1986.[78] Backpropagation can be informally described as follows:

Apply an input pattern to the neural network. After some time, certain output units of the network are informed of the values they are expected to show as their outputs. If the actual and expected values are identical, then do nothing more. However, if there is a difference then this is an error. Then change the weights of the units impinging on the output units as per the amount of error and propagate the error to earlier layers of the network. Continue error propagation and changing the weights until error signals reach the input units. Then a new input pattern is applied to the network and the process is repeated.

Rumelhart et al. showed that this procedure will always change the weights so as to reduce the differences between expected (or desired) output values and actual output values.

7.15 A "CONNECTIONIST MANIFESTO"

The one single publication that was emblematic of the new connectionism was the two-volume collection of articles published as *Parallel Distributed Processing* (1986), subtitled *Explorations in the Microstructure of Cognition*, authored by 16 scientists calling themselves the "PDP Research Group." The first volume was given to the foundations of the new connectionism (or "parallel distributed processing," PDP, as the authors preferred to call it).[79] The second volume addressed applications of the theory in psychology and biology.[80] The authors were drawn from biology, linguistics, psychology, computer science, biophysics, and mathematics (including, most notably, the Nobel Laureate Francis Crick) from a number of institutions in the United States, but epicentered at UCSD. Their adoption of the term parallel distributed processing was noteworthy in that it emphasized both the parallelism inherent in neural networks and the distributive nature of this parallelism. So influential was this book that Minsky and Papert in 1988 called it the "connectionism manifesto."[81]

A manifesto is never a textbook or a survey. Rather, it lays out the philosophy of a topic and salient principles of a discipline-in-the-making from its author's perspective. It is a call to arms, so to speak, intended to shape the minds of its readers in the direction its authors desire. It may even be polemical. *Parallel Distributed Processing* seemed to fit the manifesto bill in all these respects. Certainly, after its publication no neural network researcher could afford to ignore it.

Like any good manifesto, this too began with the whys and wherefores of neural network processing—to cajole and persuade the ignorant or the neophyte as to the worthiness of the cause; to initiate the process of conversion of nonbelievers. Thus was the mission of the first long chapter by John McClelland, of Carnegie-Mellon University, Rumelhart, and Hinton, titled, adroitly, "The Appeal of Parallel Distributed Processing."[82]

This "appeal" was followed by a chapter by the same authors on "A General Framework for Parallel Distributed Processing."[83] This, in fact, was the defining statement of this group's manifesto on connectionism. It is perhaps worthwhile for the reader to have an idea of their definition if only to compare it with the "old" connectionism of the perceptron era. Thus, they tell us:

A neural network/PDP system comprises a number of features[84]:

(a) A set of processing units (neurons).
(b) A state of activation for each unit.
(c) An output function for each unit.
(d) A pattern of connectivity among units.
(e) A propagation rule for transmitting patterns of activation through the network.
(f) An activation rule for the units.
(g) A learning rule for modifying the strengths of connections.
(h) An environment within which the system will function.

Consider, for example, a single processing unit U with three inputs from other units to it and an output from the unit to other units. Each input i to U has a value $v(i)$ and a positive or negative weight $\Omega(i)$ so that the weighted sum of input values is the *net* effect of the inputs on U:

$$net = \Sigma\ \Omega(i)\ v(i) \text{ over all } i.$$

The effect of *net* will be to produce a new activation state of U at time $t + 1$ as a function G of $net(t)$ and U's state, $a(t)$ at time t:

$$a\ (t + 1) = G\ [a(t), net(t)].$$

This is the activation rule. And the pattern of connectivity between units in the network is defined quite simply by the set of weights $\Omega\ (i, j)$ over all connected units pairs (Ui, Uj).

The output function of the unit U at time t is a function F of the activation state at time t:

$$o\ (t) = F\ [a\ (t)]$$

As for the propagation rule there are different possibilities but the simplest would be along the following lines:

If there are two types of connections in the network, inhibitory and excitatory, the net excitatory propagation effect is the set of all weighted excitatory inputs to units in the network, and the net inhibitory propagation effect is the set of all weighted inhibitory inputs to units in the network.

The learning rule can also differ (one being the backpropagation rule) but ultimately they are variants of the classic Hebbian learning rule mentioned earlier (Section 7.12).

Finally, there is the all-important environment. In the PDP model the assumption was that "at any point of time there is some probability that any of the possible set of input patterns is imposing on the input units" according to some probability distribution.[85]

7.16 (ALMOST) EGALITARIAN COMPUTING

By the end of the 1980s the new connectionism had spawned a veritable industry in multidisciplinary neural network research, no doubt stimulated by the work of the PDP Research Group. A notable signifier of the establishment of a genuine research tradition was the formation, in 1987, of the International Neural Network Society (INNS).

For computer scientists, connectionism offered an enticing computational model: The enticement stemmed from the fine-grained, parallel, distributed, and decentralized ethos that characterized connectionism. Hierarchy, the historically overarching feature of computational models and artifacts was—to some, exhilaratingly—absent. In the connectionist world no one process or processor was in command; no master, no slave. Here was a computational model that was more or less egalitarian in spirit, a community of cooperative equals, collectively embarked on a common enterprise.

An example of this quasi-egalitarianism is the connectionist approach to the "constraint satisfaction problem." Here, as the name suggests, the solution of a problem demands the satisfaction of a set of mutually interacting, often conflicting, constraints. Design problems are very frequently of this kind. The computer architect embarked on the design of a computer's ISP architecture (Chapter 6) must satisfy a plethora of constraints on each architectural feature that are imposed in part by the overall design goals, in part by one or more of its companion features, and in part by technological factors. Some constraints are mutually agreeable, some mutually hostile. But egalitarianism between constraints may not entirely rule: Some constraints may have higher priority than others. (In the case of an architecture design, the word length may be more important than instruction formats.)

The constraint satisfaction problem can be mapped onto the connectionist framework. Each unit in a neural network would represent a constraint in the form of a hypothesis, and the connection between two units would signify the relationship between the connected hypotheses. For example, there would be a positive connection between a unit pair (i, j) if the corresponding hypotheses are mutually true; the connection would be negative if the corresponding hypotheses are mutually exclusive. The positive or negative values of the connections need not be binary; they could represent extents of positivity or negativity. When such a network is activated it will eventually "settle" into a state in which as many of the constraints are satisfied subject to the priority of the most stringent constraints.

In an overview article published in 1989, David Rumelhart gave an example of a 16-unit constraint network representing a Necker cube (a well-known perceptual example in *Gestalt* psychology) and how the network would eventually settle into a state that would interpret the Necker cube from one or the other mutually exclusive perspective.[86]

A striking example of a connectionist artifact was NETtalk designed circa 1986 by Terrence Sejnowski of Johns Hopkins University and Charles Rosenberg of Princeton University.[87] This consisted of three layers of units, an array of 203 input units split up into seven equally sized groups, connected to a single "hidden layer" of 80 units that in turn were connected to 26 output units. The goal of this network was to translate a textual string of letters into a string of phonemes. That is, NETtalk was intended to "pronounce" English text; more precisely, it *learnt* to pronounce.

The input to the NETtalk system at each step of its operation consisted of a string of seven letters, each letter activating one of seven groups of input units. Of these, only the middle (fourth) letter was the target of conversion into a phoneme. The other six provided contextual information because how a letter sounds is influenced by its surrounding letters. After processing each central letter the next letter in the string would be the target of conversion. The total number of weighted connections in the network was 18,629. The hidden units were used to form internal representations that aided in mapping the letters into phonemes.

Like all neural networks, NETtalk's operation involved two phases: a learning–training phase followed by the performance phase. In the learning phase the weights were set arbitrarily, so when NETtalk received its initial input text the output was in effect phonetic gibberish. Training then involved incrementally adjusting the weights of the connections so as to minimize the "mean-squared error" based on the error or discrepancy between the values of the output units and intended output pattern— that is, backpropagation (Section 7.14) served as the learning algorithm. This required a "teacher" to provide the correct values of the output units. Each of the latter units encoded one of 26 distinct phonemic ("articulatory") features, subsets of which in turn encoded over 50 distinct phonemes.

Inevitably, the highly parallel, fine-grain, distributed nature of connectionist architectures would attract the attention of computer hardware designers. We have already seen its most prominent manifestation: Daniel Hillis's connection machine (this chapter, Sections 7.6–7.8). Another interesting manifestation of this influence was the work of Carver Mead of the California Institute of Technology. Mead, as we have noted (Chapter 5, Section 5.11), was one of the coinventors of the VLSI "structured design methodology." And the possibility offered by VLSI to implement neural networks—a "silicon nervous system"—offered an irresistible challenge. His book *Analog VLSI and Neural Systems* (1989) was published by Addison-Wesley as a volume in both their "VLSI Systems Series" and "Computation and Neural Systems Series," thus representing a marriage of circuit technology and neural computing. The book was a detailed exploration of silicon nervous systems, and among the artifacts he and his collaborators designed were a "silicon retina" and an "electronic cochlea."[88]

7.17 CONNECTIONISM THREATENING SYMBOLISM:
A KUHNIAN CONFLAGARATION?

Arguably the most contentious intellectual consequence of the rise of the new connectionism in the mid-1980s was the challenge it posed to the symbolist paradigm that had served as the bedrock of AI research for some three decades. Indeed, symbolism had extended its reach still further: The physical symbol system hypothesis, the core of the symbolist paradigm, had inspired the emergence of a computationally grounded "interdiscipline" named *cognitive science* in the late 1970s.[89] (The neologism "interdiscipline" seems appropriate because cognitive science was conceived as a science involving several "classical" disciplines, most notably, psychology, philosophy, linguistics, and computer science.)

Connectionism not only presented a threat to the symbolist paradigm in AI (as briefly mentioned in Chapter 6, Section 6.23) but also in cognitive science. But there is an important distinction between the nature of these two challenges. The aim of much of AI research is to build computational *artifacts* that manifest some form of intelligent behavior. In this project the researcher is ultimately an artificer or designer or engineer. And like artificers in other arenas, the AI researcher can freely choose whatever body of ideas that he or she believes is effective for artifact building. In contrast, the cognitive scientist's goal is to present testable models or theories about cognition—a *natural* phenomenon. The cognitive scientist, like the natural scientist, is ultimately constrained by the reality of natural phenomena. If the cognitive scientist espouses a paradigm then that paradigm must cohere with the vagaries of the natural world of cognition.

To repeat, though connectionism had made a certain mark in the late 1950s and early 1960s, AI was governed solidly by the symbolist paradigm through the 1960s, 1970s, and the first half of the 1980s. When the journal *Cognitive Science* was launched in 1977 and then, 2 years later, the Cognitive Science Society was established, thus collectively marking the recognition of a new interdisciplinary science, its computational grounding was also in the realm of the symbolic paradigm.

The physical symbol system hypothesis and its correlate the heuristic search hypothesis (Chapter 6, Section 6.5) as the computational core of the paradigms governing both AI and cognitive science was never in doubt for researchers in these fields—until the mid-1980s and especially the publication, in 1986, of *Parallel Distributed Processing*. Connectionism was reborn and almost immediately became the challenger to the symbolist paradigm.

Can we frame this situation as a Kuhnian scenario? Recall the now-classic description by philosopher of science Thomas Kuhn of how a *paradigm shift*—and thus a *revolution*—occurs in a science.[90] Did the emergent connectionist challenge to symbolism produce anything like a paradigm shift?

In the Kuhnian scenario the threat to the dominant paradigm governing a science begins with a recalcitrant problem that refuses solution according to the precepts of the governing paradigm. This anomaly, as Kuhn termed it, if sufficiently resistant to conquest, eventually gives rise to a crisis. Some immensely brave soul proposes a solution that rests on an alternative framework that contradicts in essential ways the dominant paradigm. If the evidence is there, this alternative may grow into a serious challenger to the governing paradigm. Eventually, if the challenger demonstrates greater explanatory power (in addition to explaining the anomaly) then increasingly more scientists will gravitate toward it, leading to a replacement of the governing paradigm by the challenger. Thus a paradigm shift; thus a scientific revolution. The paradigm shift is as much social and cultural as it is scientific because it entails something like a new *Weltanschauung*, a new way of viewing the world.

Did the rise of the new connectionism conform to this scenario?

In a presidential address to the Psychonomic Society[91] delivered in 1987, titled "Connectionism: Is It a Paradigm Shift for Psychology?," Walter Schneider of the University of Pittsburgh suggested, with due academic caution, that cognitive psychology was indeed in the throes of a paradigm shift.

From a social perspective, he noted the extraordinary growth in talks on connectionism at the annual Cognitive Science Society meetings—17% in 1984, 23% in 1985, 31% in 1986—and the remarkable advanced sales volume of *Parallel Distributed Processing*.[92] He quoted a well-known computer architect Dan Hammerstrom of the Oregon Graduate Institute as predicting VLSI-based computers that would emulate neural networks containing "a billion connections between millions of nodes"[93] (though there was no mention of the Connection Machine).

As for connectionism's theoretical import, Schneider wrote enthusiastically of "new types of studies" in psychology being revealed by connectionism: new modes of representation (across networks of cells), new learning concepts (supervised and unsupervised learning), and new ways of casting cognitive concepts (such as semantic nets and schemas).[94]

Schneider's mention of schemas is worth noting. As we have seen (Chapter 6, Section 6.15) schemas (and its derivatives, frames, and scripts) have a significant presence in the story being told here. But schemas are quintessential symbol systems. Connectionism, being the new kid on the block as a computational theory of the mind, being the challenger to the symbolist paradigm, was obliged to solve the same problems symbolism solved, either by showing that symbolist explanations could be *reduced* to connectionist explanations—that is, connectionism was more fundamental than symbolism—or by demonstrating the un-necessity of the symbolic level altogether, by showing that what symbolism could explain connectionism could explain "better" (in some sense).

The schema, as a significant explanatory model in both AI and cognitive science, was precisely such a candidate for the connectionists to address. Thus it was that David Rumelhart, Paul Smolensky, James McClelland, and Geoffrey Hinton, "godfathers" of the new connectionism, took up this problem.[95] Recall that in the symbolist paradigm schemas were perceived as symbol structures that represent prototypical situations or experiences, but are plastic and elastic enough to be expanded, instantiated, reshaped. The analogy was with data structures, but somehow constructed unconsciously as the outcome of learning and interacting with the world.

Rumelhart et al. had another perspective.[96] In broad terms, they proposed that the way schemas come into being is that stimuli from the environment activate a set of units (cells) that form stable subpatterns that activate one another; and, when activated, they may all identically inhibit other units. These "coalitions" of cells constitute what are called schemas.[97]

Rumelhart et al. posited that contrary to the symbolist interpretation of schemas as data structures stored in memory a schema is not the content of memory but rather an *emergent* property of an interconnected network of units.

As a test of this hypothesis they conducted a simulation experiment involving common human knowledge of different kinds of rooms. First, the experimenters identified a set of 40 "room descriptors"—features that appear in one kind of room or another, e.g., walls, ceilings, telephone, desk, chairs, ovens, books, bed, bathtub, and so on. Human subjects were then asked to imagine each of five different room types (office, living room, kitchen, bathroom, and bedroom), and identify which of the 40 descriptors would apply to each room type. A total of 16 judgments of the 40 room descriptors were collected for each of the five room types. These data, consisting of 80 room descriptors. served as the basis for constructing a network in which each unit represented a room descriptor. The weight $\Omega(i, j)$ of the connection between two units i, j were set according to some rules, for example:

(a) If the units i, j tend to be active or inactive together then $\Omega(i, j)$ will be a large positive value.

(b) If units i, j are active or inactive independently then $\Omega(i, j) = 0$.

A typical simulation run was of the following sort.

One of the units corresponding to the descriptor *oven* is set "on" ("clamped") and not allowed to change. The run begins. Eventually the system "settles" into a stable pattern of interconnected units. The stable coalition of units that emerges is precisely one that would correspond to the kitchen schema.[98] The next four runs begin with the descriptors desk, bathtub, sofa, and bed being clamped, respectively, and in each case the system settles into a prototype room most closely related to the respective clamped unit; namely, office, bathroom, living room, and bedroom schemas.

Rumelhart and his collaborators go on to describe how their connectionist model was able to accommodate other attributes of schemas, for example, the property that although some features are tightly bound in a schema, others are of a variable nature that initially have "default values" but that, in the course of new situations or experiences, may be set to some other value. For example, size was a variable room descriptor that for the room type bedroom was defaulted to the value *medium*. However, if the input to the system contained information suggesting other descriptors, then the default value of *size* may change. In one simulation run, both bed and sofa were clamped. The outcome was a room "best described as a large, fancy bedroom," with the size variable set to *large*. This setting of size in turn changed default values of other descriptors and resulted in a bedroom with a fireplace.[99]

Schneider, however, tempered his enthusiasm when he noted that in contrast to theory the empirical impact of connectionism "has been minimal and may remain limited."[100] By this he meant that an empirical phenomenon may be explained by more than one connectionist model, and cited as an example a certain well-known letter-recognition phenomenon that could be explained by two distinct connectionist models. He also made the interesting observation that although connectionist models may turn out to be effective in the realm of artificial learning, their modus operandi may not bear any resemblance to the ways in which humans learn.[101] In other words, connectionism may prove more efficacious for designing computation-based cognitive artifacts than for explaining human cognition.

Ultimately, despite the qualifications and caveats, Schneider could only answer his title question with a resounding "yes!": that connectionist models constituted "a significant paradigm shift in psychology," as momentous as the shift from behaviorism to cognitivism in the 1950s; indeed, comparable to the Chomskyan "revolution" in linguistics in the late 1950s and early 1960s.[102]

But there were others who emphatically did not share Walter Schneider's optimism. Among the most influential were the cognitive theorists Jerry Fodor and Zenon Pylyshyn of Rutgers University. In a lengthy paper published in the journal *Cognition* in 1988 they argued that as far as the architecture of cognition was concerned connectionism *could not* achieve paradigmatic status in the Kuhnian sense because there were critical features of cognition that connectionism did not (and could not) model but symbolism (or, as they called it, "classical" theories) could (and did).[103]

One such feature familiar to both linguists and computer scientists was what Fodor and Pylyshyn called the "productivity of thought." By this they meant the generative capacity of the mind: that the mind can produce an unbounded number of "thoughts" using finite resources. For example, following Chomskyan notions, a language consists of an infinite set of sentences that can be generated by the recursive application of a finite set of grammatical rules (see Chapter 1, Section 1.2). "So linguistic (and mental) representation must constitute *symbol systems* . . . so the mind cannot be a PDP"[104]— neural networks lack this productivity feature. Bolstered by further arguments, they go on to repudiate the connectionist architecture of mind.[105]

Another feature was what Fodor and Pylyshyn called the "systematicity" of cognition, that is, the idea that cognitive features form systems. For instance, in the case of language understanding, one's capacity to understand a sentence is "*intrinsically connected*" to one's ability to generate or understand certain other sentences.[106] This is intimately related to the fact that the sentences of a language have an internal compositional structure that is governed by the rules of syntax and semantics.

Systematicity is what enables an English language speaker to assume that if any one of the expressions "John loves Mary," "Mary loves John," and "the boy loves the girl" is a sentence, then they all are. And what is true of language is also true for thought: "*thought is systematic too*."[107] There is a "language of thought"—the title of a 1976 book by Fodor[108]—that allows one to comprehend that the thought that John loves Mary and the thought that Mary loves John have the same internal structure. Symbolism supports the systematicity feature of cognition; connectionism does not. "So the architecture of the mind is not a Connectionist network."[109]

Fodor and Pylyshyn were mindful of the distinction between cognitive architecture and the implementation of that architecture—and in this they were in harmony with the computer scientist's computer architecture–hardware hierarchy (see Chapter 3). Rather strangely, they made no mention of David Marr's three-level hierarchy for vision and, especially, the separation of the algorithmic and the hardware levels (Section 7.11). Instead, they made the point that perhaps "a theorist who is impressed by the virtues of Connectionism has the option of proposing PDP's as theories of implementation."[110] Indeed, citing one author, Dana Ballard,[111] they pointed to the fact that "Connectionists do sometimes explicitly take their models to be theories of implementation."[112] This would be reasonable because, after all (as we have seen in this chapter), the new connectionism inherited a legacy that reached back to Pitts, McCulloch, and von Neumann, and their idealization of the neuronal structure of the brain. A frequent argument advanced in support of connectionism has been that it is "biologically plausible." Fodor and Pylyshyn admitted that they had no "principled objection" to treating connectionism as an implementation theory.[113]

Of course, the detractors of connectionism as a governing paradigm had their own detractors. Paul Smolensky of the University of Colorado, a member of the PDP Research Group, hastened to reply.[114] So did David Chalmers of Indiana University.[115] So did Tim van Gelder of the Australian National University.[116]

We will not enter into the precise nature of their refutations. What was striking was that there was an ongoing controversy. In the Kuhnian scenario this was akin to a preparadigmatic state in which different schools of thought are warring for dominance. Or, more accurately, because symbolism had reigned as *the* paradigm for some 30 years, the situation was one in which connectionism was the intruder knocking at the door, striving to get a foothold, *desiring* to supplant symbolism, but far from achieving this.

A particularly judicious analysis of this symbolism–connectionism controversy came from Donald Norman, founder of the Institute of Cognitive Science at UCSD, the "home base" of the PDP Research Group. *Apropos* the controversy, Norman was a self-described "in-between cognitive scientist."[117] His own work, he noted, some done in collaboration with David Rumelhart, had been "in the spirit of the connectionist models." But he admitted to being troubled with some aspects of connectionism. He had no doubts about its originality and consequentiality: It had ushered in "a major rethinking of the levels" at which cognitive processes should be examined—straddling the psychological level "above" and the neuroscientific one "below." Further, it was centered on a new computational model of thinking that employed new metaphors borrowed from mathematical systems theory, such as "stability" and "global maxima."

Finally, it emphasized the role of continuous learning and adaptation in cognitive processes.[118]

Like Schneider, Norman seemed to come very close to claiming that connectionism represented a paradigm shift.[119] And yet, he admitted, there were important flaws. Connectionism seems more appropriate for *low-level* cognitive processes, those "closer to perception or to motor output."[120] When it comes to higher modes of cognition such as creativity, planning, decision making, composing music, conducting science— "Don't I need to have mental variables, symbols that I manipulate? My answer is 'yes.' I think this lack is a major deficiency in the PDP approach."[121]

Ultimately, Norman believed, connectionism failed the problem of *conscious* control and learning. This was quite different from the "preprogrammed"—thus unconscious— nature of connectionist processes, including learning.

And so Donald Norman, the "in-between cognitive scientist," acknowledged the *necessity* of multiple abstraction levels; of a hierarchy of levels—a realization that computer system designers were very familiar with (see Chapter 2) and that David Marr had brought to the cognitive scientist's and neuroscientist's attention.

In fact, by the end of the decade, symbolism continued to dominate. In *Foundations of Cognitive Science* (1989), the encyclopedic compendium of articles by various authors, the preeminence of symbolism is self-evident. There was one chapter on connectionism (by David Rumelhart[122]), but otherwise the spirit of the physical symbol system hypothesis pervaded the volume.[123] Even more startlingly, Allen Newell's William James Lectures of 1987 at Harvard, magisterially titled *Unified Theories of Cognition* (1990), published posthumously, devoted, in a book of 550 pages, a mere 2 pages to connectionism.[124]

So as the 1990s began we can only judiciously answer the question raised in this section's title as follows:

There was no evidence, circa 1990, that connectionism effected a Kuhnian paradigm shift in either cognitive science or AI. Rather, they seemed to constitute two "coexisting" paradigms—*coparadigms*—each with its own research tradition and its own practitioners quite independent of the other, each addressing quite distinct abstraction levels of cognition, but occasionally acknowledging the other's presence.

7.18   THE LANGUAGE OF BIOLOGICAL DEVELOPMENT

In 1990 a book bearing the enchanting title *The Algorithmic Beauty of Plants* was published by Springer-Verlag (surely one of the most indomitable publishers of the most recondite of scientific texts).[125] Its authors were Przemyslaw Prusinkiewicz of the University of Regina and, posthumously, Aristid Lindenmayer of the University of Utrecht, and it was in a sense the culmination of a huge body of work originating in 1968 in a remarkable paper by Lindenmayer.

Lindenmayer was a theoretical biologist and Prusenskiewicz a researcher in computer graphics, a branch of computer science. Thus the book was the fruit of one kind of interdisciplinary science straddling biology and computer science.

The book's title inevitably compels attention; it probably would not if it had been named simply "The Beauty of Plants": plants *are* beautiful to look at after all. But *algorithmic* beauty? Indeed, the association of plants with algorithms itself seem startling.

This association began in 1968 in a paper by Aristid Lindenmayer, at the time at Queens College of the City University of New York, in the *Journal of Theoretical Biology*.[126] And thus began one of most surprising—and highly creative—instances of the odd coupling of biology and computing mentioned at the beginning of this chapter.

Lindenmayer's interest lay in development biology, the study of the processes by which the cellular organization in an organism develops, differentiates, and grows. And he began his paper by noting that development biologists had "grossly neglected" the investigation of the *integrating* mechanisms that governed the "multitude of individual acts of division, differentiation and cell enlargement."[127] His objective, he then stated, was to advance a theoretical framework wherein this integrative process could be discussed and, significantly, *computed*.[128]

Lindenmayer was a biologist, not a mathematician or a computer scientist. His interest lay in constructing a biologically *valid* (not just plausible) framework that would illuminate the focus of developmental biology—how organisms develop their shape (*morphogenesis*), how cells differentiate into functionally distinct types, how cells grow. There was, then, a distinct separation of his research agenda and the agenda of cellular automata (Section 7.2) that von Neumann had initiated. Nonetheless, Lindenmayer discovered his "theoretical framework" in theoretical computer science, specifically in its branch of *finite automata theory*—concerned with the behavior of abstract artifacts called *finite-state machines* (also called sequential machines). Finite-state machines (FSMs) were, in fact, mathematical abstractions of physical digital circuits and as such had been the subject of much scrutiny by electrical engineers, computer scientists, and mathematicians throughout the 1960s (Chapter 5, Section 5.10).[129]

Lindenmayer's approach has a distinguished pedigree. In particular, the Scottish polymath D'Arcy Wentworth Thompson published in 1917 the work for which he is best remembered, a book called *On Growth and Form*, which described, in mathematical terms, aspects of the morphogenesis of plants and animals and revealed the formal beauty of living growth patterns.[130] He was thus a pioneer in what became mathematical biology.

A FSM is an abstract artifact with an input alphabet, an output alphabet, and a set of possible internal states. These states are the only way the internal structure of a FSM is known. What is of interest is its behavior, and this is defined mathematically by two functions, a next -state function δ,

$$\delta: I \times S \to S$$

and an output function λ,

$$\lambda: S \to O$$

where I and O are alphabets of input and output symbols, respectively, and S is a set of states. Thus, given an input symbol $i \in I$ and a "current" state $s \in S$ of the FSM, δ computes the "next state" $s' \in S$ of the machine as

$$\delta\,(i, s\,) = s'$$

and λ computes the output $o \in O$ as a function of the current state:

$$\lambda\,(s) = o$$

So, given an input string and an initial state of the FSM, it will compute in successive discrete moments of time a succession of states of the machine and an output string.[131]

Aristid Lindenmayer conceived the individual cells of an organism as FSMs, each cell governed by its own δ and λ functions and where a cell's inputs are "impulses" received from the outputs of its neighboring cells. The output of a cell is an abstraction of "diffusible chemical substances, or electrical impulses, or any other types of physical or chemical changes that can be propagated outside the cell."[132] The collection of all impulses to a cell from its neighbors constitute the input to the cell. As for the cell's state:

Each cell in any time unit may be present in one of a number of states that may be interpreted as certain distinguishable physiological or morphological conditions.[133]

Finally, the process governing the behavior of an individual cell was assumed to unfold in a succession of discrete time units.

Thus began Lindenmayer's highly original computational explorations in development biology. And thus was born a research area called (in his honor) *L-systems*.

But, as automata theorists knew in the 1960s—beginning, in fact, with linguist Noam Chomsky's revolutionary work on the theory of syntax in the late 1950s[134]—there is a remarkable relationship between certain classes of automata and certain types of formal (that is, artificial) languages. That is, different classes of automata have the power to recognize or generate different types of formal languages.[135] Of interest to us here is that the kind of automata Lindenmayer embraced in 1968, FSMs, have the power to recognize and generate a class of languages called regular languages. Indeed, there was an equivalence between finite automata and regular languages, a relationship proved by the mathematician Stephen Kleene of the University of Wisconsin-Madison in 1956.[136] This meant that rather than model the behavior of something by a FSM, one could model this behavior by an equivalent formal grammar and the language generated by the grammar.

In Lindenmayer's original work of 1968, biological development was conceived in terms of a linear array of FSMs. But by the early 1970s the study of L-systems had metamorphosed into a branch of formal language theory. The term developmental language came into being; and theoretical computer scientists, even had they not known before, would certainly become aware of this new research topic through a paper presented by Lindenmayer and theoretical computer scientist Grzegorg Rozenberg of the Universities of Utrecht and Antwerp in 1972 in the annual ACM Symposium on Theory of Computing (an essential conference for theorists).[137]

So what does an L-system look like? Now, in formal language theory of the kind computer scientists were thoroughly familiar with (by the end of the 1960s), a "Chomsky-style" grammar would be defined by a 4-tuple:

$$\mathbf{G} = <Vn, Vt, P, S>.$$

Here, Vn is the set of *nonterminal symbols* (denoting syntactic categories such as, in the case of a programming language, for example, "assignment statement" or, in the case of a natural language, "noun phrase"); Vt is the set of *terminal symbols*, the alphabet out of which strings signifying sentences in the language (for example, the string of letters "the cat"); S is a particular element of Vn called the "sentential symbol" (representing the syntactic category "sentence"); and P is a set of ("rewrite" or "production") *rules* the systematic application of which *generates* sentences in the language. The *language* $\mathbf{L(G)}$ corresponding to the grammar $\mathbf{G}$ is the set of all terminal symbol strings that can be generated from $\mathbf{G}$ by systematically applying the rules of P.

By the end of the 1960s a massive body of knowledge concerning Chomsky-style formal grammars had accumulated. To repeat, formal language theory was of both theoretical interest (to automata theorists) and practical interest (to programming language designers and implementers). And, as noted, the formal equivalence of particular types of Chomsky languages and particular types of automata had been identified.

It was this body of knowledge that Lindenmayer and his collaborators (such as Grzegorg Rozenberg, Arto Salomaa of the University of Turku, and Przemyslaw Prusinkiewicz) brought to the table in the 1970s to formalize L-systems as developmental languages and that was pursued by many other researchers (such as Rani Siromoney of Madras Christian College in India) over the next two decades.

An L-system is defined rather like a Chomsky grammar as

$$\mathbf{G} = <V, \omega, P>,$$

where V is an alphabet of symbols (analogous to Vt in a Chomsky grammar), ω is a word composed as a string of symbols from V, called the axiom (analogous to S in a Chomsky grammar), and P is the set of rewrite rules (as in a Chomsky grammar).

In this kind of grammar, the symbols in V represent the various *cell types* in an organism. But a crucial difference between a Chomsky grammar and a Lindenmayer grammar is that the rules in the former are applied only sequentially whereas the rules of the latter can be applied *simultaneously*.

As an example, Lindenmayer's first venture into this kind of research concerned the growth of algae. This process can be modeled by an L-system:

$$\mathbf{G}^* = <V^*, \omega^*, P^*>,$$

where $V^* = \{a, b\}$, $\omega^* = a$, and P^* is the pair of rewrite rules:

(i) $a \rightarrow ab$,

(ii) $b \rightarrow a$.

Each rule is interpreted to mean "replace the symbol on the left-hand side of the arrow with the symbol or symbol string on the right-hand side." That is, the left-hand side symbol is *rewritten* as the right-hand side symbol or string.

So given the previously mentioned grammar, successive, possibly parallel, applications of these two rules yield the developing structure as follows:

Begin with the axiom ω:	a
Apply rule (i):	$a => ab$
Apply rules (ii), (i):	=> aba
Apply rules (i), (ii), (i):	=> *abaab*
Apply rules (i), (ii), (i), (i):	=> *abaababa*
Apply rules (i), (ii), (i), (i), (ii), (i), (ii), (i)	=> abaababaabaab
. . . .	. . . . .

This can continue indefinitely, leading to increasingly larger strings, representing cellular structures. Notice that as the "algae" grows the number of possibilities of parallel rewriting increases.

Of course, plants and other organisms are not usually linear in form: They have *shape*. Development languages had to be combined with appropriate computational tools to represent the development of shapes as well as cellular growth and differentiation; that is, there had to be a means for the graphical interpretation of L-systems. In *The Algorithmic Beauty of Plants*, Przemyslaw Prusinkiewicz described the application of turtle geometry based on the LOGO-style "turtle" Seymour Papert had experimented with and described in his 1980 book *Mindstorms* (Chapter 6, Section 6.11)—and further described in a book by Papert's MIT colleagues Hal Abelson and Andrea di Sessa.[138] Perhaps it was this synergy of L-systems and turtle geometry, along with other graphics tools, that constituted the "algorithmic beauty" of the multitude of "artificial" plants described in the book.

NOTES

1. S. Dasgupta, 2014. *It Began with Babbage: The Genesis of Computer Science*. New York: Oxford University Press, pp. 159–165.

2. W. S. McCulloch and W. Pitts, 1943. "A Logical Calculus Immanent in Nervous Activity," *Bulletin of Mathematical Biophysics*, 5, pp. 115–133.

3. J. von Neumann, 1951. "The General and Logical Theory of Automata," pp. 1–41 in L. A. Jeffress (ed.), *Cerebral Mechanisms in Behavior: The Hixon Symposium*. New York: Wiley.

4. J. von Neumann, 1958. *The Computer and the Brain*. New Haven, CT: Yale University Press.

5. J. von Neumann, 1966. *Theory of Self-Reproducing Automata* (edited by A. W. Burks). Urbana, IL: University of Illinois Press.

6. A. W. Burks, 1970. "Von Neumann's Self-Reproducing Automata," pp. 3–64 in Burks (ed.), 1966.

7. Burks (ed.), 1966.

8. E. F. Codd, 1968. *Cellular Automata*. New York: Academic Press.

9. See, e.g., H. T. Kung, 1980. "Structure of Parallel Algorithms," pp. 65–117 in M. C. Yovits (ed.), *Advances in Computers, Volume 19*. New York: Academic Press, p. 86.

10. See, e.g., G. S. Almasi and A. Gottlieb, 1989. *Highly Parallel Computing*. Redwood City, CA: Benjamin/Cummings, chapter 8; H. S. Stone, 1990. *High-Performance Computer Architecture*. Reading, MA: Addison-Wesley, chapters 4, 6.

11. D. H. Lawrie, 1975. "Access and Alignment of Data in an Array Processor," *IEEE Transactions on Computers*, C-24, 12, pp. 4–12.

12. H. S. Stone, 1971. "Parallel Processing with a Perfect Shuffle," *IEEE Transactions on Computers*, C-20, 2, pp. 153–161.

13. H. T. Kung, 1979. "Let's Design Algorithms for VLSI Systems," pp. 65–90 in *Proceedings of the Caltech Conference on Very Large Scale Integration: Architecture, Design, Fabrication*. Pasadena, CA: California Institute of Technology.

14. See H. T. Kung, 1982. "Why Systolic Architectures?," *Computer*, 15, 1, pp. 37–46 and its citations.

15. Kung, 1980, p. 86.

16. H. T. Kung and C. E. Leiserson, 1979. "Systolic Arrays for VLSI," pp. 256–282 in I. S. Duff and G. W. Stewart (eds.), *Sparse Matrix Proceedings 1978*. Philadelphia: Society for Industrial and Applied Mathematics; M. J. Foster and H. T. Kung, 1980. "The Design of Special Purpose Chips," *Computer*, 13, 1, pp. 26–40.

17. Foster and Kung, 1980, pp. 26–27.

18. M. Annaratone et al., 1987. "The Warp Computer; Architecture, Implementation and Performance," Tech. Report CMU-RI-TR-87-18. Department of Computer Science/Robotics Institute, Carnegie-Mellon University, Pittsburgh, PA.

19. W. D. Hillis, 1985. *The Connection Machine*. Cambridge, MA: MIT Press.

20. W. D. Hillis, 1984. "The Connection Machine: A Computer Architecture Based on Cellular Automata," *Physica*, 10D, pp. 213–228.

21. Hillis, 1985, p. 1.

22. Ibid., p. 3.

23. Dasgupta, 2014, chapter 8.

24. Hillis, 1985, p. 5.

25. P. Hoffman, 1986. "The Next Leap in Computers," *New York Times Magazine*, December 7: http://www.ny.times.com/1986/12/07/magazine/the-next-leap-in-computers.html?. Retrieved May 23, 2016.

26. M. L. Minsky, 1985. *The Society of Mind*. New York: Simon & Schuster.

27. Hillis, 1985, p. xii.

28. G. A. Taubes, 1995. "The Rise and Fall of Thinking Machines," http://www.inc.com/magazine/19950915/2622.html. Retrieved May 23, 2016.

29. W. D. Hillis, 1988. "Intelligence as an Emergent Behavior; or The Songs of Eden," *Daedalus*, Winter, pp. 175–180, esp. p. 175.

30. Hillis, 1985, pp. 14–18.

31. Ibid., p. 38.

32. C. L. Seitz, 1985. "The Cosmic Cube," *Communications of the ACM*, 28, 1, pp. 22–33.

33. See Dasgupta, 2014, pp. 236–238, for a historical discussion of LISP.

34. Hillis, 1985, p. 47.

35. Ibid., p. 37.

36. See Dasgupta, 2014, pp. 217–215, for a historical discussion of APL.

37. Taubes, 1995.

38. J. S. Huxley, [1942] 2010. *Evolution: The Modern Synthesis* (The Definitive Edition). Cambridge, MA: MIT Press.

39. See, for example, the entry "Neo-Darwinism," pp. 196–199 in P. B. Medawar and J. S. Medawar, 1983. *Aristotle to Zoos: A Philosophical Dictionary of Biology*. Cambridge, MA: Harvard University Press.

40. Darwinism, neo-Darwinism, and evolution are, of course, the subjects of innumerable books, both technical and popular. My reference here are J. Maynard Smith, 1975. *Theory of Evolution*. Harmondsworth, UK: Penguin Books; M. Ruse, 1986. *Taking Darwin Seriously*. Oxford: Blackwell. The former was authored by a theoretical evolutionary biologist and geneticist; the latter by a philosopher of biology.

41. A. Newell and H. A. Simon, 1972. *Human Problem Solving*. Englewood Cliffs, NJ: Prentice-Hall, p. 96.

42. J. H. Holland, 1975. *Adaptation in Natural and Artificial Systems*. Ann Arbor: University of Michigan Press.

43. J. R. Sampson, 1984. *Biological Information Processing*. New York: Wiley, p. 130.

44. J. D. Bagley, 1967. "The Behavior of Adaptive Systems Which Apply Genetic and Correlation Algorithms" PhD dissertation, University of Michigan, Ann Arbor.

45. See Sampson, 1984, for an overview of this early work.

46. See, e.g., D. E. Goldberg, 1989. *Genetic Algorithms in Search, Optimization, and Machine Learning*. Reading, MA: Addison-Wesley.

47. For example, by the early 1990s, design theorists such as John Gero and Mary Lou Maher of the University of Sidney were exploring the application of genetic algorithms in computer models of the design process. See L. Alem and M. L. Maher, 1994. "A Model of Creative Design Using a Genetic Metaphor," pp. 283–294 in T. Dartnall (ed.), *Artificial Intelligence and Creativity*. Dordecht, The Netherlands: Kluwer.

48. D. Marr, 1982. *Vision*. San Francisco: W. H. Freeman.

49. Ibid., p. 19.

50. Ibid.

51. A. Newell, 1980. "Physical Symbol Systems," *Cognitive Science*, 4, pp. 135–183.

52. A. Newell, 1982. "The Knowledge Level," *Artificial Intelligence*, 18, pp. 87–127.

53. Marr, 1982, p. 25, figure 1-4.

54. Ibid., p. 32.

55. Ibid., p. 329.

56. Ibid., p. 36.

57. Ibid., p. 37.

58. D. O. Hebb, 1949. *The Organization of Behavior*. New York: Wiley.

59. A. M. Turing, 1950. "Computing Machinery and Intelligence," *Mind*, LIX, pp. 433–460.

60. W. S. McCulloch and W. Pitts, 1943. "A Logical Calculus of the Ideas Immanent in Nervous Activity," *Bulletin of Mathematical Biophysics*, 5, pp. 115–133.

61. A. M. Uttley, 1954. "The Classification of Signals in the Nervous System," *EEG Clinical Neurophysiology*, 6, pp. 479–494.

62. F. Rosenblatt, 1958. "The Perceptron: A Probabilistic Model for Information Storage and Organization in the Brain," *Psychological Review*, 65, pp. 386–408. Reprinted, pp. 92–114 in J. A. Anderson and E. Rosenfeld (eds.), 1988. *Neurocomputing*. Cambridge, MA: MIT Press. All page citations are to the reprint.

63. F. Rosenblatt, 1961. *Principles of Neurodynamics: Perceptrons and the Theory of Brain Mechanisms*. Washington, DC: Spartan Books.

64. H. D. Block, 1962. "The Perceptron: A Model for Brain Functioning I," *Review of Modern Physics*, 34, pp. 123–135.

65. Rosenblatt 1958, p. 111.

66. Ibid., p. 112.

67. Ibid., p. 113.

68. Ibid.

69. Ibid., pp. 113–114. Italics added.

70. Anderson and Rosenfeld, 1988, p. 89.

71. See, e.g., M. C. Yovits and S. Cameron (eds.), 1960. *Self-Organizing Systems: Proceedings of an Interdisciplinary Conference*. New York: Pergamon; H. von Foerster and G. W. Zopf (eds.), 1962. *Illinois Symposium on Principles of Self-Organization*. New York: Pergamon; M. C. Yovits, G. T. Jacobi, and G. D. Goldstein (eds.), 1962. *Self-Organizing Systems*. Washington, DC: Spartan Books.

72. M. L. Minsky and S. Papert, 1969. *Perceptrons*. Cambridge, MA: MIT Press.

73. M. L. Minsky, 1967. *Computation: Finite and Infinite Machines*. Englewood Cliffs, NJ: Prentice-Hall.

74. Minsky and Papert, 1969, p. 1.

75. For a representative sample of this work, see J. A. Anderson, 1972. "A Simple Neural Network Generating an Interactive Memory," *Mathematical Biosciences*, 14, pp. 197–220; T. Kohonen, 1972. "Correlating Matrix Memories," *IEEE Transactions on Computers*, C-21, pp. 353–359; L. N. Cooper, 1973. "A Possible Organization of Animal Memory and Learning," pp. 252–264 in B. Lundquist and S. Lundquist (eds.), *Proceedings of the Nobel Symposium on Collective Properties of Physical Systems*. New York: Academic Press; S. Grossberg, 1976. "Adaptive Pattern Classification and Universal Recoding: I. Parallel Development and Coding of Neural Feature Detectors," *Biological Cybernetics*, 23, pp. 121–134; D. Marr and T. Poggio, 1976. "Cooperative Computation of Stereo Disparity," *Science*, 194, pp. 283–287.

76. M. L. Minsky and S. A. Papert, 1988. *Perceptrons* (Expanded Edition). Cambridge, MA: MIT Press.

77. Ibid., p. xiv.

78. D. E. Rumelhart, G. E. Hinton, and R. J. Williams, 1986. "Learning Internal Representations by Error Propagation," pp. 318–362 in D. E. Rumelhart, J. L. McClelland, and the PDP Research Group, 1986. *Parallel Distributed Processing: Explorations in the Microstructure of Cognition. Volume 1: Foundations*. Cambridge, MA: MIT Press.

79. Ibid.

80. J. L. McCelland and D. E. Rumelhart and the PDP Research Group, 1986. *Parallel Distributed Processing: Explorations in the Microstructure of Cognition. Volume 2: Psychological and Biological Models*. Cambridge, MA: MIT Press.

81. Minsky and Papert, 1988.

82. J. L. McCelland, D. E. Rumelhart, and G. E. Hinton, 1986. "The Appeal of Parallel Distributed Processing," pp. 3–44 in Rumelhart, Hinton, and Williams et al., 1986.

83. D. E. Rumelhart, G. E. Hinton, and J. L. McClelland, 1986. "A General Framework for Parallel Distributed Processing," pp. 45–76 in McClelland, Rumelhart, and Hinton, 1986.

84. Ibid., p. 46.

85. Ibid., p. 54.

86. D. E. Rumelhart, 1989. "The Architecture of Mind: A Connectionist Approach," pp. 133–160 in M. I. Posner (ed.), *Foundations of Cognitive Science*. Cambridge, MA: MIT Press.

87. T. Sejnowski and C. Rosenberg, 1986. "NETtalk: A Parallel Network That Learns to Read Aloud." Tech. Report JHU/EEC-86/01, Johns Hopkins University, Baltimore; T. Seknowski and C. Rosenberg, 1987. "Parallel Networks That Learn to Pronounce English Texts," *Complex Systems*, 1, pp. 145–168.

88. C. A. Mead, 1989. *Analog VLSI and Neural Systems*. Reading, MA: Addison-Wesley.

89. At this time of writing, the definitive—and alarmingly encyclopedic!—history of cognitive science is Margaret Boden's (2006) provocatively titled *Mind as Machine: A History of Cognitive Science* (Volumes I and II). Oxford: Clarendon. An older, less ambitious in scope yet engaging work is H. Gardner, 1985. *The Mind's New Science*. New York: Basic Books.

90. T. S. Kuhn, [1962] 2012. *The Structure of Scientific Revolutions* (4th ed.). Chicago: University of Chicago Press.

91. For those unfamiliar with the term, *psychonomics* refers to experimental psychology, especially to experimental cognitive psychology.

92. W. Schneider, 1987. "Connectionism: Is It a Paradigm Shift for Psychology?" *Behavior Research Methods, Instruments & Computers*, 19, 2, pp. 73–83.

93. Quoted in Ibid., p. 78.

94. Ibid., p. 79.

95. D. E. Rumelhart, P. Smolensky, J. L. McClelland, and G. E. Hinton, 1986. "Schemata and Thought Processes in PDP Models," pp. 7–57 in J. L. McCelland, D. E. Rumelhart, and the PDP Research Group. *Parallel Distributed Processing: Explorations in the Microstructure of Cognition. Volume 2: Psychological and Biological Models*. Cambridge, MA: MIT Press.

96. Ibid., p. 20.

97. Ibid.

98. Ibid., p. 25.

99. Ibid., pp. 34–35.

100. Schneider, 1987, p. 79.

101. Ibid., pp. 81–82.

102. Ibid., p. 82.

103. J. A. Fodor and Z. W. Pylyshyn, 1988. "Connectionism and Cognitive Architecture, A Critical Analysis," *Cognition*, 28, pp. 3–71.

104. Ibid., p. 22.

105. Ibid., p. 23.

106. Ibid., p. 25.

107. Ibid., p. 26.

108. J. A. Fodor, 1976. *The Language of Thought*. Cambridge, MA: Harvard University Press.

109. Fodor and Pylyshyn, 1988, p. 27.

110. Ibid., p. 47.

111. D. H. Ballard, 1986. "Cortical Connections and Parallel Processing: Structure and Function," *Behavioral and Brain Sciences*, 9, pp. 167–120.

112. Ibid.

113. Fodor and Pylyshyn, 1988, p. 49.

114. P. Smolensky, 1988. "The Constituent Structure of Connectionist Mental States: A Reply to Fodor and Pylyshyn," Tech. Report, CU-CS-394-88, Department of Computer Science, University of Colorado at Boulder.

115. D. J. Chalmers, 1990. "Why Fodor and Pylyshyn Were Wrong: The Simplest Refutation," *Proceedings of the 12th Annual Conference of the Cognitive Science Society*, pp. 340–347.

116. T. van Gelder, 1990. "Compositionality: A Connectionist Variation on a Classical Theme" *Cognitive Science*, 14, 3, pp. 355–384.

117. D. A. Norman, 1986. "Reflections on Cognition and Parallel Distributed Processing," pp. 531–546 in McClelland, Rumelhart, and the PDP Research Group, Vol. 2.

118. Ibid., p. 533.

119. Ibid., p. 538.

120. Ibid.

121. Ibid., p. 541,

122. Rumelhart, 1989.

123. Posner, 1989.

124. A. Newell, 1990. *Unified Theories of Cognition*. Cambridge, MA: Harvard University Press.

125. P. Prusinkiewicz and A. Lindenmayer, 1990. *The Algorithmic Beauty of Plants*. Berlin: Springer-Verlag.

126. A. Lindenmayer, 1968. "Mathematical Models of Cellular Interaction in Development—I," *Journal of Theoretical Biology*, 18, pp. 280–315.

127. Ibid., p. 280.

128. Ibid., p. 281.

129. See, e.g., J. Hartmanis and R. E. Stearns, 1966. *Algebraic Structure Theory of Sequential Machines*. Englewood Cliffs, NJ: Prentice-Hall; M. L. Minsky, 1967. *Computation: Finite and Infinite Machines*. Englewood Cliffs, NJ: Prentice-Hall; Z. Kohavi, 1970. *Switching and Finite Automata Theory*. New York: McGraw-Hill, Part 3.

130. D. W. Thompson, [1917] 1992. *On Growth and Form*. Cambridge: Cambridge University Press.

131. In the interest of accuracy, this describes one form of the FSM called the "Moore machine": Here the output is a function of only the present state. In an alternative form, called the "Mealy machine," the output is a function of both the input and the present state. These two forms are named according to their inventors, E. F. Moore and G. H. Mealy, respectively.

132. Lindenmayer, 1968, p. 283.

133. Ibid.

134. N. Chomsky, 1956. "Three Models for the Description of Language," pp. 113–124 in *Proceedings of the Symposium on Information Theory*. Cambridge, MA: MIT; N. Chomsky, 1957. *Syntactic Structures*. The Hague: Mouton; N. Chomsky, 1959. "On Certain Formal Properties of Grammar," *Information & Control*, 2, pp. 136–167.

135. J. E. Hopcroft and J. D. Ullman, 1969. *Formal Languages and Their Relation to Automata*. Reading, MA: Addison-Wesley.

136. S. C. Kleene, 1956. "Representation of Events in Nerve Nets and Finite Automata," pp. 3–41 in *Automata Studies*. Princeton, NJ: Princeton University Press.

137. A. Lindenmayer and G. Rozenberg, 1972. "Development Systems and Languages," *Proceedings of the 4th ACM Symposium on Theory of Computing*. Denver, CO, pp. 214–221.

138. H. Abelson and A. di Sessa, 1981. *Turtle Geometry*. Cambridge, MA: MIT Press.

Epilogue

"Progress" in the Second Age of Computer Science

E.1 THE IDEA OF SCIENTIFIC PROGRESS

Did the second age of computer science show evidence of *progress*? The answer depends on what one means by this word.

In its grandest sweep the *idea of progress* is the idea that history is a triumphant record of human development; of the improvement of the human condition.[1] The American historian Robert Nisbet in his magisterial survey of 1980 traced the origins of this idea to antiquity.[2] Others hold that this idea is of much more recent origin. For the English historian John Bury in his classic (1932) study it was first seriously considered in the 16th century.[3] Bruce Mazlish, an American historian of the late 20th century, locates it in the 17th century, stimulated at least in part by the Scientific Revolution.[4]

But regardless of its origins, the idea of progress in the grand sense—that history itself is the story of continuous human development—has evoked much controversy, especially in the 20th century, many scholars, especially postmodernists, denying, even despising the very idea.[5] The title of the volume edited by Bruce Mazlish and Leo Marx—*Progress: Fact or Illusion?* (1996)—speaks to its contentious and contested nature.

Even in the realm of science—of specific concern here—the story is complicated. On the one hand, most practicing scientists scarcely harbor any doubts on this matter. For what they see is that despite the many false alleys along which science has wandered across the centuries, despite the waxing and waning of theories and models, the history of science, at least since the "early modern period" (16th and 17th centuries), is a history of both steady and sometimes breathtaking growth of scientific *knowledge*. For most scientists this growth of knowledge *is* progress;[6] a necessary and sufficient condition for something to be deemed a science. To deny either the possibility or the actuality of progress in science is to deny science's raison d'être. Perhaps this is why Nobel Laureate biologist Peter Medawar poured scorn on those who would deny the idea of progress in science.[7]

On the other hand, there are the doubters, skeptics, unbelievers among those who, if not scientists, are concerned with what science is about.[8] Careful examination by historians and philosophers of science has shown that identifying progress in science is, in many ways, a formidable, even elusive problem. The most influential of these scholars such as philosophers Karl Popper, Thomas Kuhn, Larry Laudan, and Paul Thagard do not doubt that science makes progress; rather they have debated on *how* science makes progress; or *what is it* about science that makes it inherently progressive.[9]

E.2   KNOWLEDGE-CENTRIC PROGRESS

"The function of science is to produce knowledge." So wrote physicist–philosopher John Ziman in 2000.[10] No surprise, then, that to the scientist growth of knowledge is manifest evidence of scientific progress.[11] Let us call this the idea of knowledge-centric progress. Still, this view is not without its problems, perhaps most vividly exemplified by Thomas Kuhn's concept of the paradigm shift that produces a scientific revolution.[12] For this very notion suggests that a new paradigm that overthrows and replaces the old may well constitute a body of knowledge that is (in Kuhn's famous and controversial term) "incommensurable" with the knowledge constituting the ousted paradigm.[13] That is, the core elements of the two paradigms, the old and the new, offer such fundamentally different views of the world, they cannot be compared. So how can one claim that a paradigm shift constitutes growth of knowledge if the two paradigms are incommensurable?

But paradigm shifts—and scientific revolutions—are relatively rare events in any given science. Within a paradigm scientists practice what Kuhn called "normal science": They solve problems generated by and constrained by the paradigm's dominating *diktat*. And, as Kuhn noted, the conditions (both social and intellectual) under which normal science is practiced are such that growth of knowledge *inside that paradigm* is ensured; hence progress is ensured.[14] The textbooks that students rely on are repositories of the state of knowledge in the prevalent paradigm at any point of historical time. Successive editions of a textbook reflect the growth of knowledge resulting from normal scientific practice.

As for what Kuhn called "extraordinary science"—the science that produces paradigm shifts—Kuhn argued that the very fact that a revolution occurs and a new paradigm comes into place to replace the old can hardly be seen as anything other than progress.[15] In practical terms the relevant scientific community will "renounce" most of the books and papers that had embodied the overthrown paradigm; new textbooks will be published reflecting the new paradigm—and this repudiation of past knowledge and embracing of new "present" knowledge is perceived as progress.

E.3 PROBLEM-CENTRIC PROGRESS

There is another way of understanding the nature of scientific progress and that is by recognizing that scientists are, first and foremost, *problem solvers;* that if the business of science is the production of knowledge the means to do so is by solving problems. The practicing scientist *qua* scientist is not concerned with the grand sweep of history; in fact not much with history at all. His or her worldview is considerably more modest, more localized:

Scientists identify a problem, make solving the problem their goal, and the extent to which they succeed in meeting this goal marks the extent of their contributions to progress in their particular domains.

So the growth of knowledge scientists almost instinctively associate with scientific progress is the outcome of how effective their problem solutions are.

Let us call this view problem-centric progress. In fact, such philosophers as Popper, Kuhn, and Laudan have all recognized and adopted the problem-centric position, with the implicit understanding that problem-centrism and knowledge-centrism are complementary views of scientific progress.

Still, problem-centrism offers a different perspective that theorists of progress have explored. Larry Laudan began his book *Progress and Its Problems* (1977) with the bald statement "Science is essentially a problem-solving activity."[16] (Compare this with Ziman's announcement quoted in Section E.2.) This premise, for Laudan, shaped the very nature of his theory of scientific progress. He tells us that the "acid test" for a scientific theory is not so much whether the theory is true but whether "it provides satisfactory solutions to important problems."[17] He went on to propose a concept he termed problem-solving effectiveness that, roughly speaking, means the extent to which a theory solves important empirical (real-world) problems posed in a science *minus* the ambiguities and anomalies the theory might generate.[18] This led Laudan to propose that progress in a science occurs if and only if successive theories in that science demonstrate a growth in problem-solving effectiveness.[19]

E.4 LINKING PROGRESS TO CREATIVITY

Laudan devoted his entire book to the exploration of his proposition and its implications. But appealing though it is, his problem-centric theory of progress poses a difficulty: It demands *comparison* between scientific theories. The difficulty lies in that it makes sense to compare solutions only if they address exactly the same problem.

But this may not always be the situation, in which case something like Kuhn's incommensurability issue will arise. The solutions may well be incommensurable, and thus noncomparable, because they are solutions to different problems.

In the realm of computer architecture, for example, we have seen that the problem of constructing a taxonomy produced several distinct solutions (Chapter 3, Section 3.18). The Hwang–Briggs taxonomy purported to be an improvement on the Flynn scheme; Skillicorn's system was intended to be an improvement on the Hwang–Briggs scheme; the present author's "molecular" taxonomy was intended to improve on Skillicorn's system. But did this "succession of theories" (in Laudan's words) manifest increasing problem-solving effectiveness? Did this succession indicate progress? We are hard-pressed to answer "yes" to these questions because we can judge each taxonomy only according to *what problem it was trying to solve*. And Flynn's problem was not the same as Hwang and Brigg's problem; the latter was not the same as Skillicorn's problem; and Skillicorn's problem was not identical to this author's problem. They were incommensurable solutions.

The reader will also observe that knowledge-centrism and problem-centrism offer complementary approaches to the problem of understanding scientific progress. To John Ziman, the business of science is knowledge production;[20] to Larry Laudan, science is a problem-solving activity.[21] But not all knowledge produced is necessarily significant to making progress, nor are all problems necessarily significant to making progress. The knowledge must be both original and valuable; the problem must have significant relevance.

What seems to be explicitly lacking in both problem-centrism (in Laudan's sense) and knowledge-centrism (in Ziman's sense) is the recognition that for something (problem solution or knowledge) to seriously contribute to progress in a discipline demands the exercise of *creativity*. In other words, the idea of progress in science (and, in fact, in art, technology, and the humanities) must be linked to creativity. Let us call this view creativity-centric progress.

E.5 CREATIVITY-CENTRIC PROGRESS IN THE ARTIFICIAL SCIENCES

For most philosophers and historians of science, "science" means the natural sciences whose domains are natural phenomena. The knowledge-centric and problem-centric views of progress as enunciated by such scholars as Kuhn, Popper, Laudan, Ziman, and Thagard pertain to the natural sciences. But the domain of computer science—our particular interest in this book—is composed of computational artifacts (Prologue, Section P.2). Computer science is an artificial science. So let us confine this discussion to the realm of the artificial.

We begin by noting a fundamental ontological distinction between natural and artificial problems: The natural world comprises entities *that have no purpose*; that is, their creation and existence entail no prior purpose, no goals, no needs. They just *are*, and the natural scientist's task is to uncover the laws governing their existence. In contrast, the artificial world comprises artifacts—things that are *consciously* conceived and produced by an earthly maker—an *artificer*—in response to some goals, needs, or desires. Their creations presuppose some *purpose* on the part of their artificers.

So whereas a problem in natural science is to explain purposeless natural phenomena, a problem in artificial science begins with purpose. We can then say this:

A problem in artificial science is a set of goals (or needs) to create an artifact that demonstrably satisfies these goals. A solution in artificial science is an artifact created in response to the problem.

However, not all problems demand creativity on the part of artificers. Something more is demanded:

A problem *of interest* in artificial science is a problem that (a) has never been encountered before; that is, its goals are of a kind never previously conceived; or (b) the goals, though previously identified, have never led to an artifact's satisfying them; that is, no solution is known to exist for the problem; or (c) solutions exist for the problem but none are deemed satisfactory according to some artificer's or the relevant community's standards.

And thus: A solution to a problem of interest is a *creative solution*. In other words, a solution to a problem of interest demanding creativity on the part of some artificer is a creative solution. Such a solution is both an *artifact* that is original (because it is a solution to a problem of interest) and it generates one or more new items of knowledge associated with the artifact. For example, the Pascal programming language was both a particular original (abstract) artifact that could be used as a tool by programmers and a body of new knowledge about programming languages of a certain kind.

We may now define creativity-centered progress thus:

(1) Progress in the context of a *particular* problem of interest over some given period or epoch is judged by the effectiveness of the creative solutions to that problem of interest.
(2) Progress in a *given domain* of an artificial science over some period or epoch is judged by the effectiveness of its constituent creative solutions.

Some caveats are in order:

One. Whether something is a creative solution and its effectiveness as a creative solution are matters of *judgment*. Depending on the nature of the problem of interest, the judgment may be personal or communal, objective or subjective. Effectiveness is judged by taking into account such factors as the completeness of the creative solution and whether the solution generated new problems, and the extent of agreement within the community of judges.

Two. Who does the judging is clearly important. Whether a solution is deemed creative will be critically dependent not only on the nature of the problem of interest but also on the knowledge, experience, and even taste of the judge(s).[22]

Three. Judgment of progress can be limited to a single problem of interest that has evoked one or more creative solutions, or it can apply to an entire domain or branch of an artificial science, or indeed to an entire artificial science.

Four. Creativity—and thus progress—is a matter of *historical* judgment. That is, it will entail the judges' familiarity with the history of the problem of interest, their knowledge of the artificer's personal history, and their understanding that creative solutions constitute progress over historical periods (or epochs).[23]

Five. Creativity-centric progress, unlike Laudan's problem-centric progress, is not judged relative to some distinctly identified *previous* epoch. (This is to avoid the incommensurability issue discussed earlier.) Historicity is intraepochal rather than interepochal. Indeed, creativity-centrism does not recognize any kind of *relative* progress, whether intraepochal or interepochal.

Six. Finally, there is no imputation of "goodness" or "badness" to creativity-centric progress. Judgment of progress does not entail judgment of a "better" situation than what prevailed before.

E.6   JUDGING PROGRESS IN THE SECOND AGE OF COMPUTER SCIENCE

Armed with this idea of creativity-centered progress let us return to the question posed at the very beginning of this Epilogue. Did the second age of computer science evince any evidence of progress?

Recall the principal themes characterizing this age (Prologue, Sections P.6–P.10): (a) the quest for scientificity in real-world computing; (b) the desire for human-centered computing; (c) the search for a unified theory of intelligence; (d) liberating computing from the shackles of sequentiality; and (e) revolting against the von Neumann core of the computational paradigm. To explore the question of progress in the second age, let us consider a few key episodes from the preceding chapters representative of these themes.

E.6.1 The Scientificity Theme

The Algol 60 project of the 1960s had the goal of universality as an expression of this theme (Chapter 1, Section 1.1). However, after it was designed and defined, universality was found wanting in some respects, for instance, in its bias toward numerical computation. Moreover, its ambiguities and internal inconsistencies led to difficulties in implementing the full language, another blow against universality.

Hence the Algol 68 project—to create a successor to Algol 60 that was "more universal" and shorn of Algol 60's ambiguities and inconsistencies. Universality was addressed in Algol 68 in a number of ways. One was the unification of statements and expressions, another was by inventing the **mode** concept, thus affording the user the ability to construct an unlimited number of data types, a third by facilitating both numerical and nonnumerical computations.

To the extent that Algol 68 met these goals one could claim that it had advanced progress toward the goal of universality. But the design of this language generated new problems of its own. Most visible was its complexity. Although arguably much new knowledge was embedded in the language, this very knowledge—increasing its complexity—seriously hindered its understandability and learnability. The size of the Algol 68 report was a manifestation of this complexity. Thus the effectiveness of the language as a creative solution to the problem of creating a universal, machine-independent, algorithmic language was seriously undermined—at least in the perception of some. On the other hand the first implementers of the language at the Royal Radar Establishment (RRE) reported that they had little difficulty in teaching Algol 68 to their user community. Again, it must be noted that this first implementation, called Algol 68R, was not that of the full language.

As noted, the creativity-centric model does not recognize relative progress because of the problem of incommensurabilty. So the question of whether Algol 68 marked progress relative *to* Algol 60 is meaningless. We may ask this instead: Did Algol 68 solve *its* problem of interest more effectively than Algol 60 solved *its* problem of interest? According to some criteria the answer is "yes." Others (such as Dijkstra, Hoare, and Wirth) would claim that Algol 68 constituted a regression in the design of universal programming languages.

E.6.2 Combining the Themes of Scientificity and Human-Centered Computing

Pascal, yet another successor to Algol 60, was born from the need for human-centered computing (Chapter 1, Section 1.7). For Nicklaus Wirth, its designer, this meant a quest for simplicity (Chapter 1, Section 1.8) as a means (a subgoal) to that larger end. But there were other goals. One, also a subgoal of the human-centered computing theme, was to facilitate a certain style of program design and composition generically called structured programming (Chapter 1, Sections 1.9, 1.11). The other originated in the desire for scientificity: By the time Wirth had embarked on the development of Pascal this goal had led to the idea of programs as mathematical entities. A mathematical logic for characterizing program behavior had emerged through the work of Robert Floyd and C. A. R. Hoare in the late 1960s (Chapter 1, Sections 1.12,1.13; Chapter 5, Section 5.7).

So Pascal as a computational artifact was a response to three (closely related) goals: linguistic simplicity, thereby aiding learnability and understandability; a means for structured programming; and as a mean for thinking mathematically about programs and programming. And as discussed in Chapter 1, there is sufficient evidence in support of the claim that *in the context of these goals* Pascal manifested progress as a response to both the scientificity and human-centered computing objectives.

E.6.3 The Human-Centered Computing Theme, Revisited

The concern for human-centered computing—"Dijkstra's discontent" (Chapter 2, Section 2.1)—also spawned several conceptual problems under the rubric of programming methodology. Their solutions, the subjects of Chapter 2, were techniques.

Structured programming was one, but there were several others such as information hiding, abstract specification, the concept of abstract data type, a relational model of data organization, the class concept, and object orientedness.

These techniques were abstract artifacts. Their effectiveness as creative solutions lay in that they constituted *new* knowledge (primarily but not exclusively procedural) that has proved to be highly consequential in the realm of programming methodology, software engineering, and programming language design. It is fair to say that the inventions of these techniques clearly manifested progress in the domain of software design and development because of their effectiveness as programming artifacts.

E.6.4 The Parallelism Theme

Consider now the theme of liberating computing from the shackles of sequentiality by exploiting parallelism (in short, the "parallelism theme"). This problem was certainly not new: It had been addressed in the second half of the 1960s by such scientists as Dijkstra, who had suggested the concepts of critical regions and semaphores as early solutions (Chapter 4, Section 4.3). However, in the 1970s the parallelism theme was refined, yielding several closely related but distinct problems of interest. A plethora of abstract computational artifacts emerged in response to these new problems, some as techniques for constructing parallel programs (e.g., monitors, conditional critical regions, shared class, concurrent statement), some as programming languages (e.g., Concurrent Pascal, Glypnir, Actus, and Edison), all within the constraints imposed by the von Neumann computing style (Chapter 4, Sections 4.4–4.10).

If we take these solutions *collectively* as a class of parallel programming artifacts, they clearly convey a sense of progress in addressing the overlying parallelism theme—both as new knowledge and as practical techniques. Viewed over a period from the mid-1960s through the duration of the second age it seems uncontroversial to claim that these artifacts were effective as creative solutions to the problem of liberating computing from the shackles of sequentiality.

E.6.5 Combining the Parallelism and Human-Centered Computing Themes

The parallelism theme was addressed in a different way when combined with the desire for human-centered computing. The argument recognized the reality of bounded rationality (Chapter 6, Section 6.5): that people are not adept at thinking in parallel; they are naturally sequential thinkers. Thus the user of a computing system at any particular abstraction level (as application programmer or systems programmer or microprogrammer) should be shielded from the perils of parallelism. The user thinks sequentially, and the task of parallel processing is delegated to some combination of hardware, architecture, and software resources. The techniques—automatic parallelism detection and optimization strategies—and knowledge generated in the second age in response to this goal were impressive creative solutions. Progress was undoubtedly manifested.

But if we compare the two "ideologies"—hiding parallelism from the user versus forcing the user to cope with parallel thinking—in terms of their creative solutions, the issue of progress becomes cloudy. The goals were different, the problems of interest were different, and the resulting solutions were incommensurable.

E.6.6 Combining the Parallelism and the Subversion Themes

What if the parallelism theme is combined with the theme of subverting the von Neumann style (in short, the "subversion theme")? In fact, at least two research programs in the second age addressed precisely this combination, one being data-flow computing (Chapter 4, Sections 4.16, 4.17), the other the Connection Machine project (Chapter 7, Section 7.6).

Both were concerned with the production of massively parallel processing power; both sought to revolt and reject the classical von Neumann style. But in other important respects they differed in their origins. The Connection Machine was inspired by brain architecture on the one hand and the architecture of cellular automata on the other. Data-flow computing originated in formal, graph-theoretic models of computation. The resulting solutions were thus vastly different.

For the sake of brevity let us focus on data flow largely because it attracted much greater attention as a serious challenger to the von Neumann style and prompted research in all aspects of computing—hardware, architecture, programming, and languages—in Europe, North America, and Asia.

Data flow as a computing style arose in response to a desire to exploit the "natural" parallelism between operations in a computation constrained only by date availability for instructions to execute. Parallelism would then be "data driven" and inherently asynchronous rather than "control driven" as in von Neumann computing (Chapter 4, Sections 4.16, 4.17). This goal of course feeds into the combination of the larger themes of parallelism and subversion.

Different interpretations of this natural parallelism goal generated, in turn, two alternative possibilities called, respectively, "static" and "dynamic" data flow. The various projects undertaken in America, France, England, and Japan espoused one or the other of these styles.

Ultimately though, the data-flow movement turned out to be *unprogressive* in the following sense. A critical subgoal for achieving natural, asynchronous, data-driven parallelism was that there would be no central memory to hold data objects that would have to be accessed to fetch data for instructions to operate on. Accessing memory is part of the essence of von Neumann computing and gives rise to the so-called von Neumann bottleneck that impedes efficient computing.[24] In fact, in data flow there would be (ideally) no data memory, hence no data objects. Rather, data values produced by instructions would "flow" to other instructions needing those values.

This crucial goal of no central data memory could not be satisfied because the contents of large data structures (such as arrays or linked lists) could not be efficiently passed around as might simple scalar values. Rather, data structures would have to be stored in some central memory and accessed as in the von Neumann machine. The effectiveness of computational artifacts as creative solutions was seriously undermined because of this constraint. One cannot convincingly make a case that the data-flow style of computing manifested any kind of progress.

E.6.7   Unified Intelligence Theme

As a final set of examples, consider the quest for a unified science of intelligence (in short, "unified intelligence"). The goal here was to arrive at a theory of intelligence that would encompass both natural and artificial intelligence. So the kind of solution sought was a theory, a piece of propositional knowledge: The theory was the desired artifact.

A seminal creative solution to the problem was the heuristic search hypothesis originally proposed by Allen Newell and Herbert Simon in the late 1950s. The 1960s witnessed the further development of heuristic search and its use by Newell, Simon, and other investigators across a range of human and machine intelligence. By the beginning of the second age, Newell and Simon could encapsulate the power and breadth of heuristic search as a theory of intelligence in their tome *Human Problem Solving* (1972).

Yet its deployment also revealed its limitation,s such as the realization on the part of many researchers that "universal" or "domain-independent" (or "weak") heuristics such as MEA or "generate-and-test" were insufficient; rather domain-specific (or "strong") heuristics were necessary in many problem situations. The effectiveness of weak heuristic search as a creative solution to the universal intelligence was thus undermined. Thus one could assert that domain-independent heuristic search was only a *partial* solution—thus a manifestation of only partial progress—in the quest for unified intelligence.

Another project in the unified intelligence theme worth considering here is Terry Winograd's SHRDLU system (Chapter 6, Section 6.12). Winograd's goal was to construct a simulated robot (a liminal artifact) that would understand natural language in the context of a blocks world and carry out tasks in that world accordingly. Within the limits of this goal its effectiveness as a creative solution was spectacular. Among its contributions to the universal intelligence theme was demonstrating the systemic nature of language understanding (Chapter 6, Section 6.14), the significance of procedural semantics for eliciting meaning (Chapter 6, Section 6.13), and the precise way in which knowledge about the world and linguistic knowledge interact in understanding and doing things. SHRDLU embodied such knowledge and can be said to unequivocally represent progress in the realm of language understanding in limited world domains.

On the other hand, taking a wider-angled view, the question of progress was far less conclusive. If the goal was to come up with a creative solution to the general problem of universal intelligence, SHRDLU failed. In this larger sense and within the confines of the second age of computer science, SHRDLU's contribution to progress in natural language understanding (let alone universal intelligence) was pronouncedly muted.

E.7 THE KNOWLEDGE-CENTRIC VIEW, REVISITED

Scientists—natural or unnatural—are fundamentally problem solvers. The problem-centric view was predicated on this assumption. But, as we have noted, scientists conventionally refer to the growth of knowledge as a "measure" of progress in their respective fields. The creativity-centric view attempts to establish a common ground between these alternative perspectives. A solution to a problem of interest is judged as progressive to the extent that it is an effective creative solution. At the same time each such creative solution generates new knowledge *about* the problem of interest.

In any given epoch the knowledge accumulated over the course of that epoch is the cumulative outcome of creative problem solving in the relevant science. The growth of knowledge in a science (or some domain of a science) over the epoch is then taken to be an objective record of progress in that domain over that epoch. Thus progress in the second age of computer science would be objectively reflected in the growth of computational knowledge over the period (approximately) 1970–1990.

But some might object that not all such knowledge is *relevant* or *useful* or *consequential*. All knowledge is not equal. Our brief and selective exploration of creativity-centric progress in the second age indicates that sometimes the knowledge generated is of a "negative" kind because certain problems were not satisfactorily solved or because there were unanticipated consequences. For example, the realization that the data-flow style is not entirely conducive to creating an alternative to von Neumann computing; or that the microworld approach espoused in the SHRDLU project is not entirely suited as a foundation for generalized natural language understanding theories or models; or that the undue complexity of Algol 68 is not conducive to the design of a universal programming language. Skeptics and critics might claim that such "negative" knowledge should not count as signifiers of progress.

On the other hand, as the saying goes, we learn from our failures. "Negative" knowledge may turn out to be as valuable as "positive" knowledge, for it constitutes knowledge of what not to do or what to avoid. Sometimes negative knowledge leads to new goals, new problems of interest not previously recognized. The Algol 68 experience led to the quest for simplicity in language design. The SHRDLU experience led to the search for general, theoretical frameworks for organizing and representing knowledge, such as frames, naïve physics, and semantic networks. The data-flow experience revealed that in some important, practical sense the von Neumann style cannot be entirely dispensed with. And, though not mentioned as one of the examples in this Epilogue, the growth in the complexity of computer architectures *became* negative knowledge in the minds of some computer scientists and designers. The RISC movement was generated as a reaction to such perceived negative knowledge (Chapter 3, Section 3.16).

More interestingly (and perplexingly), the relevance or consequence of a piece of knowledge may be "discovered" much later than when it was first generated. A classic instance of this was the invention of microprogramming in 1951. This did not evoke much interest for well over a decade, until IBM realized its significance for their specific goals for the System/360 series of computers in the early 1960s.[25] The remarkable growth of interest in user microprogramming and universal host machines evinced in the second age and the various problems of interest thus engendered (Chapter 3, Sections 3.7–3.13) were grounded in what, circa 1960, might have seemed rather irrelevant knowledge to some computer designers.

The fact is, no knowledge, no matter how obscure, esoteric, or recondite it might seem at the time of its production, should be condemned, because one never knows *when* that knowledge might become significant.

In sum, from the creativity-centric perspective, each of the five themes that dominated the second age of computer science showed *both* evidence of progress *and* evidence of lack of progress (or even regression). But from a strictly knowledge-centric perspective we can claim that *all* creative solutions, regardless of how effective they are at the time of their emergence, have the *potential* as manifestations of progress. In this ambiguity lies the ambivalent nature of the idea of scientific progress.

NOTES

1. B. Mazlish and L. Marx, 1996. "Introduction," pp. 1–7 in B. Mazlish and L. Marx (eds.), 1996. *Progress: Fact or Illusion?* Ann Arbor, MI: University of Michigan Press.

2. R. Nisbet, [1980] 1994. *History of the Idea of Progress* (2nd ed.). New Brunswick, NJ: Transaction Publishers.

3. J. B. Bury, [1932] 1955. *The Idea of Progress.* New York: Dover Books.

4. B. Mazlish, 1996. "Progress: A Historical and Critical Perspective," pp. 27–44 in Mazlish and Marx (eds.), 1996.

5. Nisbet, [1980] 1994; Mazlish, 1996.

6. See, e.g., H. Bondi, 1975. "What is Progress in Science?," pp. 1–10 in R. Harré (ed.), 1975. *Problems of Scientific Revolutions: Progress and Obstacles to Progress in the Sciences.* Oxford: Clarendon; J. L. Monod, 1975. "On the Molecular Theory of Evolution," pp. 11–24 in Harré, 1975; J. Ziman, 2000. *Real Science; What Is It, and What It Means.* Cambridge: Cambridge University Press. Bondi and Ziman were physicists; Monod, a Nobel Laureate, was a molecular biologist.

7. P. B. Medawar, 1973. *The Hope of Progress.* New York: Knopf Doubleday.

8. For a sampler of oppositional views see S. Harding, 1998. *Is Science Multicultural?* Bloomington: Indiana University Press; N. Koertege (ed.), 1998. *A House Built on Sand.* New York: Oxford University Press; Z. Sardar, 2000. *Thomas Kuhn and the Science Wars.* Cambridge: Icon Books; A. Sokal, 2008. *Beyond the Hoax: Science, Philosophy and Culture.* New York: Oxford University Press. For a concise review of the controversy, see G. Holton, 1996. "Science and Progress Revisited," pp. 9–26 in Mazlish and Marx (eds.), 1996.

9. See, e.g., K. R. Popper, 1975. "The Rationality of Scientific Revolutions," pp. 72–101 in Harré, 1975; T. S. Kuhn, 2012. *The Structure of Scientific Revolutions* (4th ed.). Chicago: University of Chicago Press; L. Laudan, 1977. *Progress and Its Problems.* Berkeley, CA: University of California Press; P. R. Thagard, 1988. *Computational Philosophy of Science.* Cambridge, MA: MIT Press.

10. Ziman, 2000, p. 83.

11. See, e.g., ibid., chapters 4 and 9.

12. Kuhn, 2012.

13. Ibid., pp. 147–149.

14. Ibid., pp. 164–165.

15. Ibid., pp. 165–166.

16. Laudan, 1977, p. 11.

17. Ibid., pp. 13–14.

18. Ibid., p. 68.

19. Ibid.

20. Ziman, 2000, pp. 56, 83.

21. Laudan, 1977, p. 11.

22. For a discussion of judgment of creativity see S. Dasgupta, 2011. "Contesting (Simonton's) Blind Variation, Selective Retention Theory of Creativity," *Creativity Research Journal*, 23, 2, pp. 166–182, esp. pp. 170–172.

23. This is the phenomenon of the *historicity of creativity* that I have discussed elsewhere. See S. Dasgupta, 2013. "The Historicity of Creativity," pp. 61–72 in W. J. Gonzalez (ed.), 2013. *Creativity, Innovation and Complexity in Science*. La Coruña: Netbiblio; S. Dasgupta, 2016. "From *The Sciences of the Artificial* to Cognitive History," pp. 60–70 in R. Frantz and L. Marsh (eds.), 2016. *Minds, Models and Milieux: Commemorating the Centennial of the Birth of Herbert Simon*. Basingstoke, UK: Palgrave Macmillan.

24. Arguably, the most well-known criticism of the von Neumann bottleneck was by John Backus, in his 1977 Turing Award lecture. See J. W. Backus, [1978] 1987. "Can Programming Be Liberated From the von Neumann Style?," pp. 63–30 in Anon. (ed.), 1987. *ACM Turing Award Lectures: The First Twenty Years 1966–1985*. New York: ACM Press/Reading, MA: Addison-Wesley.

25. S. Dasgupta, 2014. *It Began with Babbage: The Genesis of Computer Science*. New York: Oxford University Press, chapters 12 and 15.

Bibliography

H. Abelson and A. di Sessa, 1981. *Turtle Geometry*. Cambridge, MA: MIT Press.

W. B. Ackerman, 1982. "Data Flow Languages," *Computer*, 15, 2, pp. 15–25.

ACM Curriculum Committee, 1968. "Curriculum 68," *Communications of the ACM*, 11, 3, pp. 151–197.

D. A. Adams, 1968. "A Computational Model with Data Flow Sequencing," Tech. Report CS 117. Stanford University.

U. Agüero and S. Dasgupta, 1987. "A Plausibility Driven Approach for Computer Architecture Design," *Communications of the ACM*, 30, 11, pp. 922–932.

A. V. Aho, J. E. Hopcroft, and J. D. Ullman, 1974. *The Design and Analysis of Computer Algorithms*. Reading, MA: Addison-Wesley.

A. V. Aho and J. D. Ullman, 1973. *The Theory of Parsing, Translation and Compiling, Volume 2*. Englewood Cliffs, NJ: Prentice-Hall.

S. Alagic and M. A. Arbib, 1978. *The Design of Well-Structured and Correct Programs*. New York: Springer-Verlag.

S. Alem and M. L. Maher, 1994. "A Model of Creative Design Using a Genetic Metaphor," pp. 283–294 in Dartnall, 1994.

C. Alexander, 1964. *Notes on the Synthesis of Form*. Cambridge, MA: Harvard University Press.

G. S. Almasi and A. Gottlieb, 1989. *Highly Parallel Computing*. Redwood City, CA: Benjamin/ Cummings.

M. Amamiya, M. Tokesue, R. Hasegawa et al., 1986. "Implementation and Evaluation of a List Processing-Oriented Data Flow Machine," pp. 10–19 in *Proceedings of the 13th Annual International Symposium on Computer Architecture*. Los Alamitos, CA: IEEE Computer Society Press.

G. M. Amdahl, G. A. Blaauw, and F. P. Brooks, Jr., 1964. "Architecture of the IBM System/360," *IBM Journal of Research & Development*, 8, 2, pp. 87–101.

U. Amman, 1977. "On Code Generation in a Pascal Compiler," *Software – Practice & Experience*, 7, 3, pp. 391–423.

D. W. Anderson, F. J. Sparacio, and R. M. Tomasulo, 1967. "The IBM System/360 Model 91: Machine Philosophy and Instruction Handling," *IBM Journal*, January, pp. 8–24.

H. Anderson, D. Dieks, W. J. Gonzalez, T. Uebel, and G. Wheeler (eds.), 2013. *New Challenges to Philosophy of Science*. Heidelberg: Springer.

J. A. Anderson, 1972. "A Simple Neural Network Generating an Interactive Memory," *Mathematical Biosciences*, 14, pp. 197–220.

J. A. Anderson and E. Rosenfeld (eds.), 1988. *Neurocomputing*. Cambridge, MA: MIT Press.

J. R. Anderson, 1983. *The Architecture of Cognition*. Cambridge, MA: Harvard University Press.

T. Anderson and B. Randell (eds.), 1979. *Computing Systems Reliability*. Cambridge: Cambridge University Press.

G. R. Andrews and F. B. Schneider, 1983. "Concepts and Notations for Concurrent Programming," *ACM Computing Surveys*, 15, 1, pp. 3–44.

M. Annaratone et al., 1987. "The Warp Computer: Architecture, Implementation and Performance," Tech. Report CMU-RI-TR-87-18. Department of Computer Science/Robotics Institute, Carnegie-Mellon University, Pittsburgh, PA.

Anon., 1981. *The Programming Language Ada Reference Manual* (Proposed Standard Document, United States Department of Defense). Berlin: Springer-Verlag.

Anon., 1987. *ACM Turing Award Lectures: The First Twenty Years 1966–1985*. New York: ACM Press/Reading, MA: Addison-Wesley.

K. Appel, 1984. "The Use of the Computer in the Proof of the Four-Color Theorem," *Proceedings of the American Philosophical Society*, 178, 1, pp. 35–39.

K. R. Apt, 1981. "Ten Years of Hoare Logic," *ACM Transactions on Programming Languages and Systems*, 3, 4, pp. 431–483.

M. A. Arbib and S. Alagic, 1979. "Proof Rules for Gotos," *Acta Informatica*, 11, pp. 139–148.

Arvind, K. P. Gostelow, and W. Plouffe, 1978. "An Asynchronous Programming Language and Computing Machine," Tech. Report 114a. Department of Information & Computer Science, University of California, Irvine.

Arvind and R. A. Ianucci, 1986. "Two Fundamental Issues in Multiprocessing," CSG Memo 226-5. Laboratory for Computer Science, MIT, Cambridge, MA.

Arvind and R. E. Thomas, 1981. "I-Structures: An Efficient Data Structure for Functional Languages," Tech. Report LCS/TM-178. Laboratory for Computer Science, MIT, Cambridge, MA.

E. A. Ashcroft and Z. Manna, 1971. "Formalization of Properties of Parallel Programs," pp. 17–41 in D. M. Michie (ed.), *Machine Intelligence 6*. Edinburgh: Edinburgh University Press.

E. A. Ashcroft and W. W. Wadge, 1982. "Rx for Semantics," *ACM Transactions on Programming Languages and Systems*, 4, 2, pp. 283–294.

I. Asimov, 1989. "Foreword," pp. vii-viii in Boyer, 1989.

A. J. Ayer, [1936] 1971. *Language, Truth and Logic*. Harmondsworth, UK: Penguin Books.

F. T. Baker and H. D. Mills, 1973. "Chief Programmer Teams," *Datamation*, 19, 12, pp. 58–61.

R. C. Backhouse, 1986. *Program Construction and Verification*. London: Prentice-Hall International.

J. W. Backus, [1978] 1987. "Can Programming Be Liberated from the von Neumann Style?," pp. 63–80 in Anon., 1987.

J. W. Backus, R. W. Beeber, S. Best, R. Goldberg, L. M. Halbit, H. C. Herrick, R. A. Nelson, D. Sayre, P. B. Sheridan, H. Stern, I. Ziller, R. A. Hughes, and R. Nutt, 1957. "The FORTRAN

Automatic Coding System," *Proceedings of the Western Joint Computer Conference*, Los Angeles, CA, pp. 188–197.

J. L. Baer and B. Koyama, 1979. "On the Minimization of the Width of the Control Memory of Microprogrammed Processors," *IEEE Transactions on Computers*, C-28, 4, pp. 310–316.

J. D. Bagley, 1967. "The Behavior of Adaptive Systems Which Apply Genetic and Correlation Algorithms," PhD dissertation, University of Michigan, Ann Arbor.

D. H. Ballard, 1986. "Cortical Connections and Parallel Processing: Structure and Function," *Behavioral and Brain Sciences*, 9, 1, pp. 67–90.

M. R. Barbacci, 1981. "The Instruction Set Processor Specifications (ISPS): The Notation and Its Applications," *IEEE Transactions on Computers*, C-30, 1, pp. 24–40.

M. R. Barbacci (ed.), 1985. Special Issue on Hardware Description Languages, *Computer*, 18, 2.

M. R. Barbacci, G. E. Barnes, R. G. Cattell, and D. P. Siewiorek, 1979. "The ISPS Computer Description Language," Department of Computer Science, Carnegie-Mellon University, Pittsburgh, PA.

G. H. Barnes, R. M. Brown, M. Kato et al., 1968. "The ILLIAC IV Computer," *IEEE Transactions on Computers*, C-18, 8, pp. 746–757.

A. Barr and E. A. Feigenbaum (eds.), 1981. *The Handbook of Artificial Intelligence, Volume 1*. Stanford, CA: HeurisTech Press/Los Altos, CA: Kaufman.

D. W. Barron, J. N. Buxton, D. F. Hartley, E. Nixon, and C. Strachey, 1963. "The Main Features of CPL," *Computer Journal*, 6, 2, pp. 134–143.

D. R. Barstow, 1985. "Domain-Specific Automatic Programming," *IEEE Transactions on Software Engineering*, SE-11, 11, pp. 1321–1336.

D. R. Barstow, H. E. Shrobe, and E. Sandwall (eds.), 1984. *Interactive Programming Environments*. New York: McGraw-Hill.

J. J. Bartik, 2013. *Pioneer Programmer* (J. T. Richman and K. D. Todd, eds.). Kirkville, MO:_ Truman State University Press.

F. C. Bartlett, 1932. *Remembering*. Cambridge: Cambridge University Press.

R. S. Barton, 1961. "A New Approach to the Functional Design of a Digital Computer," *Proceedings of the Western Joint Computer Conference*, Los Angeles, CA, pp. 393–396.

R. S. Barton, 1970. "Ideas for Computer Systems Organization: A Personal Survey," pp. 7–16 in J. T. Tou (ed.), 1970. *Software Engineering, Volume I*. New York: Academic Press.

K. E. Batcher, 1974. "STARAN Parallel Processor System," pp. 405–410 in *Proceedings of the National Computer Conference, Volume 43*. Montvale, NJ: AFIPS Press.

K. E. Batcher, 1980. "Design of a Massively Parallel Processor," *IEEE Transactions on Computers*, C-29, 9, pp. 836–840.

F. L. Bauer and H. Wössner, 1972. "The 'Plankalkul' of Konrad Zuse: A Forerunner of Today's Programming Languages," *Communications of the ACM*, 15, pp. 678–685.

C. G. Bell, R. Cady, H. McFarland, B. A. Delagi, J. F. O'Loughlin, R. Noonan, and W. A. Wulf, 1970. "A New Architecture for Minicomputers – the DEC PDP-11," *Proceedings of the AFIPS Spring Joint Computer Conference*, 36, pp. 657–675.

C. G. Bell and A. Newell, 1970. "The PMS and ISP Description System for Computer Structures," *Proceedings of the AFIPS Spring Joint Computer Conference*, 36, pp. 351–374.

C. G. Bell and A. Newell, 1971. *Computer Structures: Readings and Examples*. New York: McGraw-Hill.

A. J. Bernstein, 1966. "Analysis of Programs for Parallel Processing," *IEEE Transactions on Computers*, EC-5, 10, pp. 757–763.

A. Birman, 1974. "On Proving Correctness of Microprograms," *IBM Journal of Research & Development*, 9, 5, pp. 250–266.

G. Birtwhistle and P. A. Subrahamanyam (eds.), 1988. *VLSI Specification, Verification and Synthesis*. Boston, MA: Kluwer Academic Press.

D. Bjørner (ed.), 1980. *Abstract Software Specifications*. Berlin: Springer-Verlag.

G. A. Blaauw and F. P. Brooks, Jr., 1964. "The Structure of SYSTEM/360. Part I – Outline of Logical Structure," *IBM Systems Journal*, 3, 2 & 3, pp. 119–136.

G. A. Blaauw and W. Händler (eds.), 1981. *International Workshop on Taxonomy in Computer Architecture*. Nuremburg, Germany: Friedrich Alexander Universität.

H. D. Block, 1962. "The Perceptron: A Model for Brain Functioning I," *Review of Modern Physics*, 34, pp. 123–135.

D. G. Bobrow and A. M. Collins (eds.), 1975. *Representation and Understanding: Studies in Cognitive Science*. New York: Academic Press.

D. G. Bobrow and T. Winograd, 1977. "An Overview of KRL: A Knowledge Representation Language," *Cognitive Science*, 1, 1, pp. 3–46.

M. Boden (ed.), 1990. *The Philosophy of Artificial Intelligence*. Oxford: Oxford University Press.

M. A. Boden, 1991. *The Creative Mind*. New York: Basic Books.

M. A. Boden, 2006. *Mind as Machine: A History of Cognitive Science* (two volumes). Oxford: Clarendon.

C. Böhm and G. Jacopini, 1966. "Flow Diagrams, Turing Machines and Languages With Only Two Formation Rules," *Communications of the ACM*, 9, 5, pp. 366–371.

H. Bondi, 1975. "What is Progress in Science?," pp. 1–10 in Harré, 1975.

J. T. Bonner, 1988. *The Evolution of Complexity by Natural Selection*. Princeton, NJ: Princeton University Press.

D. Borrione (ed.), 1987. *From HDL Descriptions to Guaranteed Circuit Designs*. Amsterdam: North-Holland.

W. Bouknight, S. A. Denenberg, D. E. McIntyre et al., 1972. "The ILLIAC IV System," *Proceedings of the IEEE*, 60, 4, pp. 369–388.

C. B. Boyer, 1989. *A History of Mathematics* (2nd ed.). Revised by U. C. Merzbach. New York: Wiley.

R. S. Boyer and J. S. Moore, 1981. *The Correctness Problem in Computer Science*. New York: Academic Press.

R. J. Brachman and H. J. Levesque (eds.), 1985. *Readings in Knowledge Representation*. San Mateo, CA: Morgan Kaufmann.

R. J. Brachman and B. C. Smith (ed.), 1980. "Special Issue on Knowledge Representation," *SIGART Newsletter*, 70, February.

P. W. Bridgeman, 1927. *The Logic of Modern Physics*. New York: Macmillan.

P. Brinch Hansen, 1972a. "Structured Multiprogramming," *Communications of the ACM*, 15, 7, pp. 574–578.

P. Brinch Hansen, 1972b. "An Outline of a Course on Operating Systems Principles," pp. 29–36 in Hoare and Perrott, 1972.

P. Brinch Hansen, 1973. *Operating Systems Principles*. Englewood Cliffs, NJ: Prentice-Hall.

P. Brinch Hansen, 1975. "The Programming Language Concurrent Pascal," *IEEE Transactions on Software Engineering*, 1, 2, pp. 195–202.

P. Brinch Hansen, 1976. "The SOLO Operating System," *Software – Practice and Experience*, 6, 2, pp. 141–200.

P. Brinch Hansen, 1977. *The Architecture of Concurrent Programs*. Englewood Cliffs, NJ: Prentice-Hall.

P. Brinch Hansen, 1981. "The Design of Edison," *Software – Practice and Experience*, 11, 4, pp. 363–396.

P. Brinch Hansen, 1982. *Programming a Personal Computer*. Englewood Cliffs, NJ: Prentice-Hall.

P. Brinch Hansen, 1987. "Joyce – A Programming Language for Distributed Systems," *Software – Practice and Experience*, 17, 1, pp. 29–50.

P. Brinch Hansen, 1993. "Monitors and Concurrent Pascal: A Personal History," *SIGPLAN Notices*, 28, 3, pp. 1–35.

P. Brinch Hansen, 1996. *The Search for Simplicity: Essays in Parallel Programming*. Los Alamitos, CA: IEEE Computer Society Press.

D. C. Brown and B. Chandrasekaran, 1986. "Knowledge and Control for a Mechanical Design Expert System," *Computer*, 19, 7, pp. 92–110.

H. Brown, C. Tong, and G. Foyster, 1983. "Palladio: An Exploratory Environment for Circuit Design," *Computer*, 16, 12, pp. 41–56.

B. G. Buchanan, 1988. "AI as an Experimental Science," pp. 209–250 in Fetzer (ed.), 1988.

B. G. Buchanan and E. A. Feigenbaum, 1978. "DENDRAL and Meta-DENDRAL: Their Application Dimension," *Artificial Intelligence*, 11, pp. 5–24.

B. G. Buchanan and E. H. Shortcliffe, 1984. *Rule-Based Expert Systems: The MYCIN Experiments of the Stanford Heuristic Programming Project*. Reading, MA: Addison-Wesley.

J. K. Buckle, 1978. *The ICL 2900 Series*. London: Macmillan.

A. W. Burks, 1970. "Von Neumann's Self-Reproducing Automata," pp. 3–64 in Burks (ed.), 1970.

A. W. Burks (ed.), 1970. *Essays on Cellular Automata*. Urbana: University of Illinois Press.

A. W. Burks, H. H. Goldstine, and J. von Neumann, [1946] 1971. "Preliminary Discussion of the Logical Design of an Electronic Computing Instrument," pp. 92–119 in Bell and Newell, 1972.

Burroughs Corporation, 1961. *The Descriptor – A Definition of the B5000 Information Processing System*. Detroit, MI: Burroughs Corporation.

Burroughs Corporation, 1969. *B6500 System Reference Manual (001)*. Detroit, MI: Burroughs Corporation.

J. B. Bury, [1932] 1955. *The Idea of Progress*. New York: Dover.

C. Butler, 2002. *Postmodernism: A Very Short Introduction*. Oxford: Oxford University Press.

W. C. Carter, W. H. Joyner, and D. Brand, 1978. "Microprogram Verification Considered Necessary," pp. 657–664 in *Proceedings of the National Computer Conference*. Arlington, VA: AFIPS Press.

D. J. Chalmers, 1990. "Why Fodor and Pylyshyn Were Wrong: The Simplest Refutation," *Proceedings of the 12th Annual Conference of the Cognitive Science Society*, pp. 340–347.

K. M. Chandy and J. Misra, 1988. *Parallel Program Design: A Foundation*. Reading, MA: Addison-Wesley.

E. Charniak and D. McDermott, 1985. *Introduction to Artificial Intelligence*. Reading, MA: Addison-Wesley.

N. Chomsky, 1956. "Three Models for the Description of Language," pp. 113–124 in *Proceedings of the Symposium on Information Theory*. Cambridge, MA: MIT.

N. Chomsky, 1957. *Syntactic Structures*. The Hague: Mouton.

N. Chomsky, 1959. "On Certain Formal Properties of Grammar," *Information and Control*, 2, pp. 136–167.

Y. Chu, 1971. *Computer Organization and Microprogramming*. Englewood Cliffs, NJ: Prentice-Hall.

W. A. Clark, 1980. "From Electron Mobility to Logical Structure: A View of Integrated Circuits," *ACM Computing Surveys*, 12, 3, pp. 325–356.

E. F. Codd, 1968. *Cellular Automata*. New York: Academic Press.

E. F. Codd, 1970. "A Relational Model of Data for Large Shared Data Banks," *Communications of the ACM*, 16, 6, pp. 377–387.

K. M. Colby, 1973. "Simulation of Belief Systems," pp. 251–282 in Schank and Colby, 1973.

A. M. Collins and E. F. Loftus, 1975. "A Spreading Activation Theory of Semantic Processing," *Psychological Review*, 82, pp. 407–428.

R. P. Colwell, C. Y. Hitchcock, E. D. Jensen et al., 1985. "Computers, Complexity and Controversy," *Computer*, 18, 9, pp. 8–20.

R. W. Conway and D. G. Gries, 1973. *Introduction to Programming: A Structured Approach Using PL/1 and PL/C*. Cambridge, MA: Winthrop Publishers.

R. W. Cook and M. J. Flynn, 1970. "System Design of a Dynamic Microprocessor," *IEEE Transactions on Computers*, C-19, 3, pp. 213–222.

S. Cook, 1971. "The Complexity of Theorem Proving Procedures," *Proceedings of the 3rd ACM Symposium on Theory of Computing*, Shaker Heights, OH, pp. 151–158.

L. N. Cooper, 1973. "A Possible Organization of Animal Memory and Learning," pp. 252–264 in Lundquist and Lundquist, 1973.

K. J. W. Craik, [1943] 1967. *The Nature of Explanation*. Cambridge: Cambridge University Press.

N. Cross (ed.), 1984. *Developments in Design Methodology*. Chichester, UK: Wiley.

I. F. Currie, S. G. Bond, and J. D. Morison, 1970. "ALGOL 68-R," pp. 21–34 in Peck (ed.), 1970.

H. B. Curry and R. Feys, 1958. *Combinatory Logic, Volume 1*. Amsterdam: North-Holland.

O-J. Dahl, 2001. "The Birth of Object Orientation: The Simula Language. http://www.olejohandahl.info/old/birth-of-oo.pdf.

O-J. Dahl, E. W. Dijkstra, and C. A. R. Hoare, 1972. *Structured Programming*. New York: Academic Press.

O-J. Dahl and C. A. R. Hoare, 1972. "Hierarchical Program Structures," pp. 175–220 in Dahl, Dijkstra, and Hoare, 1972.

O-J. Dahl, B. Myhrhaug, and K. Nygaard, 1968. *The SIMULA 67 Common Base Language*. Oslo: Norweigian Computing Centre.

W. Damm, 1985. "Design and Specification of Microprogrammed Computer Architectures," pp. 3–10 in *Proceedings of the 18th Annual Microprogramming Workshop*. Los Alamitos, CA: IEEE Computer Society Press.

W. Damm, 1988. "A Microprogramming Logic," *IEEE Transactions on Software Engineering*, 14, 5, pp. 559–574.

W. Damm and G. Doehman, 1985. "Verification of Microprogrammed Computer Architectures in the AADL/S* System: A Case Study," pp. 61–73 in *Proceedings of the 18th Annual Microprogramming Workshop*. Los Alamitos, CA: IEEE Computer Society Press.

W. Damm, G. Doehman, K. Merkel, and M. Sichelsschmidt, 1986. "The AADL/S* Approach to Firmware Verification," *IEEE Software*, 3, 4, pp. 27–37.

T. Dartnell (ed.), 1994. *Artificial Intelligence and Creativity*. Dordecht, The Netherlands: Kluwer.

S. R. Das, D. K. Banerji, and A. Chattopadhyay, 1973. "On Control Memory Minimization in Microprogrammed Digital Computers," *IEEE Transactions on Computers*, C-22, 9, pp. 845–848.

S. Dasgupta, 1977. "Parallelism in Loop-Free Microprograms," *Information Processing 77* (Proceedings of the IFIP Congress 1977). Amsterdam: North-Holland, pp. 745–750.

S. Dasgupta, 1979. "The Organization of Microprogram Stores," *ACM Computing Surveys*, 12, 3, pp. 295–324.

S. Dasgupta, 1980. "Some Aspects of High-Level Microprogramming," *ACM Computing Surveys*, 12, 3, pp. 295–324.

S. Dasgupta, 1982. "Computer Design and Description Languages," pp. 91–155 in Yovits, 1982.

S. Dasgupta, 1990. "A Hierarchical Taxonomic System for Computer Architectures," *Computer*, 22, 3, pp. 64–74.

S. Dasgupta, 1991. *Design Theory and Computer Science*. Cambridge: Cambridge University Press.

S. Dasgupta, 1996. *Technology and Creativity*. New York: Oxford University Press.

S. Dasgupta, 1997. "Technology and Complexity," *Philosophica*, 59, 1, pp. 113–140.

S. Dasgupta, 2003a. "Multidisciplinary Creativity: The Case of Herbert A. Simon," *Cognitive Science*, 27, pp. 683–707.

S. Dasgupta, 2003b. "Innovation in the Social Sciences: Herbert Simon and the Birth of a Research Tradition," pp. 458–470 in Shavinina, 2003.

S. Dasgupta, 2011. "Contesting (Simonton's) Blind-Variation, Selective-Retention Theory of Creativity," *Creativity Research Journal*, 23, 2, pp. 166–182.

S. Dasgupta, 2014. *It Began with Babbage: The Genesis of Computer Science*. New York: Oxford University Press.

S. Dasgupta, 2016. "From *The Sciences of the Artificial* to Cognitive History," pp. 60–70 in Frantz and Marsh, 2016.

S. Dasgupta and B. D. Shriver, 1985. "Developments in Firmware Engineering," pp. 101–176 in Yovits, 1985.

S. Dasgupta and J. Tartar, 1976. "The Identification of Maximal Parallelism in Straight-Line Microprograms," *IEEE Transactions on Computers*, C-25, 10, pp. 986–992.

S. Davidson, 1986. "Progress in High Level Microprogramming," *IEEE Software*, 3, 4, pp. 18–26.

A. L. Davis, 1978. "The Architecture and System Method of DDM1: A Recursively Structured Data Driven Machine," pp. 210–215 in *Proceedings of the 5th Annual Symposium on Computer Architecture*. New York: ACM/IEEE, New York.

R. Davis, B. G. Buchanan, and E. A. Shortcliffe, 1977. "Production Rules as Representation for a Knowledge-Based Consultative Program," *Artificial Intelligence*, 8, 1, pp. 15–45.

J. de Bakker, 1980. *The Mathematical Theory of Program Correctness*. London: Prentice-Hall International.

R. De Millo, R. J. Lipton, and A. J. Perlis, 1979. "Social Processes and Proofs of Theorems and Programs," *Communications of the ACM*, 22, 5, pp. 271–280.

P. J. Denning, 1970. "Virtual Memory," *Computing Surveys*, 2, 3, pp. 153–190.

J. B. Dennis, 1974. "First Version of a Data Flow Procedural Language," pp. 362–376 in *Proceedings Colloque sue le Programmation* (Lecture Notes on Computer Science, Volume 19). Berlin: Springer-Verlag.

J. B. Dennis, 1979. "The Varieties of Data Flow Computers," Computation Structures Group Memo 183–1, MIT, Cambridge, MA.

J. B. Dennis, 1986. "Data Flow Ideas and Future Supercomputers," pp. 78–96 in Metropolis, Sharp, Worlton, and Ames, 1986.

J. B. Dennis, S. H. Fuller, W. B. Ackerman, R. J. Swan, and K-S. Wang, 1979. "Research Directions in Computer Architecture," pp. 514–556 in Wegner, 1979.

J. B. Dennis, C. K. C. Leung, and D. P. Misunas, 1979. "A Highly Parallel Processor Using a Data Flow Machine Language," Computational Structures Group Memo 134-1, MIT, Cambridge, MA.

J. B. Dennis and D. P. Misunas, 1974. "A Preliminary Architecture for a Basic Data Flow Computer," pp. 126–132 in *Proceedings of the 2nd Annual Symposium on Computer Architecture*. New York: ACM/IEEE, New York.

F. DeRemer and H. Kron, 1975. "Programming-in-the-Large versus Programming-in-the-Small," *ACM SIGPLAN Notices*, 10, 6, pp. 114–121.

E. W. Dijkstra, 1965a. "Programming Considered as a Human Activity," pp. 213–217 in *Proceedings of the 1965 IFIP Congress*. Amsterdam: North-Holland.

E. W. Dijkstra, 1965b. "Cooperating Sequential Processes," Tech. Report EWD 123. Mathematics Department, Technische Universiteit Eindhoven, Eindhoven.

E. W. Dijkstra, 1968a. "Goto Statement Considered Harmful" (Letter to the Editor). *Communications of the ACM*, 11, pp. 147–148.

E. W. Dijkstra, 1968b. "The Structure of the 'THE' Multiprogramming Operating System," *Communications of the ACM*, 11, 5, pp. 341–346.

E. W. Dijkstra, 1972. "Notes on Structured Programming," pp. 1–82 in Dahl, Dijkstra, and Hoare, 1972.

E. W. Dijkstra, 1976. *A Discipline of Programming*. Englewood Cliffs, NJ: Prentice-Hall.

E. W. Dijkstra, [1969] 1979. "Structured Programming," pp. 43–50 in Yourdon, 1979.

E. W. Dijkstra, 2001. "What Led to 'Notes on Structured Programming," EWD 1308-0.

J. E. Donahue, 1976. *Complementary Definitions of Programming Language Semantics*. New York: Springer-Verlag.

J. R. Donaldson, 1973. "Structured Programming," *Datamation*, 19, 12, pp. 52–54.

R. W. Doran, 1979. *Computer Architecture: A Structured Approach*. New York: Academic Press.

H. E. Dreyfus, 1979. *What Computers Can't Do* (revised ed.). New York: Harper & Row.

I. S. Duff and G. W. Stewart (eds.), 1979. *Sparse Matrix Proceedings 1978*. Philadelphia: Society for Industrial and Applied Mathematics.

C. Eastman (ed.), 1975. *Spatial Synthesis in Computer-Aided Building Design*. New York: Wiley.

J. Eatwell, M. Milgate, and P. Newman (eds.), 1987. *The New Palgrave: A Dictionary of Economics*, *Volume 1*. London: Macmillan.

A. H. Eden, 2007. "Three Paradigms of Computer Science," *Minds and Machines*, 17, 2, pp. 135–167.

G. W. Ernst and A. Newell, 1969. *GPS: A Case Study in Generality and Problem Solving*. New York: Academic Press.

A. D. Falkoff, K. E. Iverson, and E. H. Sussenguth, 1964. "A Formal Description of SYSTEM/360," *IBM Systems Journal*, 3, 2 & 3, pp. 198–262.

E. A. Feigenbaum, B. G. Buchanan, and J. Lederberg, 1971. "On Generality and Problem Solving: A Case Study Using the DENDRAL Program," pp. 165–190 in Meltzer and Michie, 1971.

J. H. Fetzer, 1988. "Program Verification: The Very Idea," *Communications of the ACM*, 37, 9, pp. 1048–1063.

J. H. Fetzer (ed.), 1988. *Aspects of Artificial Intelligence*. Dordecht, The Netherlands: Kluwer.

N. V. Findler (ed.), 1979. *Associative Networks: Representation and Use of Knowledge by Computers*. New York: Academic Press.

D. A. Fisher, 1978. "DoD's Common Programming Language Effort," *Computer*, 11, 3, pp. 24–33.

J. A. Fisher, 1981. "Trace Scheduling: A Technique for Global Microcode Compaction," *IEEE Transactions on Computers*, C-30, 7, pp. 478–490.

J. A. Fisher, D. Landskov, and B. D. Shriver, 1981. "Microcode Compaction: Looking Backward and Looking Forward," pp. 95–102 in *Proceedings, National Computer Conference*. Montvale, NJ: AFIPS Press.

R. W. Floyd, 1961. "A Descriptive Language for Symbol Manipulation," *Journal of the ACM*, 8, 4, pp. 579–584.

R. W. Floyd, 1967. "Assigning Meaning to Programs," pp. 19–32 in Anon., 1967. *Mathematical Aspects of Computer Science*, Volume XIX. Providence, RI: American Mathematical Society.

M. J. Flynn, 1966. "Very High Speed Computing Systems," *Proceedings of the IEEE*, 54, 12, pp. 1901–1912.

M. J. Flynn, 1974. "Trends and Problems in Computer Organization," *Information Processing 74*. (Proceedings of the 1974 IFIP Congress.) Amsterdam: North-Holland.

M. J. Flynn and R. F. Rosin, 1971. "Microprogramming: An Introduction and Viewpoint," *IEEE Transactions on Computers*, C-20, 7, pp. 727–731.

J. A. Fodor, 1976. *The Language of Thought*. Cambridge, MA: Harvard University Press.

J. A. Fodor and Z. W. Pylyshyn, 1988. "Connectionism and Cognitive Architecture: A Critical Analysis," *Cognition*, 28, pp. 3–71.

C. C. Foster, 1970. *Computer Architecture*. New York: Van Nostrand-Rheinhold.

C. C. Foster, 1972. "A View of Computer Architecture," *Communication of the ACM*, 15, 7, pp. 557–565.

M. J. Foster and H. T. Kung, 1980. "The Design of Special Purpose Chips," *Computer*, 13, 1, pp. 26–40.

R. Franz and L. Marsh (eds.), 2016. *Minds, Models and Milieux: Commemorating the Centennial of the Birth of Herbert Simon*. Basingstoke, UK: Palgrave Macmillan,

U. O. Gagliardi, 1973. "Report of Workshop 4 – Software Related Advances in Computer Hardware," in *Proceedings of a Symposium on the High Cost of Software*. Menlo Park, CA: Stanford Research Institute.

H. Gardner, 1985. *The Mind's New Science*. New York: Basic Books.

M. R. Garey and D. S. Johnson, 1979. *Computers and Intractability: A Guide to the Theory of NP-Completeness*. San Francisco: W. H. Freeman.

M. R. Gensereth and N. J. Nilsson, 1987. *Logical Foundations of Artificial Intelligence*. Los Altos, CA: Morgan Kaufmann.

J. S. Gero and R. D. Coyne, 1987. "Knowledge-Based Planning as a Design Paradigm," pp. 339–373 in Yoshika and Waterman, 1987.

W. Giloi, 1983. "Towards a Taxonomy of Computer Architecture Based on the Machine Data Type View," pp. 6–15 in *Proceedings of the 10th Annual International Symposium on Computer Architecture*. Los Alomitos, CA: IEEE Computer Society Press.

A. Goldberg, 1984. "The Influence of an Object-Oriented Language on the Programming Environment," pp. 141–174 in Barstow, Shrobe, and Sandwall, 1984.

D. E. Goldberg, 1989. *Genetic Algorithms in Search, Optimization and Machine Learning*. Reading, MA: Addison-Wesley.

M. J. C. Gordon, 1979. *The Denotational Description of Programming Languages*. Berlin: Springer-Verlag.

M. J. C. Gordon, 1986. "Why Higher-Order Logic Is a Good Formalism for Specifying and Verifying Hardware," pp. 153–177 in Milne and Subhramanayam, 1986.

M. J. C. Gordon, 1988. "HOL – A Proof Generating System for Higher Order Logic," pp. 73–128 in Birtwhistle and Subhramanayam, 1988.

M. J. C. Gordon and T. Melham, 1987. "Hardware Verification Using Higher-Order Logic," pp. 43–67 in Borrione, 1987.

A. Grasselli and U. Montanari, 1970. "On the Minimization of Read-Only Memories in Microprogrammed Digital Computers," *IEEE Transactions on Computers*, C-19, 11, pp. 1111–1114.

D. G. Gries, 1974. "On Structured Programming" (Letter to the Editor). *Communications of the ACM*, 17, 11, pp. 655–657.

D. G. Gries (ed.), 1978. *Programming Methodology. A Collection of Articles by Members of IFIP WG. 2.3*. New York: Springer-Verlag.

D. G. Gries, 1981. *The Science of Programming*. New York: Springer-Verlag.

S. Grossberg, 1976. "Adaptive Pattern Classification and Universal Recoding: I. Parallel Development and Coding of Neural Feature Detectors," *Biological Cybernetics*, 23, pp. 121–134.

J. R. Gurd and W. Bohm, 1988. "Implicit Parallel Processing: SISAL on the Manchester Dataflow Computer," pp. 175–205 in G. Paul and G. S. Almasi (eds.), 1988. *Parallel Systems and Computation*. Amsterdam: North-Holland.

J. R. Gurd and C. C. Kirkham, 1986. "Data Flow: Achievements and Prospects," pp. 61–68 in H-J. Kugler (ed.), 1986. *Information Processing 86* (Proceedings of the 1986 IFIP Congress). Amsterdam: North-Holland.

J. R. Gurd, C. C. Kirkham, and I. Watson, 1985. "The Manchester Prototype Dataflow Computer," *Communications of the ACM*, 28, 1, pp. 34–52.

J. R. Gurd, I. Watson, and J. R. Glauert, 1978. "A Multilayered Data Flow Computer Architecture," Internal Report, Department of Computer Science, University of Manchester.

H. W. Gschwind, 1967. *Design of Digital Computers*. New York: Springer-Verlag.

J. V. Guttag, 1975. "The Specification and Application to Programming of Abstract Data Types," PhD dissertation, Department of Computer Science, University of Toronto.

J. V. Guttag, 1977. "Abstract Data Types and the Development of Data Structures," *Communication of the ACM*, 20, 6, pp. 396–404.

J. V. Guttag, E. Horowitz, and D. R. Musser, 1978. "The Design of Data Type Specifications," pp. 60–79 in Yeh, 1978.

H. Hagiwara, S. Tomita et al., 1980. "A Dynamically Microprogrammable Computer with Low-Level Parallelism," *IEEE Transactions on Computers*, C-29, 7, pp. 577–595.

W. Haken, K. Appel, and J. Koch, 1977. "Every Planar Map is Four Colorable," *Illinois Journal of Mathematics*, 21, 84, pp. 429–467.

A. C. D. Haley, 1962. "The KDF.9 Computer System," *Proceedings of the 1962 Fall Joint Computer Conference*. Washington, DC: Spartan Books.

M. A. K. Halliday, 1967–1968. "Notes on Transitivity and Theme in English," Parts 1–3, *Journal of Linguistics*, 3(1), pp. 37–81; 3(2), pp. 199–244; 4(2), pp. 179–215.

V. C. Hamachar, Z. G. Vranesic, and S. G. Zaky, 1982. *Computer Organization* (2nd ed.). New York: McGraw-Hill.

W. Händler, 1977. "The Impact of Classification Schemes on Computer Architecture," *Proceedings of the 1977 International Conference on Parallel Processing*, pp. 7–15.

W. Händler, 1981. "Standards, Classification and Taxonomy: Experiences with ECS," pp. 39–75 in Blaauw and Händler, 1981.

R. E. M. Harding, 1942. *An Anatomy of Inspiration* (2nd ed.). Cambridge: Heffer & Sons.

S. Harding, 1998. *Is Science Multicultural?* Bloomington: Indiana University Press.

G. H. Hardy, [1940]1969. *A Mathematician's Apology* (with a foreword by C. P. Snow). Cambridge: Cambridge University Press.

R. Harré (ed.), 1975. *Problems of Scientific Revolutions: Progress and Obstacles to Progress in the Sciences*. Oxford: Clarendon.

J. Hartmanis and R. E. Stearns, 1965. "On the Computational Complexity of Algorithms," *Transactions of the American Mathematical Society*, 177, pp. 285–306.

J. Hartmanis and R. E. Stearns, 1966. *Algebraic Structure Theory of Sequential Machines*. Englewood Cliffs, NJ: Prentice-Hall.

A. C. Hartmann, 1977. *A Concurrent Pascal Compiler for Minicomputers.* (Lecture Notes in Computer Science 50.) Berlin: Springer-Verlag.

J. Haugeland, 1985. *Artificial Intelligence: The Very Idea.* Cambridge, MA: MIT Press.

J. E. Hayes, D. Michie, and L. I. Mikulich (eds.), 1979. *Machine Intelligence 9.* Chichester, UK: Ellis Horwood.

P. J. Hayes, 1979a. "The Logic of Frames," pp. 46–61 in Metzing, 1979.

P. J. Hayes, 1979b. "The Naïve Physics Manifesto," pp. 240–270 in Michie, 1979.

P. J. Hayes, 1985. "The Second Naïve Physics Manifesto," pp. 1–36 in Hobbs and Moore, 1985.

D. O. Hebb, 1949. *The Organization of Behavior.* New York: Wiley.

J. L. Hennessy, N. Jouppi, and S. Przybylski, 1982. "MIPS: A Microprocessor Architecture," pp. 17–22 in *Proceedings of the 15th Annual Workshop on Microprogramming.* Los Angeles, CA: IEEE Computer Society Press.

J. L. Hennessy, N. Jouppi, and S. Przybylski, 1983. "Design of a High-Performance VLSI Processor," pp. 33–54 in *Proceedings of the 3rd Caltech Conference on VLSI.* Pasadena, CA: California Institute of Technology.

C. Hewitt, 1969. "PLANNER: A Language for Proving Theorems in Robots," pp. 295–301 in *Proceedings of the International Joint Conference on Artificial Intelligence* (IJCAI). Bedford, MA: Mitre Corporation.

W. D. Hillis, 1984. "The Connection Machine: A Computer Architecture Based on Cellular Automata," *Physica,* 10D, pp. 213–228.

W. D. Hillis, 1985. *The Connection Machine.* Cambridge, MA: MIT Press.

W. D. Hillis, 1988. "Intelligence as an Emergent Behavior: Or The Songs of Eden," *Daedalus,* Winter, pp. 175–180.

C. A. R. Hoare, 1962. "Quicksort," *Computer Journal,* 5, 1, pp. 10–15.

C. A. R. Hoare, 1965. "Record Handling," *ALGOL Bulletin,* 21 (Nov.), pp. 39–69.

C. A. R. Hoare, 1969. "An Axiomatic Basis of Computer Programming," *Communications of the ACM,* 12, 10, pp. 576–580, 583.

C. A. R. Hoare, 1971. "Proof of a Program: FIND," *Communications of the ACM,* 14, 1, pp. 39–45.

C. A. R. Hoare, 1972a. "Notes on Data Structuring," pp. 83–174 in Dahl, Dijkstra, and Hoare, 1972.

C. A. R. Hoare, 1972b. "Towards a Theory of Parallel Programming," pp. 61–71 in Hoare and Perrot, 1972.

C. A. R. Hoare, 1973. "A Structured Paging System," *The Computer Journal,* 16, 8, pp. 209–214.

C. A. R. Hoare, 1974. "Monitors: An Operating System Structuring Concept," *Communications of the ACM,* 17, 10, pp. 550–557.

C. A. R. Hoare, 1981. "The Emperor's Old Clothes," *Communication of the ACM,* 24, 2, pp. 75–83.

C. A. R. Hoare, 1986. *The Mathematics of Programming.* Oxford: Clarendon.

C. A. R. Hoare, 1986. *Communicating Sequential Processes.* London: Prentice-Hall International.

C. A. R. Hoare, 1987. "An Overview of Some Formal Methods of Program Design," *Computer,* 20, 9, pp. 85–91.

C. A. R. Hoare and R. H. Perrot (eds.), 1972. *Operating Systems Techniques.* New York: Academic Press.

C. A. R. Hoare and N. Wirth, 1973. "An Axiomatic Definition of the Programming Language PASCAL," *Acta Informatica,* 2, pp. 335–355.

J. R. Hobbs and R. C. Moore (eds.), 1985. *Formal Theories of the Commonsense World.* Norwood, NJ: Ablex.

P. Hoffman, 1986. "The Next Leap in Computers," *New York Times Magazine*, December 7. www.nytimes.com/1986/12/07/magazine/the-next-leap-in-computers.html. Retrieved November 28, 2017.

J. H. Holland, 1975. *Adaptation in Natural and Artificial Systems*. Ann Arbor: University of Michigan Press.

G. Holton, 1996. "Science and Progress Revisited," pp. 9–26 in Mazlish and Marx (eds.), 1996.

J. E. Hopcroft and J. D. Ullman, 1969. *Formal Languages and Their Relation to Automata*. Reading, MA: Addison-Wesley.

M. E. Hopkins, 1972. "A Case for the GOTO," *Proceedings of the 25th National ACM Conference*, pp. 787–790.

R. M. Hord, 1982. *The ILLIAC IV: The First Supercomputer*. Rockville, MD: Computer Science Press.

J. J. Horning and B. Randell, 1973. "Process Structuring," *ACM Computing Surveys*, 5, 1, pp. 5–30.

S. S. Husson, 1970. *Microprogramming: Principles and Practices*. Englewood Cliffs, NJ: Prentice-Hall.

J. S. Huxley, [1942] 2010. *Evolution: The Modern Synthesis* [The Definitive Edition]. Cambridge, MA: MIT Press.

K. Hwang and F. Briggs, 1984. *Computer Architecture and Parallel Processing*. New York: McGraw-hill.

IBM, 1964. "IBM System/360 Principles of Operation," Form A22-6821-0. New York: IBM Corporation.

J. K. Iliffe, 1972. *Basic Machine Principles* (2nd ed.). London: Macdonald/New York: American Elsevier.

D. H. H. Ingalls, 1978. "The Smalltalk-76 Programming System: Design and Implementation," *Proceedings of the 5th Symposium on the Principles of Programming Languages*, pp. 9–16.

Intel Corporation, 1981. *The iAPX-432 GDP Architecture Reference Manual*. Santa Clara, CA: Intel Corporation.

K. E. Iverson, 1962. *A Programming Language*. New York: Wiley.

L. W. Jackson and S. Dasgupta, 1974. "The Identification of Parallel Microoperations," *Information Processing Letters*, 2, 6, pp. 180–184.

T. Jayasri and D. Basu, 1976. "An Approach to Organizing Microinstructions Which Minimizes the Width of Control Store Words," *IEEE Transactions on Computers*, C-25, 5, pp. 514–521.

L. Jeffress (ed.), 1951. *Celebral Mechanisms in Behavior: The Hixon Symposium*. New York: Wiley.

K. Jensen and N. Wirth, 1975. *PASCAL User Manual and Report* (2nd ed.). New York: Springer-Verlag.

D. Johnson et al., 1980. "Automatic Partitioning of Programs in Multiprocessor Systems," pp. 175–178 in *Proceedings of IEEE COMPCON*. New York: IEEE Press.

C. B. Jones, 1980. *Software Development: A Rigorous Approach*. London: Prentice-Hall International.

C. B. Jones, 1986. *Systematic Software Development Using VDM*. London: Prentice-Hall International.

J. J. Joyce, 1988. "Formal Verification and Implementation of a Microprocessor," pp. 129–157 in Birtwhistle and Subhramanayam 1988.

W. H. Joyner, W. C. Carter & G. B. Leeman, 1976. "Automated Proofs of Microprogramming Correctness," pp. 51–55 in *Proceedings of the 9th Annual Microprogramming Workshop*. New York: ACM/IEEE.

R. M. Karp, 1972. "Reducibility Among Combinatorial Problems," pp. 85–104 in Miller and Thatcher, 1972.

R. M. Karp and R. E. Miller, 1966. "Properties of a Model for Parallel Computation: Determinacy, Termination, Queuing," *SIAM Journal of Applied Mathematics*, 14, 6, pp. 1390–1411.

M. G. H. Katevenis, 1985. *Reduced Instruction Set Computer Architecture for VLSI*. Cambridge, MA: MIT Press.

A. Kay, 1993. "The Early History of Smalltalk," *ACM SIGPLAN Notices*, 28, 3, pp. 1–54.

A. Kay and A. Godelberg, 1977. "Personal Dynamic Media," *Computer*, 10, 3, pp. 31–41.

B. Kernighan and D. M. Ritchie, 1993. *The C Programming Language*. Englewood Cliffs, NJ: Prentice-Hall.

S. C. Kleene, 1956. "Representation of Events in Nerve Nets and Finite Automata," pp. 3–41 in McCarthy and Shannon, 1956.

R. L. Kleir and C. V. Ramamoorthy, 1971. "Optimization Strategies for Microprograms," *IEEE Transactions on Computers*, C-20, 7, pp. 783–795.

D. E. Knuth, 1968a. *The Art of Computer Programming. Volume 1. Fundamental Algorithms*. Reading, MA: Addison-Wesley.

D. E. Knuth, 1968b. "The Semantics of Context Free Languages," *Mathematical Systems Theory*, 2, 2, pp. 127–145.

D. E. Knuth, 1973. *The Art of Computer Programming. Volume 3. Sorting and Searching*. Reading, MA: Addison-Wesley.

D. E. Knuth, 1974. "Structured Programming *With* **goto** Statements," *Computing Surveys*, 6, 12, pp. 261–301.

D. E. Knuth, 1992. *Literate Programming*. Stanford, CA: Center for the Study of Language and Information.

D. E. Knuth, 1996. *Selected Writings on Computer Science*. Stanford, CA: Center for the Study of Language and Information.

D. E. Knuth and R. W. Floyd, 1971. "Notes on Avoiding **go to** Statements," *Information Processing Letters*, 1, pp. 23–31.

N. Koertege (ed.), 1998. *A House Built on Sand*. New York: Oxford University Press.

P. M. Kogge, 1981. *Architecture of Pipelined Computers*. New York: McGraw-Hill.

Z. Kohavi, 1970. *Switching and Finite Automata Theory*. New York: McGraw-Hill.

T. Kohonen, 1972. "Correlating Matrix Memories," *IEEE Transactions on Computers*, C-21, pp. 353–359.

P. Kornerup and B. D. Shriver, Sr., 1975. "An Overview of the MATHILDA System," *ACM SIGMICRO Newsletter*, 5, 4, pp. 25–53.

C. H. A. Koster, 1970. "Syntax Directed Parsing of ALGOL 68 Programs," Tech. Rept. MR 115. Mathematics Centrum, University of Amsterdam.

D. J. Kuck, 1976. "Parallel Processing of Ordinary Programs," pp. 119–179 in Rubinoff and Yovits, 1976.

D. J. Kuck, 1977. "A Survey of Parallel Machine Organization and Programming," *ACM Computing Surveys*, 9, 1, pp. 29–60.

D. J. Kuck, 1978. *The Structure of Computers and Computation, Volume 1*. New York: Wiley.

D. J. Kuck, R. H. Kuhn, B. Leasure, and M. J. Wolfe, 1980. "The Structure of an Advanced Retargetable Vectorizer," pp. 957–974 in K. Hwang (ed.), 1980. *Supercomputers: Design and Applications Tutorial*. Silver Spring: IEEE Computer Society Press.

D. J. Kuck, R. H. Kuhn, D. Padua et al., 1981. "Dependence Graphs and Compiler Organization," *Proceedings of the 8th Annual ACM Symposium on Principles of Programming Languages*. Williamsburg, VA, pp. 207–218.

D. J. Kuck and R. A. Stokes, 1980. "The Burroughs Scientific Processor (BSP)," *IEEE Transactions on Computers*, C-31, 5, pp. 363–376.

T. S. Kuhn, [1962] 2012. *The Structure of Scientific Revolutions* (4th ed.). Chicago: University of Chicago Press.

H. T. Kung, 1979. "Let's Design Algorithms for VLSI Systems," pp.65–90 in *Proceedings of the Caltech Conference on Very Large Scale Integration: Architecture, Design, Fabrication*. Pasadena, CA: California Institute of Technology.

H. T. Jung, 1980. "Structure of Parallel Algorithms," pp. 65–117 in Yovits, 1980.

H. T. Kung, 1982. "Why Systolic Architectures?," *Computer*, 15, 1, pp. 37–46.

H. T. Kung and C. E. Leiserson, 1979. "Systolic Arrays for VLSI," pp. 256–282 in Duff and Stewart.

D. LaCapra, 1983. *Rethinking Intellectual History: Text, Context, Language*. Ithaca, NY: Cornell University Press.

I. Lakatos, 1976. *Proofs and Refutations*. Cambridge: Cambridge University Press.

D. Lamire, 2013. "Should Computer Scientists Run Experiments?," http://lamire.me/blog/archives/2013/07/10/should-computer-scientists-run-experiments.

P. J. Landin, 1964. "The Mechanical Evaluation of Expressions," *Computer Journal*, 6, 4, pp. 308–320.

G. G. Langdon, Jr., 1974. *Logic Design: A Review of Theory and Practice*. New York: Academic Press.

B. Latour and S. Woolgar, 1986. *Laboratory Life: The Construction of Scientific Facts* (2nd ed.). Princeton, NJ: Princeton University Press.

W. W. Lattin, J. A. Bayliss, D. L. Budde et al., 1981. "A 32b VLSI Micro-Mainframe Computer Ssytem," *Proceedings of the 1981 IEEE International Solid State Circuits Conference*, pp. 110–111.

L. Laudan, 1977. *Progress and Its Problems*. Berkeley, CA: University of California Press.

S. H. Lavington, 1998. *A History of Manchester Computers* (2nd ed.). Swindon, UK: British Computer Society.

D. H. Lawrie, 1975. "Access and Alignment of Data in an Array Processor," *IEEE Transactions on Computers*, C-24, 12, pp. 4–12.

D. H. Lawrie, T. Layman, D. Baer et al., 1975. "Glypnir – A Programming Language for ILLIAC IV," *Communications of the ACM*, 18, 3, pp. 157–163.

H. W. Lawson, 1968. "Programming Language-Oriented Instruction Streams," *IEEE Transactions on Computers*, C-17, pp. 743–747.

H. W. Lawson and B. Malm, 1973. "A Flexible Asynchronous Microprocessor," *BIT*, 13, pp. 165–176.

H. W. Lawson and B. K. Smith, 1971. "Functional Characteristics of a Multilingual Processor," *IEEE Transactions on Computers*, C-20, 7, pp. 732–742.

G. B. Leeman, 1975. "Some Problems in Certifying Microprograms," *IEEE Transactions on Computers*, C-24, 5, pp. 545–553.

G. B. Leeman, W. C. Carter, and A. Birman, 1974. "Some Techniques for Microprogram Validation," *Information Processing 74* (Proceedings of the 1974 IFIP Congress). Amsterdam: North-Holland, pp. 76–80.

A. Lindenmayer, 1968. "Mathematical Models of Cellular Interaction in Development– I.," *Journal of Theoretical Biology*, 18, pp. 280–315.

A. Lindenmayer and G. Rozenberg, 1972. "Development Systems and Languages," *Proceedings of the 4th ACM Symposium on Theory of Computing*, Denver, CO, pp. 214–221.

C. H. Lindsey and S. G. van der Meulin, 1971. *Informal Introduction to ALGOL 68*. Amsterdam: North-Holland.

B. Liskov, 1980. "Modular Program Construction Using Abstractions," pp. 354–389 in Bjørner, 1980.

B. Liskov and V. Berzins, 1979. "An Appraisal of Program Specifications," pp. 276–301 in Wegner, 1979.

B. Liskov and S. Gillies, 1974. "Programming With Abstract Data Types," *SIGPLAN Notices*, 9, 4, pp. 50–59.

B. Liskov, A. Snyder, R. Atkinson, and C. Shaffert, 1977. "Abstraction Mechanisms in CLU," *Communications of the ACM*, 20, 8, pp. 564–576.

A. E. Lovejoy, [1936] 1964. *The Great Chain of Being*. Cambridge, MA: Harvard University Press.

P. E. Lucas, 1972. "On the Semantics of Programming Languages and Software Devices," pp. 41–58 in Rustin, 1972.

P. E. Lucas and K. Walk, 1969. "On the Formal Description of PL/1," *Annual Review of Automatic Programming*, 6, 3, pp. 105–182.

B. Lundquist and S. Lundquist (eds.), 1973. *Proceedings of the Nobel Symposium on Collective Properties of Physical Systems*. New York: Academic Press.

A. S. Maida and S. C. Schapiro, 1982. "Intensional Concepts in Propositional Semantic Networks," *Cognitive Science*, 6, 4, pp. 291–330.

B. J. Mailloux, 1967. "On the Implementation of ALGOL 68," PhD dissertation, University of Amsterdam.

L. March, 1977. "A Boolean Description of a Class of Built Forms," pp. 41–73 in March (ed.), 1977.

L. March (ed.), 1977. *The Architecture of Form*. Cambridge: Cambridge University Press.

D. Marr, 1982. *Vision*. San Francisco: W. H. Freeman.

D. Marr and T. Poggio, 1976. "Cooperative Computation of Stereo Disparity," *Science*, 194, pp. 283–287.

L. H. Martin and J. Sørensen (ed.), 2011. *Past Minds: Studies in Cognitive Historiography*. London: Equinox.

J. Maynard Smith, 1975. *Theory of Evolution*. Harmondsworth, UK: Penguin.

B. Mazlish, 1996. "Progress: A Historical and Critical Perspective," pp. 27–44 in Mazlish and Marx (eds.), 1996.

B. Mazlish and L. Marx, 1996. "Introduction," pp. 1–7 in Mazlish and Marx (eds.), 1996.

B. Mazlish and L. Marx (eds.), 1996. *Progress: Fact or Illusion?* Ann Arbor: University of Michigan Press.

E. Mayr, 1982. *The Growth of Biological Thought*. Cambridge, MA: Belknap Press of Harvard University Press.

J. McCarthy, 1963. "Towards a Mathematical Science of Computation," pp. 21–28 in *Proceedings of the 1962 IFIP Congress*. Amsterdam: North-Holland.

J. McCarthy, 1979. "First Order Theories of Individual Concepts and Propositions," pp. 129–147 in Hayes, Michie, and Mikulich 1979.

J. McCarthy, 1981. "History of LISP," pp. 173–185 in Wexelbat, 1981.

J. McCarthy and P. J. Hayes, 1969. "Some Philosophical Problems From the Standpoint of Artificial Intelligence," pp. 463–502 in Michie and Meltzer, 1969.

J. McCarthy and C. E. Shannon (eds.), 1956. *Automata Studies*. Princeton, NJ: Princeton University Press.

E. J. McCluskey, 1965. *Introduction to the Theory of Switching Circuits*. New York: McGraw-Hill.

D. D. McCracken, 1973. "Revolution in Programming: An Overview," *Datamation*, 19, 12, pp. 50–52.

W. S. McCulloch and W. Pitts, 1943. "A Logical Calculus Immanent in Nervous Activity," *Bulletin of Mathematical Biophysics*, 5, pp. 115–133.

R. M. McKeag and R. Macnaghten (eds.), 1980. *The Construction of Programs*. Cambridge: Cambridge University Press.

W. M. McKeeman, J. J. Horning, and D. B. Wortman, 1970. *A Compiler Generator*. Englewood Cliffs, NJ: Prentice-Hall.

J. L. McClelland, D. E. Rumelhart and the PDP Research Group, 1986. *Parallel Distributed Processing: Explorations in the Microstructure of Cognition. Volume 2: Psychological and Biological Models*. Cambridge, MA: MIT Press.

J. L. McClelland, D. E. Rumelhart, and G. E. Hinton, 1986. "The Appeal of Parallel Distributed Processing," pp. 3–44 in Rumelhart, McClelland, and the PDP Research Group, 1986.

D. W. McShea, 1997. "Complexity in Evolution: A Skeptical Assessment," *Philosophica*, 59, 1, pp. 79–112.

C. A. Mead, 1989. *Analog VLSI and Neural Systems*. Reading, MA: Addison-Wesley.

C. A. Mead and L. Conway, 1980. *Introduction to VLSI Systems*. Reading, MA: Addison-Wesley.

P. B. Medawar and J. S. Medawar, 1983. *Aristotle to Zoos: A Philosophical Dictionary of Biology*. Cambridge, MA: Harvard University Press.

P. M. Melliar-Smith, 1979. "System Specifications," pp. 19–65 in Anderson and Randell, 1979.

B. Meltzer and D. Michie (eds.), 1971. *Machine Intelligence 6*. Edinburgh: Edinburgh University Press.

N. Metropolis, D. H. Sharp, W. J. Worlton, and K. R. Ames (ed.), 1986. *Frontiers of Supercomputing*. Berkeley, CA: University of California Press.

D. Metzing (ed.), 1979. *Frame Conceptions and Text Understanding*. Berlin: de Gruyter.

D. Michie (ed.), 1979. *Expert Systems in the Micro-Electronic Age*. Edinburgh: Edinburgh University Press.

D. Michie and B. Meltzer (eds.), 1969. *Machine Intelligence 4*. Edinburgh: Edinburgh University Press.

Microdata Corporation, 1970. *Microprogramming Handbook* (2nd ed.). Sanata Ana, CA: Microdata Corporation.

E. F. Miller and G. E. Lindamood, 1973. "Structured Programming: A Top-Down Approach," *Datamation*, 19, 12, pp. 55–57.

R. E. Miller and J. W. Thatcher (eds.), 1972. *Complexity of Computer Computations*. New York: Plenum.

R. Milner, 1971. "An Algebraic Definition of Simulation Between Programs," *Proceedings of the 2nd International Joint Conference on Artificial Conference* (IJCAI-71).

R. Milner, 1980. *A Calculus of Communicating Systems*. Berlin: Springer-Verlag.

G. J. Milne and P. A. Subhramanayam (eds.), 1986. *Formal Aspects of VLSI Design*. Amsterdam: North-Holland.

M. L. Minsky, 1967. *Computation: Finite and Infinite Machines*. Englewood Cliffs, NJ: Prentice-Hall.

M. L. Minsky, 1975. "A Framework for Representing Knowledge," pp. 211–277 in Winston (ed.), 1975.

M. L. Minsky, 1985. *The Society of Mind*. New York: Simon & Schuster.

M. L. Minsky and S. Papert, 1969. *Perceptrons*. Cambridge, MA: MIT Press.

M. L. Minsky and S. Papert, 1972. *Artificial Intelligence: A Progress Report*. Artificial Intelligence Laboratory, MIT, Cambridge, MA.

M. L. Minsky and S. Papert, 1988. *Perceptrons* (expanded edition). Cambridge, MA: MIT Press.

W. J. Mitchell, 1990. *The Logic of Architecture*. Cambridge, MA: MIT Press.

J. Monod, 1975. "On the Molecular Theory of Evolution," pp. 11–24 in Harré, 1975.

C. Montangero, 1974. "An Approach to Optimal Specification of Read-Only Memories in Microprogrammed Digital Computers," *IEEE Transactions on Computers*, C-23, 4, pp. 375–389.

G. E. Moore, 1965. "Cramming More Components Onto Integrated Circuits," *Electronics*, April 19, pp. 114–117.

G. E. Moore, 1979. "VLSI: Some Fundamental Challenges," *IEEE Spectrum*, April, pp. 30–37.

D. Morris and R. N. Ibbett, 1979. *The MU5 Computer System*. London: Macmillan.

T. Moto-oka (ed.), 1982. *Fifth Generation Computer Systems*. Amsterdam: North-Holland.

R. A. Mueller and J. Varghese, 1995. "Knowledge-Based Code Selection in Retargetable Microcode Synthesis," *IEEE Design and Test*, 2, 3, pp. 44–55.

L. Mumford, 1967. *Technics and Civilization*. New York: Harcourt Brace Jovanovich.

S. Muroga, 1982. *VLSI System Design*. New York: Wiley.

G. J. Myers, 1982. *Advances in Computer Architecture* (2nd ed.). New York: Wiley.

Nanodata Corporation, 1979. *The QM-1 Hardware Level User's Manual*. Williamsburg, NY: Nanodata Corporation.

P. Naur, 1969. "Programming by Action Clusters," *BIT*, 9, pp. 250–258.

P. Naur et al. (eds.), 1962–1963. "Revised Report on the Algorithmic Language ALGOL 60," *Numerische Mathematik*, 4, pp. 420–453.

N. Nersessian, 1995. "Opening the Black Box: Cognitive Science and the History of Science," *Osiris* (2nd series), 10, pp. 194–211.

C. J. Neuhauser, 1977. "Emmy System Processor – Principles of Operation" Computer Systems Laboratory Technical Note TH-114. Stanford University, Stanford, CA.

A. Newell, 1980, "Physical Symbol Systems," *Cognitive Science*, 4, pp. 135–183.

A. Newell, 1982. ""The Knowledge Level," *Artificial Intelligence*, 18, pp. 87–127.

A. Newell, 1990. *Unified Theories of Cognition*. Cambridge, MA: Harvard University Press.

A. Newell, A. J. Perlis, and H. A. Simon, 1967. "What Is Computer Science?," *Science*, 157, pp. 1373–1374.

A. Newell, C. J. Shaw, and H. A. Simon, 1958. "Elements of a Theory of Human Problem Solving," *Psychological Review*, 65, pp. 151–166.

A. Newell, C. J. Shaw, and H. A. Simon, 1960. "A Variety of Intelligent Learning in a General Problem Solver," pp. 153–189 in Yovits and Cameron, 1960.

A. Newell and H. A. Simon, 1956. "The Logic Theory Machine: A Complex Information Processing System," *IRE Transactions on Information Theory*, IT-2, pp. 61–79.

A. Newell and H. A. Simon, 1972. *Human Problem Solving*. Englewood Cliffs, NJ: Prenctice-Hall.

A. Newell and H. A. Simon, 1976. "Computer Science as Empirical Inquiry: Symbols and Search," *Communications of the ACM*, 19, 3, pp. 113–126.

R. B. M. Nichols, 1942. "The Birth of a Poem," pp. 104–126 in Harding 1942.

N. J. Nilsson, 1980. *Principles of Artificial Intelligence*. Palo Alto, CA: Tioga Publishing Company.

R. Nisbet, [1980] 1994. *History of the Idea of Progress* (2nd ed.). New Brunswick, NJ: Transaction Publishers.

D. A. Norman, 1986. "Reflections on Cognition and Parallel Distributed Processing," pp. 531–546 in McClelland, Rumelhart, and the PDP Research Group.

K. Nygaard and O-J. Dahl, 1981. "The Development of the SIMULA Languages," pp. 439–480 in Wexelblat, 1981.

A. Opler, 1967. "Fourth Generation Software," *Datamation*, 13, 1, pp. 22–24.

E. I. Organick, 1973. *Computer Systems Organization: The B5700/6700 Series*. New York: Academic Press.

E. I. Organick and J. A. Hinds, 1978. *Interpreting Machines: Architecture and Programming of the B1700/B1800 Series.* New York: North-Holland

S. Owicki and D. G. Gries, 1976. "An Axiomatic Proof Technique for Parallel Programs," *Acta Informatica,* 6, pp. 319–340.

S. Owicki and L. Lamport, 1982. "Proving Liveness Properties of Concurrent Programs," *ACM Transactions on Programming Languages and Systems,* 4, 3, pp. 455–495.

A. Padegs, 1964. "Channel Design Considerations," *IBM Systems Journal,* 3, 2 & 3, pp. 165–180.

S. Papert, 1980. *Mindstorms: Children, Computers and Powerful Ideas.* New York: Basic Books.

S. Papert, 1984. "Microworlds: Transforming Education," ITT Key Issues Conference, Annenberg School of Communication, University of Southern California, Los Angeles, CA.

D. L. Parnas, 1972. "A Technique for Software Module Specification With Examples," *Communications of the ACM,* 15, 5, pp. 330–336.

D. L. Parnas, 1972. "On the Criteria to be Used in Decomposing Systems Into Modules," *Communications of the ACM,* 15, 12, pp. 1053–1058.

J. R. Partington, [1957] 1989. *A Short History of Chemistry* (3rd ed.). New York: Dover.

L. M. Patnaik, R. Govindarajan, and N. S. Ramadoss, 1986. "Design and Performance Evaluation of EXMAN: An Extended MANchester Data Flow Computer," *IEEE Transaction on Computers,* C-35, 3, pp. 229–244.

D. A. Patterson, 1976. "STRUM: A Structured Microprogram Development System for Correct Firmware," *IEEE Transactions on Computers,* C-25, 10, pp. 974–985.

D. A. Patterson, 1985. "Reduced Instruction Set Computers," *Communications of the ACM,* 28, 1, pp. 8–21.

D. A. Patterson and D. Ditzel, 1980. "The Case for a Reduced Instruction Set Computer," *Computer Architecture News,* 8, 6, pp. 25–33.

D. A. Patterson and R. Piepho, 1982. "RISC Assessment: A High Level Language Experiment," pp. 3–8 in *Proceedings of the 9th Annual International Symposium on Computer Architecture.* Los Angeles, CA: IEEE Computer Society Press.

D. A. Patterson and C. Sequin, 1980. "Design Considerations for Single-Chip Computers of the Future," *IEEE Transactions on Computers,* C-29, 2, pp. 108–116.

D. A. Patterson and C. Sequin, 1981. "RISC-1: Reduced Instruction Set Computer," pp. 443–458 in *Proceedings of the 8th Annual International Symposium on Computer Architecture.* New York: IEEE Computer Society Press.

D. A. Patterson and C. Sequin, 1982. "A VLSI RISC," *Computer,* 15, 9, pp. 8–21.

J. E. L. Peck, 1970. "Preface," pp. v–vi in Peck (ed), 1970.

J. E. L. Peck (ed.), 1970. *Algol 68 Implementation.* Amsterdam: North-Holland.

J. E. L. Peck, 1978. "The Algol 68 Story," *ACS Bulletin,* November, pp. 4–6.

A. J. Perlis, 1981. "Transcripts and Presentation," pp. 139–147 in Wexelblat, 1981.

R. H. Perrott, 1979. "A Language for Array and Vector Processors," *ACM Transactions on Programming Languages and Systems,* 1, 2, pp. 177–195.

R. H. Perrott, 1980. "Languages for Parallel Computers," pp. 255–282 in McKeag and MacNaughtan, 1980.

C. A. Petri, 1962. "Kommunication mit Automaten," PhD dissertation, University of Bonn.

J. Piaget, 1955. *The Child's Construction of Reality.* New York: Routledge & Kegan Paul.

J. Piaget, 1976. *The Child and Reality.* Harmondsworth, UK: Penguin.

V. Pitchumani and E. P. Stabler, 1984. "An Inductive Assertion Method for Register Transfer Level Design Verification," *IEEE Transactions on Computers,* C-32, 12, pp. 1073–1080.

K. R. Popper, 1965. *Conjectures and Refutations: The Growth of Scientific Knowledge.* New York: Harper & Row.

K. R. Popper, 1968. *The Logic of Scientific Discovery.* New York: Harper & Row.

K. R. Popper, 1972. *Objective Knowledge.* Oxford: Clarendon.

K. R. Popper, 1975. "The Rationality of Scientific Revolutions," pp. 72–101 in Harré, 1975.

M. I. Posner (ed.), 1989. *Foundations of Cognitive Science.* Cambridge, MA: MIT Press.

E. Post, 1943. "Formal Reductions of the General Combinatorial Decision Problem," *American Journal of Mathematics,* 65, pp. 197–268.

P. Prusinkiewicz and A. Lindenmayer, 1990. *The Algorithmic Beauty of Plants.* Berlin: Springer-Verlag.

M. O. Rabin and D. S. Scott, 1959. "Finite Automata and Their Decision Problems," *IBM Journal,* 3, 2, pp. 114–125.

G. Radin, 1982. "The IBM 801 Minicomputer," pp. 39–47 in *Proceedings of the ACM Symposium on Architectural Support for Programming Languages and Operating Systems.* New York: ACM.

C. V. Ramamoorthy and H. F. Li, 1977. "Pipeline Architecture," *ACM Computing Surveys,* 9, 1, pp. 61–102.

C. V. Ramamoorthy & M. Tsuchiya, 1974. "A High Level Language for Horizontal Microprogramming," *IEEE Transactions on Computers,* C-23, 8, pp. 791–801.

B. Randell & L. J. Russell, 1964. *ALGOL 60 Implementation.* New York: Academic Press.

S. F. Reddaway, 1973. "DAP – Distributed Array Processor," *Proceedings of the First Annual Symposium on Computer Architecture,* Gainesville, FL.

W. R. Reitman, 1964. "Heuristic Decision Procedures, Open Constraints, and the Structure of Ill-Defined Problems," pp. 282–315 in Shelley and Bryan, 1964.

W. R. Reitman, 1965. *Cognition and Thought.* New York: Wiley.

M. Richards, 1969. "BCPL – A Tool for Compiler Writing and Systems Programming," *Proceedings AFIPS Spring Joint Computer Conference,* 34, pp. 557–566.

M. Richards and C. Whitby-Strevens, 1981. *BCPL, The Language and Its Compiler.* Cambridge: Cambridge University Press.

R. K. Richards, 1955. *Arithmetic Operations in Digital Computers.* Princeton, NJ: D. Van Nostrand.

J. P. Riganati and P. B. Schneck, 1984. "Supercomputing," *Computer,* 17, 10, pp. 97–113.

D. M. Ritchie, 1993. "The Development of the C Language," *ACM SIGPLAN Notices,* 28, 3, pp. 201–208.

H. W. J. Rittel and M. M. Weber, 1973. "Dilemmas in a General Theory of Planning," *Policy Sciences,* 4, pp. 155–169.

D. N. Robinson, 1995. *An Intellectual History of Psychology* (3rd ed.). Madison: University of Wisconsin Press.

J. E. Rodriguez, 1969. "A Graph Model for Parallel Computation," Report MAC-TR-64. Project MAC, MIT, Cambridge, MA.

N. Roe (ed.), 2005. *Romanticism.* Oxford: Oxford University Press.

S. Rosen, 1969. "Electronic Computers: A Historical Survey," *Computing Surveys,* 1, 1, pp. 7–36.

N. Rosenberg and W. G. Vincenti, 1978. *The Britannia Bridge: The Generation and Diffusion of Technological Knowledge.* Cambridge, MA: MIT Press.

F. Rosenblatt, 1958. "The Perceptron: A Probabilistic Model of Information Storage and Organization in the Brain," *Psychological Review,* 65, pp. 386–408.

F. Rosenblatt, 1961. *Principles of Neurodynamics: Perceptrons and Theory of Brain Mechanisms.* Washington, DC: Spartan Books.

P. S. Rosenbloom, 2010. *On Computing: The Fourth Great Scientific Domain*. Cambridge, MA: MIT Press.

R. F. Rosin, 1969a. "Supervisory and Monitor Systems," *Computing Surveys*, 1, 1, pp. 37–54.

R. F. Rosin, 1969b. "Contemporary Concepts of Microprogramming and Emulation," *Computing Surveys*, 1, 4, pp. 197–212.

R. F. Rosin, G. Frieder, and R. H. Eckhouse, Jr., 1972. "An Environment for Research in Microprogramming and Emulation," *Communications of the ACM*, 15, 8, pp. 748–760.

H. H. Ross, 1974. *Biological Systematics*. Reading, MA: Addison-Wesley.

N. Rothenberg and W. G. Vincenti, 1978. *The Britannia Bridge: The Generation and Diffusion of Technological Knowledge*. Cambridge, MA: MIT Press.

J. A. Rowson, 1980. "Understanding Hierarchical Design," Tech. Report (PhD dissertation). California Institute of Technology, Pasadena, CA.

M. Rubinoff and M. C. Yovits (eds.), 1976. *Advances in Computers, Volume 15*. New York: Academic Press.

J. Rumbaugh, 1977. "A Data Flow Multiprocessor," *IEEE Transactions on Computers*, C-26, 2, pp. 138–146.

J. L. Rumelhart, 1989. "The Architecture of Mind: A Connectionist Approach," pp. 133–160 in Posner, 1989.

D. E. Rumelhart, G. E. Hinton, and J. L. McClelland, 1986. "A General Framework for Parallel Distributed Processing," pp. 45–76 in Rumelhart, McClelland, and the PDP Research Group, 1986.

D. E. Rumelhart, G. E. Hinton, and R. J. Williams, 1986. "Learning Internal Representations by Error Propagation," pp. 318–362 in Rumelhart, McClelland, and the PDP Research Group, 1986.

D. E. Rumelhart, J. L. McClelland and the PDP Research Group, 1986. *Parallel Distributed Processing: Explorations in the Microstructure of Cognition. Volume 1: Foundations*. Cambridge, MA: MIT Press.

D. E. Rumelhart, P. Smolensky, J. L. McClelland & G. E. Hinton, 1986. "Schemata and Thought Processes in PDP Models," pp. 7–57 in McClelland, Rumelhart, and the PDP Research Group, 1986.

M. Ruse, 1973. *The Philosophy of Biology*. London: Hutchinson University Library.

M. Ruse, 1986. *Taking Darwin Seriously*. Oxford: Blackwell.

B. Russell, [1919] Reprint n.d. *Introduction to Mathematical Philosophy*. New York: Touchstone/ Simon & Schuster.

R. Rustin (ed.), 1972. *Formal Semantics of Programming Languages*. Englewood Cliffs, NJ: Prentice-Hall.

H. Rutihauser, 1967. *Description of ALGOL 60*. Berlin: Springer-Verlag.

A. B. Salisbury, 1976. *Microprogrammable Computer Architectures*. New York: Elsevier.

G. R. Salton, 1968. *Automatic Information Organization and Retrieval*. New York: McGraw-Hill.

J. R. Sampson, 1984. *Biological Information Processing*. New York: Wiley.

Z. Sardar, 2000. *Thomas Kuhn and the Science Wars*. Cambridge: Icon Books.

R. C. Schank, 1973. "Identification of Conceptualization Underlying Natural Language," pp. 187–247 in Schank and Colby, 1973.

R. C. Schank and R. P. Abelson, 1975. "Scripts, Plans, and Knowledge," pp. 151–157 in *Proceedings of the International Joint Conference on Artificial Intelligence, Volume 1*. San Francisco: Morgan Kaufmann.

R. C. Schank and R. P. Abelson, 1977. *Scripts, Plans, Goals and Understanding.* Hillsdale, NJ: Lawrence Erlbaum.

R. C. Schank and K. M. Colby (eds.), 1973. *Computer Models of Thought and Language.* San Francisco: W. H. Freeman.

D. Scherlis and D. S. Scott, 1983. "First Steps Towards Inferential Programming," pp. 199–212 in *Information Processing 83* (Proceedings of the IFIP Congress). Amsterdam: North-Holland.

W. Schneider, 1987. "Connectionism: Is It a Paradigm Shift for Psychology?," *Behavior Research Methods, Instruments & Computers*, 19, 2, pp. 73–83.

L. K. Schubert, 1976. "Extending the Expressive Power of Semantic Networks," *Artificial Intelligence*, 7, pp. 163–198.

L. K. Schubert, 1990. "Semantic Nets are in the Eye of the Beholder," Tech. Report 346. Department of Computer Science, University of Rochester, Rochester, NY.

S. J. Schwartz, 1968. "An Algorithm for Minimizing Read-Only Memories for Machine Control," pp. 28–33 in *Proceedings of the 10th Annual Symposium on Switching and Automata Theory.* New York: IEEE.

D. S. Scott and C. Strachey, 1971. "Toward a Mathematical Semantics for Computer Languages," Tech. Monograph PRG-6, August. Oxford University Computing Laboratory.

J. R. Searle, 1980. "Minds, Brains and Programs," *Behavioral and Brain Sciences*, 3, pp. 417–424.

J. R. Searle, 1985. *Minds, Brains and Science.* London: BBC Publications.

C. L. Seitz, 1985. "The Cosmic Cube," *Communications of the ACM*, 28, 1, pp. 22–33.

T. Sejnowski and C. Rosenberg, 1986. "NETtalk: A Parallel Network That Learns to Read Aloud," Tech. Report JHU/EEC-86/01. Baltimore, MD: Johns Hopkins University,.

T. Sejnowski and C. Rosenberg, 1987. "Parallel Networks That Learn to Pronounce English Texts," *Complex Systems*, 1, pp. 145–168.

M. Shahdad, R. Lipsett, E. Marschner et al., 1985. "VHSIC Hardware Description Language," *Computer*, 18, 2, pp. 94–104.

L. V. Shavinina (ed.), 2003. *The International Handbook on Innovation.* Amsterdam: Elsevier Science.

M. W. Shelley and G. L. Bryan (ed.), 1964. *Human Judgements and Optimality.* New York: Wiley.

R. E. Shostak, 1983. "Formal Verification of Circuit Designs," pp. 13–30 in Uehara and Barbacci, 1983.

D. P. Siewiorek, C. G. Bell & A. Newell, 1982. *Computer Structures: Principles and Examples.* New York: McGraw-Hill.

C. B. Silo, Jr., J. H. Pugley, and B. A. Jeng, 1981. "Control Memory Width Optimization Using Multiple-Valued Circuits," *IEEE Transactions on Computers*, C-30, 2, pp. 148–153.

H. A. Simon, 1957a. "Rationality in Administrative Decision Making," pp. 196–206 in Simon, 1957b.

H. A. Simon, 1957b. *Models of Man.* New York: Wiley.

H. A. Simon, 1962. "The Architecture of Complexity," *Proceedings of the American Philosophical Society*, 60, pp. 468–482.

H. A. Simon, 1973. "The Structure of Ill-Structured Problems," *Artificial Intelligence*, 4, pp. 181–200.

H. A. Simon, 1975. "Style in Design," pp. 287–309 in Eastman, 1975.

H. A. Simon, 1976. *Administrative Behavior* (3rd ed.). New York: Free Press.

H. A. Simon, 1987. "Bounded Rationality," pp. 266–268 in Eatwell, Milgate, and Newman, 1987.

H. A. Simon, 1996. *The Sciences of the Artificial* (3rd ed.). Cambridge, MA: MIT Press.

H. A. Simon, 1987. "Bounded Rationality," pp. 266–268 in Eatwell, Milgate, and Newman, 1987.

D. B. Skillicorn, 1988. "A Taxonomy for Computer Architecture," *Computer*, 21, 11, pp. 46–57.

J. R. Slagle, 1971. *Artificial Intelligence: The Heuristic Programming Approach*. New York: McGraw-Hill.

B. C. Smith, 1982. "Reflections and Semantics in a Procedural Language," PhD dissertation. Tech. Report MIT/CCS/TR-272, MIT, Cambridge, MA.

B. C. Smith [1982] 1985. "Prologue," pp. 31–40 in Brachman & Levesque 1985.

P. Smolensky, 1988. "The Constituent Structure of Connectionist Mental States: A Reply to Fodor and Pylyshyn," Tech. Report CU-CS-394-88. Department of Computer Science, University of Colorado, Boulder, CO.

R. T. Snodgrass, 2010. "*Ergalics*: A Natural Science of Computation," Version Feb. 12. http://www.cs.arizona.edu/projects/focal/erga.

L. Snyder, 1984. "Supercomputers and VLSI: The Effect of Large-Scale Integration on Computer Architectures," pp. 1–33 in Yovits, 1984.

A. Sokal, 2008. *Beyond the Hoax: Science, Philosophy and Culture*. New York: Oxford University Press.

N. Sounderajan, 1984. "A Proof Technique for Parallel Programs," *Theoretical Computer Science*, 31, 1 & 2, pp. 13–29.

J. P. Steadman, 1977. "Graph-Theoretic Representation of Architectural Arrangements," pp. 94–115 in March (ed.), 1977.

K. Stevens, 1975. "CFD – A Fortran-like Language for the ILLIAC IV," *SIGPLAN Notices*, 3, pp. 72–80.

H. S. Stone, 1971. "Parallel Processing with a Perfect Shuffle," *IEEE Transactions on Computers*, C-20, 2, pp. 153–161.

H. S. Stone, 1980. *Introduction to Computer Architecture*. Chicago: SRA.

H. S. Stone, 1990. *High Performance Computer Architecture*. Reading, MA: Addison-Wesley.

J. E. Stoy, 1977. *Denotational Semantics: The Scott-Strachey Approach to Programming Language Theory*. Cambridge, MA: MIT Press.

W. D. Strecker and D. W. Clark, 1980. "Comments on 'The Case for the Reduced Instruction Set Computer' by Patterson and Ditzer," *Computer Architecture News*, 8, 6, pp. 34–38.

G. J. Sussman, 1973. *A Computer Model of Skill Acquisition*. New York: American Elsevier.

I. C. Sutherland, [1963] 2003. "Sketchpad: A Man-Machine Graphical Communication System," Tech. Report 575. Computer Laboratory, University of Cambridge. http://www.cl.cam.ac.uk/TechReport.

E. E. Swartzlander (ed.), 1976. *Computer Design Development: Principal Papers*. Rochelle Park, NJ: Hayden.

J. C. Syre, D. Comte, G. Durrieu et al., 1977. "LAU System – A Data Driven Software/Hardware System Based on Single Assignment," pp. 347–351 in M. Filmeier (ed.), 1977. *Parallel Computers-Parallel Mathematics*. Amsterdam: North-Holland.

A. S. Tanenbaum, 1984. *Structured Computer Organization*. Englewood Cliffs, NJ: Prentice-Hall.

G. A. Taubes, 1995. "The Rise and Fall of Thinking Machines," http://www.inc.com/magazine/19950915/2622.html. Retrieved May 23, 2016.

C. P. Thacker, E. M. McCreight, B. W. Lampson, R. F. Sproull, and D. R. Boggs, 1979. "Alto: A Personal Computer," Palo Alto: Xerox Corporation. Reprinted, pp. 549–580 in Siewiorek, Bell, and Newell, 1982.

P. R. Thagard, 1988. *Computational Philosophy of Science*. Cambridge, MA: MIT Press.

D. W. Thompson, [1917] 1992. *On Growth and Form*. Cambridge: Cambridge University Press.

J. E. Thornton, 1964. "Parallel Operation in the Control Data 6600," pp. 33–40 in *Proceedings of the AFIPS Fall Joint Computer Conference, Volume 26, Part II*. Washington, DC: Spartan Books.

K. J. Thurber, 1976. *Large Scale Computer Architecture*. Rochelle Park, NJ: Hayden.

W. F. Tichy, 1988. "Should Computer Scientists Experiment More?," *Computer*, 21, 5, pp. 32–40.

S. Timoshenko, [1953] 1983. *History of Strength of Materials*. New York: Dover.

M. Tokoro, 1978. "A Technique for Global Optimization of Microprograms," *Proceedings of the 11th Microprogramming Workshop*, Pacific Grove, CA, pp. 41–50.

M. Tokoro, E. Tamura, and T. Takizuke, 1981. "Optimization of Microprograms," *IEEE Transactions on Computers*, C-30, 4, pp. 491–504.

S. Tomita, K. Shibayama et al., 1983. "A User-Microprogrammable Local Host Computer With Low-Level Parallelism," pp. 151–159 in *Proceedings of the 10th Annual International Symposium on Computer Architecture*. Los Angeles, CA: IEEE Computer Society Press.

M. Tsuchiya and M. J. Gonzales, 1976. "Towards Optimization of Horizontal Microprograms," *IEEE Transactions on Computers*, C-25, 10, pp. 992–995.

A. B. Tucker and M. J. Flynn, 1971. "Dynamic Microprogramming: Processor Organization and Programming," *Communications of the ACM*, 14, 4, pp. 240–250.

A. M. Turing, 1936. "On Computable Numbers with an Application to the *Entscheidungsproblem*," *Proceedings of the London Mathematical Society*, 2, pp. 230–236.

A. M. Turing, 1950. "Computing Machinery and Intelligence," *Mind*, LIX, pp. 433–460.

T. Uehara and M. R. Barbacci (eds.), 1983. *Computer Hardware Description Languages and Their Applications* (CHDL-83). Amsterdam: North-Holland.

A. M. Uttley, 1954. "The Classification of Signals in the Nervous System," *EEG Clinical Neurophysiology*, 6, pp. 479–494.

T. van Gelder, 1990. "Compositionality: A Connectionist Variation on a Classical Theme," *Cognitive Science*, 14, pp. 355–381.

A. van Wijngaarden, B. J. Mailloux, J. E. L. Peck, and C. H. A. Koster, 1969. "Report on the Algorithmic Language ALGOL 68," *Numerische Mathematik*, 14, pp. 79–218.

A. van Wijngaarden, B. J. Mailloux, J. E. L. Peck, C. H. A. Koster, M. Sintsoff, C. H. Lindsey, L. G. L. T. Meertens, and R. G. Frisker, 1975. "Revised Report on the Algorithmic Language Algol 68," *Acta Informatica*, 5, pp. 1–234.

A. van Wijngaarde, B. J. Mailloux, J. E. L. Peck, C. H. A. Koster, M. Sintsoff, C. H. Lindsey, L. G. L. T. Meertens, and R. G. Frisker, 1976. *Revised Report on the Algorithmic Language Algol 68*. Berlin: Springer-Verlag.

Varian Data Corporation, 1975. *Varian Microprogramming Guide*. Irvine, CA: Varian Data Corporation.

W. G. Vincenti, 1990. *What Engineers Know and How They Know It*. Baltimore, MD: Johns Hopkins University Press.

H. von Foerster and G. W. Zopf (ed.), 1962. *Illinois Symposium on Principles of Self-Organization*. New York: Pergamon.

J. von Neumann, 1945. "First Draft of a Report on EDVAC," pp. 355–364 in B. Randell (ed.), 1975. *Origins of Digital Computers*. New York: Springer-Verlag.

J. von Neumann, 1951. "The General and Logical Theory of Automata," pp. 1–41 in Jeffress, 1951.

J. von Neumann, 1958. *The Computer and the Brain*. New Haven, CT: Yale University Press.

J. von Neumann, 1966. *Theory of Self-Reproducing Automata* (A. W. Burks, ed.). Urbana, IL: University of Illinois Press.

E. G. Wallach, 1972. "Emulation: A Survey," *Honeywell Computer Journal*, 6, 4, pp. 287–297.

H. Weber, 1967. "A Microprogrammed Implementation of EULER on IBM System 360/30," *Communications of the ACM*, 10, pp. 579–588.

P. Wegner, 1972. "The Vienna Definition Language," *ACM Computing Surveys*, 4, 1, pp. 5–67.

P. Wegner (ed.), 1979. *Research Directions in Software Technology*. Cambridge, MA: MIT Press.

J. Weizenbaum, 1976. *Computer Power and Human Reason*. San Francisco: W. H. Freeman.

R. L. Wexelblat (ed.), 1981. *History of Programming Languages*. New York: Academic Press.

D. J. Wheeler, 1949. "Automatic Computing With the EDSAC," PhD dissertation, University of Cambridge.

B. L. Whorf, 1964. *Language, Thought and Reality*. Cambridge, MA: MIT Press.

M. V. Wilkes, 1969. "The Growth of Interest in Microprogramming: A Literature Survey," *Computing Surveys*, 1, 3, pp. 139–145.

M. V. Wilkes, [1951] 1976. "The Best Way to Design an Automatic Calculating Machine," pp. 266–270 in Swartzlander, 1976.

M. V. Wilkes, 1975. *Time-Sharing Computer Systems* (3rd ed.). London: MacDonald & Jane's/ New York: American Elsevier.

M. V. Wilkes and R. M. Needham, 1979. *The Cambridge CAP Computer and Its Operating System*. New York: North-Holland.

M. V. Wilkes, D. J. Wheeler, and S. Gill, 1951. *Preparation of Programmes for an Electronic Digital Computer*. Cambridge, MA: Addison-Wesley.

W. T. Wilner, 1972. "Design of the Burroughs B1700," *Proceedings of the AFIPS Fall Joint Computer Conference*. Montvale, NJ: AFIPS Press.

K. V. Wilson, 1980. *From Associations to Structure*. Amsterdam: North-Holland.

T. Winograd, 1972. *Understanding Natural Language*. New York: Academic Press.

T. Winograd, 1973. "A Procedural Model of language Understanding," pp. 152–186 in Schank and Colby, 1973.

T. Winograd, 1980. "What Does It Mean to Understand Language?," *Cognitive Science*, 4, pp. 209–241.

P. H. Winston, 1975. "Learning Structural Descriptions From Examples," pp. 157–210 in Winston (ed.), 1975.

P. H. Winston, 1977. *Artificial Intelligence*. Reading, MA: Addison-Wesley.

P. H. Winston (ed.), 1975. *The Psychology of Computer Vision*. New York: McGraw-Hill.

N. Wirth, 1968. "PL360, A Programming Language for the 360 Computers," *Journal of the ACM*, 15, 1, pp. 34–74.

N. Wirth, 1971a. "The Programming Language PASCAL," *Acta Informatica*, 1, 1, pp. 35–63.

N. Wirth, 1971b. "The Design of a PASCAL Compiler," *Software -- Practice & Experience*, 1, pp. 309–333.

N. Wirth, 1971c. "Program Development by Stepwise Refinement," *Communications of the ACM*, 14, 4, pp. 221–227.

N. Wirth, 1973. *Systematic Programming: An Introduction*. Englewood Cliffs, NJ: Prentice-Hall.

N. Wirth, 1976. *Algorithms + Data Structures = Programs*. Englewood Cliffs, NJ: Prentice-Hall.

N. Wirth, 1977. "Modula: A Language for Modular Programming," *Software - Practice & Experience*, 7, 1, pp. 3–35.

N. Wirth, [1984] 1987. "From Programming Language Design to Computer Construction," pp. 179–190 in Anon., 1987.

N. Wirth and C. A. R. Hoare, 1966. "A Contribution to the Development of ALGOL," *Communications of the ACM*, 9, pp. 413–432.

N. Wirth and H. Weber, 1966a. "EULER: A Generalization of ALGOL and Its Formal Definition, Part I," *Communications of the ACM*, 9, 1, pp. 13–25.

N. Wirth and H. Weber, 1966b. "EULER: A Generalization of ALGOL and Its Formal Definition, Part II," *Communications of the ACM*, 9, 2, pp. 89–99.

W. G. Wood, 1978. "On the Packing of Microoperations into Microinstructions," *Proceedings of the 11th Microprogramming Workshop*, Pacific Grove, CA, pp. 51–55.

M. Woodger, 1978. "A History of IFIP WG 2.3 (Programming Methodology)," pp. 1–5 in Gries, 1978.

W. A. Woods, 1975. "What's in a Link? Foundations for Semantic Networks," pp. 35–82 in Bobrow and Collins, 1975.

P. M. Woodward, 1970. "Introduction," p. 28 in Currie, Bond, and Morison, 1970.

W. A. Wulf, 1971. "Programming Without the GOTO," pp. 408–413 in *Proceedings of the 1971 IFIP Congress, Volume 1*. Amsterdam: North-Holland.

W. A. Wulf, 1972. "A Case Against the GOTO," *Proceedings of the 25th National ACM Conference*, pp. 791–797.

W. A. Wulf, R. L. London, and M. Shaw, 1976. "An Introduction to the Construction and Verification of Alphard Programs," *IEEE Transactions on Software Engineering*, 2, 4, pp. 253–264.

W. A. Wulf, D. B. Russell, and A. N. Haberman, 1971. "BLISS: A Language for Systems Programming," *Communications of the ACM*, 14, 12, pp. 780–790.

S. S. Yau, A. Schowe and M. Tsuchiya, 1974. "On Storage Optimization of Horizontal Microprograms," *Proceedings of the 7th Annual Microprogramming Workshop*, Palo Alto, CA, pp. 98–106.

R. T. Yeh (ed.), 1978. *Current Trends in Programming Methodology, Volume 2*. Englewood Cliffs, NJ: Prentice-Hall.

H. Yoshika and E. A. Waterman (ed.), 1987. *Design Theory for CAD*. Amsterdam: Elsevier Science.

E. N. Yourdon (ed.), 1979. *Classics in Software Engineering*. New York: Yourdon Press.

E. N. Yourdon (ed.), 1982. *Writings of the Revolution*. New York: Yourdon Press.

M. C. Yovits (ed.), 1980. *Advances in Computers, Volume 19*. New York: Academic Press.

M. C. Yovits (ed.), 1982. *Advances in Computers, Volume 21*. New York: Academic Press.

M. C. Yovits (ed.), 1984. *Advances in Computers, Volume 23*. New York: Academic Press.

M. C. Yovits (ed.), 1985. *Advances in Computers, Volume 24*. New York: Academic Press.

M. C. Yovits and S. Cameron (eds.), 1960. *Self-Organizing Systems*. New York: Pergamon.

M. C. Yovits, G. T. Jacobi, and G. D. Goldstein (eds.), 1962. *Self-Organizing Systems*. Washington, DC: Spartan Books.

S. N. Zillies, 1980. "An Introduction to Data Algebra," pp. 248–272 in Bjørner, 1980.

J. Ziman, 2000. *Real Science: What Is It and What It Means*. Cambridge: Cambridge University Press,

Index

Note: The *italicized* entries refer to books and journals.